INTRODUCTION TO
Probability with *Mathematica*®
Second Edition

TEXTBOOKS in MATHEMATICS

Series Editor: Denny Gulick

PUBLISHED TITLES

ABSTRACT ALGEBRA: AN INTERACTIVE APPROACH
William Paulsen

COMPLEX VARIABLES: A PHYSICAL APPROACH WITH APPLICATIONS AND MATLAB®
Steven G. Krantz

ESSENTIALS OF TOPOLOGY WITH APPLICATIONS
Steven G. Krantz

INTRODUCTION TO ABSTRACT ALGEBRA
Jonathan D. H. Smith

INTRODUCTION TO MATHEMATICAL PROOFS: A TRANSITION
Charles E. Roberts, Jr.

INTRODUCTION TO PROBABILITY WITH MATHEMATICA®, SECOND EDITION
Kevin J. Hastings

LINEAR ALBEBRA: A FIRST COURSE WITH APPLICATIONS
Larry E. Knop

MATHEMATICAL AND EXPERIMENTAL MODELING OF PHYSICAL AND BIOLOGICAL PROCESSES
H. T. Banks and H. T. Tran

TEXTBOOKS in MATHEMATICS

INTRODUCTION TO
Probability with
Mathematica®
Second Edition

Kevin J. Hastings

Knox College
Galesburg, Illinois, U.S.A.

CRC Press
Taylor & Francis Group
Boca Raton London New York

CRC Press is an imprint of the
Taylor & Francis Group, an **informa** business

A CHAPMAN & HALL BOOK

CRC Press
Taylor & Francis Group
6000 Broken Sound Parkway NW, Suite 300
Boca Raton, FL 33487-2742

First issued in paperback 2019

© 2010 by Taylor & Francis Group, LLC
CRC Press is an imprint of Taylor & Francis Group, an Informa business

No claim to original U.S. Government works

ISBN-13: 978-1-4200-7938-8 (hbk)
ISBN-13: 978-0-367-38519-4 (pbk)

Library of Congress Cataloging-in-Publication Data

Hastings, Kevin J., 1955-
    Introduction to probability with Mathematica / Kevin J. Hastings. -- 2nd ed.
        p. cm. -- (Textbooks in mathematics)
    Includes bibliographical references and index.
        ISBN 978-1-4200-7938-8 (hardcover)
        1. Probabilities--Data processing--Textbooks. 2. Mathematica (Computer file)--Textbooks. I. Title. II. Series.

QA273.19.E4H37 2010
519.20285'536--dc22   2009030737

**Visit the Taylor & Francis Web site at
http://www.taylorandfrancis.com**

**and the CRC Press Web site at
http://www.crcpress.com**

To my parents, for their wisdom and for helping me to understand the value of doing your best

Preface

When *Introduction to Probability with Mathematica*® was written in the late 1990s, I had noticed what other probabilists and statisticians had also noticed. Students were having difficulty understanding the main concepts of probability: event, sample space, random variable, distribution, and expectation. This weakened their foundation for the study of statistics and other subjects that relied on probability, such as operations research and finance. In addition, real problem solving, with ill-specified problems, large disorganized data sets, and important and intrinsically interesting applications were substantially missing from mathematical statistics courses. Students would emerge from such courses with some understanding of theory, a collection of techniques for preprogrammed situations, and not a great deal of experience in exploring a problem from the model formulation stage completely through to the conclusion. Good writing was also not especially emphasized. The second edition of this book does not claim to fully address all of these problems, but its aim is to make progress on several fronts, as the first edition did.

It is now rather firmly established that technology can be used as a tool to let students visualize concepts, to make computations easily, and to encourage open-ended exploration in realistic, non-contrived problems. Students experience an active learning style in which they can experiment, make conjectures, and trace the consequences of those conjectures. For an author who is cognizant of these ideas when writing, the end result is a text that is involving, that develops the habits of going deeper, questioning assumptions, and wrestling with ambiguities, and in the end leaves students with real understanding of probabilistic modeling and the analysis of data, and the ability to put that understanding to use.

One of the main pedagogical tools used in the first edition was simulation. I recognized then how powerful a tool simulation could be to help students understand randomness and sampling concepts better. Simulation also was used to gain insight into probabilistic models that are not analytically easy to solve.

Simulation programs are easy to write in *Mathematica* and, in addition, with a little extra support provided by the *Mathematica* package and notebooks that I have constructed to accompany this text, students will be able to easily create their own simulations from templates, as well as use *Mathematica* in other kinds of problem-solving situations. The notebook environment, with its excellent symbol processing capability and smooth integration of graphics, numerics, and text, yields another valuable pedagogical opportunity: to encourage students to do professional quality mathematical writing of which they can be legitimately proud. The use of *Mathematica* provides a consistent platform for the study and exposition of probability theory.

The time has come for a new edition of *Introduction to Probability with Mathematica*® for a few reasons, one of which is very pragmatic. Starting with version 6.0, *Mathematica* changed some important functions and syntax and, in addition provided a few more features that can be exploited. Because of these changes, the *Mathematica* notebooks into which the first edition of the book was divided no longer worked seamlessly, and in a few cases notebooks didn't work at all. An update of the book is necessary to support current and future users who update their version of the program. Version 7.0 was introduced just as I was completing work on the second edition, and that also forced a few last-minute changes.

But also, years of experience have revealed important ways in which the presentation could be modified to be even more effective in the new edition.

- Chapter 6 on applications really intends to be a chapter on stochastic processes and their applications, and it should be billed that way and reinforced. I have transferred some of the earlier material on Poisson processes to a new section in Chapter 6, expanded the section on Markov chains to include a study of absorbing chains, and included a new section on Brownian motion.
- The first edition did not touch in any significant way on order statistics, which are important in several topic areas in mathematical statistics. A new section in Chapter 4 has been added.
- Another area of mathematical statistics that was not addressed very well was multivariate random variables, and transformations of multivariate normal random variables, particularly with regard to quadratic forms. Such material sets up the study of regression and analysis of variance in the second course in the usual sequence. So a new section on transformations of multivariate normal random variables has been included into Chapter 4.
- To give more attention to elementary data analysis and to ground the theory of the normal distribution in reality better, more example data and the graphical technique of the normal quantile plot is covered in Section 4.1.
- More attention has been given to conditional expectation in Section 2.6, a topic which has increased in importance with the rise to prominence of financial mathematics.
- There are new visualization and animation features in *Mathematica* version 6.0 and higher that will increase the interactive nature and visual appeal of some of the existing examples and facilitate new and interesting animations. For example, when a new probability density is introduced, a user-controlled manipulation showing parameter dependence can be done.
- Throughout the book, particularly in Chapters 2 and 4 on discrete and continuous distributions, I have added some well-chosen problems drawn from Actuarial Exam P. This course does serve that audience, and these problems should give the students a head start on preparing for that professional credential.
- A new appendix giving a basic introduction to *Mathematica* has been included. The end-of-section lists of *Mathematica* commands used in the section have been eliminated. They are unnecessary in the presence of the complete index to the commands in the appendices.
- And finally, it is always a good idea to refresh examples and exercises, and add a few more. In teaching from the book a number of times, I noticed that a few of the existing examples, even the technological ones of which I was so fond, didn't advance learning especially well, and so those have been retooled or eliminated. New data sets have been introduced, particularly in the section on the bivariate normal distribution.

A road map through the text is fairly straightforward. The instructor should cover all of Chapter 1 on discrete probability. Those who prefer a less simulation-oriented approach may skip lightly over Section 1.3 without major damage, but in my opinion this is the most interesting and fun part of the book. Chapter 2 is a fairly standard treatment of the common discrete distributions, but with heavy use of technology. On first reading, the material on conditional expectation in Section 2.6 can be minimized (students ought to at least know the definition), but the section will be very helpful later for students going on to study finance. Chapter 3 is an important transition to continuous probability, but the first section contains some material on higher level ideas like σ-algebras that is usually difficult for students at the first pass, so handle with care. In Chapter 4 on continuous distributions, certainly Sections 4.1, 4.2, 4.3, 4.5, and 4.6 are must-do sections. Section 4.4 on order statistics may be omitted if time is pressing. The final section of Chapter 4 requires a bit of linear algebra. Courses that have this as a prerequisite ought not to be afraid to cover this material, which helps students a great deal when studying regression and analysis of variance in mathematical statistics. But courses that do not assume linear algebra should probably pass by Section 4.7. Chapter 5 examines some of the history of probability while covering the Laws of Large Numbers and the Central Limit Theorem, and so it should not be missed. The treatment of stochastic processes and applications in Chapter 6 is optional; an instructor can include more of it or less of it depending on time. The sections are nearly independent of one another, although it is hard to imagine studying queues in Section 6.3 without first doing Poisson processes in Section 6.2. Among these sections, surely 6.2 is the most important, filling in the theory of arrival times that was not touched in Section 2.4 where the Poisson process is first introduced. The instructor may find the Chapter 6 material to be fertile ground for student projects and/or independent reading. Students intending to go on to operations research or finance will particularly benefit by reading and doing these sections.

The second edition of *Introduction to Probability with Mathematica*®, like the first edition, is available in the traditional print copy and comes with a CD with executable notebooks, one per text section. It is my intention for the students to really interact with the electronic book, but I also think that the paper copy stands on its own merits as a nice treatment of probability. Having the electronic version ought to greatly benefit the instructor as well in creating class notes, demonstrations, and projects. I continue to believe that the book is an important contribution to the field, valued particularly by mathematics departments that are becoming more and more technologically oriented.

I have updated the complete solutions manual to include the new problems and to be compatible with *Mathematica* versions 6 and higher. This is available upon request from the publisher.

This is now my third project with Chapman & Hall/CRC Press and I must thank the helpful folks there, including Executive Editor Bob Stern, Project Coordinator Jessica Vakili, and technical assistants and reviewers for their help in getting the book out. I should also mention Maryka Baraka and others at Wolfram Research for their *Mathematica* advice. And, lastly, I continue to marvel at the patience and encouragement given to me by my wife Gay Lynn and daughter Emily during my authoring experiences. They are the best.

Kevin J. Hastings

Supplementary Resources Disclaimer

Additional resources were previously made available for this title on CD. However, as CD has become a less accessible format, all resources have been moved to a more convenient online download option.

You can find these resources available here: https://www.routledge.com/9781420079388

Please note: Where this title mentions the associated disc, please use the downloadable resources instead.

Note on the Electronic Version and the KnoxProb7`Utilities` Package

The CD that accompanies this book contains a *Mathematica* notebook for each text section. Copyright law forbids you from distributing these notebooks without the express permission of the publisher, just as with the print version. The print text was made without color, by publisher specification, but color effects are included into the electronic version. Otherwise the electronic notebooks are identical to the print version. When a notebook is first loaded you can select the Evaluate Initialization Cells command under the Evaluation menu to reproduce the output. The look of the notebooks in the front end will depend on what style sheet you are using; I have applied the Book/Textbook stylesheet to these files.

Also on the disk are two packages: the KnoxProb7`Utilities` package for users of version 7 of *Mathematica*, and KnoxProb6`Utilities` for users of version 6. Move the appropriate folder into the AddOns\ExtraPackages subdirectory of your main *Mathematica* directory in order to allow the electronic book to run. The electronic files are written for version 7, so the Needs commands in the notebooks should be modified to load Knox-Prob6`Utilities` instead of KnoxProb7`Utilities` for *Mathematica* 6 users. The main differences between the two Utilities versions are: (1) the Histogram command that I had written for the version 6 package is eliminated in the version 7 package because Wolfram Research included its own suitable Histogram command into the kernel in version 7; and (2) the ProbabilityHistogram command had to be rewritten for version 7 due to the elimination of the BarCharts package that it had relied on. Both packages are updates of those in the KnoxProb`Utilities` package that was available for the first edition of the book, adapting to the major changes in the structure of *Mathematica* that happened as it went into version 6.0. Among the most important changes for our purposes was that many statistical commands that had previously been contained in external packages were moved into the kernel, so that these packages no longer needed to be loaded. The electronic book has been subjected to a lot of testing, and it should operate well.

Table of Contents

CHAPTER 1

DISCRETE PROBABILITY

1.1 The Cast of Characters

Why study probability? All around us is uncertainty and we must understand its nature in order to conduct ourselves in the world. We join the shortest waiting line in hopes that it has the best chance of allowing us to be served quickly. We invest our savings in assets that we believe are most likely to yield high returns. We buy the battery that we think is most likely to last the longest. The study of probability also lets you understand the mechanics of games of chance like roulette, poker, and blackjack; what is predictable and what is not, what to bet on (usually nothing) and what not to bet on (usually everything). In all of these cases we do not know for certain what will happen before the phenomenon is observed (we don't know how much time we will wait in line, what our opponent's blackjack hand will be, etc.). But some common-sense assumptions, background knowledge, or a long history of observed data from similar experiments can enable us to quantify likelihoods. Several of these examples have actually given birth to significant subfields of probability or related areas of study, for example, queueing theory (i.e., the study of waiting lines) and mathematical finance, which are introduced in Chapter 6. Probability also provides the theoretical basis for statistics, which is the art of using data obtained randomly to draw conclusions about social and scientific phenomena.

In addition to all this, probability exhibits many of the aesthetically appealing properties that attract people to mathematics in general. It proceeds logically from a few simple axioms to build an abstract structure which is straightforward, yet powerful. The cast of characters of probability is both foreign and tangible, which implies that you must work a little to get to know them, but there is hope in succeeding in the acquaintance.

This text will be somewhat different from others in the field. While the main ideas of probability have not changed much in at least fifty years, the world has, and learning in mathematics has changed with it. Many things are technologically possible today of which mathematics education should take advantage. This book is a living, interactive text which asks you to do activities as you read carefully to help you understand and discover the ideas of probability in new ways, on your own terms. The complete text is available electronically in the form of *Mathematica* notebooks, one for each section. The notebooks are executable, so that you can reenter existing commands, modify them, or write your own supplemental commands to take the study in the direction you wish. You will be building probability yourself in a kind of workshop, with *Mathematica* and your own insight and love of experimentation as your most powerful tools. I caution you against depending too much on the machine, however. Much of the time, traditional pencil, paper, and brainpower are what you need most to learn. What the computer does, however, is open up new problems, or let you get new insights into concepts. So we will be judiciously blending new technology with traditionally successful pedagogy to enable you to learn effectively. Consistent themes will be the idea of taking a sample randomly from a larger universe of objects in order to get information about that universe, and the computer simulation of random phenomena to observe patterns in many replications that speak to the properties of the phenomena.

The cast of characters in probability whose personalities we will explore in depth in the coming chapters include the following six principal players: (1) Event; (2) Sample Space; (3) Probability Measure; (4) Random Variable; (5) Distribution; and (6) Expectation. The rest of this section will be a brief intuitive introduction to these six.

Most of the time we are interested in assessing likelihoods of *events*, that is, things that we might observe as we watch a phenomenon happen. The event that a national chain's bid to acquire a local grocery store is successful, the event that at least 10 patients arrive to an emergency room between 6:00 and 7:00 AM, the event that a hand of five poker cards forms a straight, the event that we hold the winning ticket in the lottery, and the event that two particular candidates among several are selected for jury duty are just a few examples. They share the characteristic that there is some uncertain experiment, sampling process, or phenomenon whose result we do not know "now". But there is a theoretical "later" at which we can observe the phenomenon taking place, and decide whether the event did or didn't occur.

An *outcome* (sometimes called a *simple event*) is an event that cannot be broken down into some combination of other events. A *composite event* is just an event that isn't simple, that is, it has more than one outcome. For instance the event that we are dealt a poker hand that makes up a straight is composite because it consists of many outcomes which completely specify the hand, such as 2 through 6 of hearts, 10 through ace of clubs, etc.

This is where the *sample space* makes its entry: the sample space is the collection of all possible outcomes of the experiment or phenomenon, that is all possible indecomposable results that could happen. For the poker hand example, the sample space would consist of all possible five-card hands. In specifying sample spaces, we must be clear about assumptions; for instance we might assume that the order in which the cards were dealt does not matter, and we cannot receive the same card twice, so that the cards are dealt without replacement.

Activity 1 How can you characterize the sample space for the grocery store example? Make up your own random phenomenon, informally describe the sample space, and give examples of events.

Activity 2 In the emergency room example above, is the event that was described an outcome or a composite event? If that event is composite, write down an example of an outcome relating to that phenomenon.

In random phenomena, we need a way of measuring how likely the events are to occur. This may be done by some theoretical considerations, or by experience with past experiments of the same kind. A *probability measure* assigns a likelihood, that is a number between 0 and 1, to all events. Though we will have to be more subtle later, for now we can intuitively consider the extremes of 0 and 1 as representing impossibility and complete certainty, respectively.

Much of the time in probability problems, probability measures are constructed so that the probability of an event is the "size" of the event as a proportion of the "size" of the whole sample space. "Size" may mean different things depending on the context, but the two most frequent usages are cardinality, that is number of elements of a set, and length (or area in two dimensions). For example, if you roll a fair die once, since there are six possible faces that could land on top, the sample space is {1, 2, 3, 4, 5, 6}. Since three of the faces are odd, it seems reasonable that the probability that an odd number is rolled is 3/6. Size means length in the following scenario. Suppose that a real number is to be randomly selected in the interval [0, 4]. This interval is the sample space for the random phenomenon. Then the probability that the random real number will be in the interval [1, 3] is $2/4 = 1/2$, which is the length of the interval [1, 3] divided by the length of the whole sample space interval.

The first two chapters will concentrate on discrete probability, in which counting the number of elements in sets is very important. We will leave much of the detail for later sections, except to remind you of the intuitively obvious *multiplication principle* for counting. If outcomes of a random experiment consist of two stages, then the total number of outcomes of the experiment is the number of ways the first stage can occur times the number of ways that the second can occur. This generalizes to multiple stage experiments. For instance, if a man has 3 suits, 6 shirts, and 4 ties, and he doesn't care about color coordination and the like, then he has $3 \times 6 \times 4 = 72$ possible outfits.

It is possible to experimentally estimate a probability by repeating the experiment over and over again. The probability of the event can be estimated by the number of times that it occurs among the replications divided by the number of replications. Try this exercise. In the KnoxProb7`Utilities` package that you should have installed is a *Mathematica* command called

DrawIntegerSample[a, b, n]

which outputs a sequence of n numbers drawn at random, without reusing numbers previously in the sequence, from the positive integers in the range a, ... , b. (You should be sure that the package has been loaded into the ExtraPackages subdirectory within the *Mathematica* directory you are using on your computer.) First execute the command to load the package, then try repeating the experiment of drawing samples of two numbers from $\{1, ... , 5\}$ a large number of times. I have shown one such sample below. Keep track of how frequently the number 1 appears in your samples. Empirically, what is a good estimate of the probability that 1 is in the random sample of size two? Talk with a group of your classmates, and see if you can set up a good theoretical foundation that supports your conclusion. (Try to specify the sample space of the experiment.) We will study this type of experiment more carefully later in this chapter. Incidentally, recall that *Mathematica* uses braces { } to delineate sequences as well as unordered sets. We hope that the context will make it clear which is meant; here the integer random sample is assumed to be in sequence, so that, for example, $\{1, 3\}$ is different from $\{3, 1\}$.

```
Needs["KnoxProb7`Utilities`"]
```

```
DrawIntegerSample[1, 5, 2]
```

$\{1, 3\}$

We are halfway through our introduction of the cast of characters. Often the outcomes of a random phenomenon are not just numbers. They may be sequences of numbers as in our sampling experiment, or even non-numerical data such as the color of an M&M that we randomly select from a bag. Yet we may be interested in encoding the simple events by single numbers, or extracting some single numerical characteristic from more complete information. For example, if we are dealt a five card hand in poker, an outcome is a full specification of all five cards, whereas we could be interested only in the number of aces in the hand, which is a numerical valued function of the outcome. Or, the sample space could consist of subjects in a medical experiment, and we may be interested in the blood sugar level of a randomly selected subject. Such a function for transforming outcomes to numerical values is called a *random variable*, the fourth member of our cast of characters. Since the random variable depends on the outcome, and we do not know in advance what will happen, the value of the random variable is itself uncertain until after the experiment has been observed.

> **Activity 3** Use DrawIntegerSample again as below to draw samples of size five from a population of size 50. Suppose we are interested in the random variable which gives the minimum among the members of the sample. What values of this random variable do you observe in several replications?

```
outcome = DrawIntegerSample[1, 50, 5]
X = Min[outcome]
```

The cell below contains a command I have called SimulateMinima for observing a number *m* of values of the minimum random variable from Activity 3. Read the code carefully to understand how it works. An integer sample is drawn from the set {1,2,...,popsize}, and then its minimum is taken. The Table command assembles a list of *m* of these minima. To get an idea of the pattern of those sample minima, we plot a graph called a *histogram* which tells the proportion of times among the *m* replications of the experiment that the sample minimum fell into each of a given number of categories. The Histogram command is in the kernel in *Mathematica* 7. Its exact syntax is

Histogram[listofdata, binspecifications, barheightspecifications]

where the latter two arguments are optional. (See the help documentation, the note in the preface, and the appendix on *Mathematica* commands for version 6 of *Mathematica* and the KnoxProb6`Utilities` package.) The binspecifications argument can be set to be a number for a desired number of categories, or it can be set to a list {*d w*} for a desired category width. The third argument can be set to "Count," "Probability," or "ProbabilityDensity," respectively, to force the bar heights to be freqencies, relative frequencies, or relative frequencies divided by category widths. One output cell is shown in Figure 1 using 100 replications, six histogram rectangles, and using samples of size 5 from {1, ... , 50}. You should reexecute several times to see whether your replications give similar histograms. Should they be identical to this one? Why or why not? (If you want to suppress the list of minima and get only the histogram, put a semicolon between the SimulateMinima and Histogram commands.)

```
SimulateMinima[m_, popsize_, sampsize_] :=
    Table[Min[
    DrawIntegerSample[1, popsize, sampsize]], {i, 1, m}]
```

```
mins = SimulateMinima[100, 50, 5]
Histogram[mins, 6, ChartStyle → Gray]
```

{7, 6, 4, 13, 6, 8, 2, 10, 6, 9, 2, 4, 2, 5, 1, 2, 11, 5,
3, 2, 22, 2, 8, 17, 6, 6, 13, 4, 2, 6, 15, 32, 3, 7, 15,
21, 5, 17, 2, 4, 8, 12, 7, 13, 1, 6, 28, 26, 14, 12, 3,
7, 6, 1, 12, 11, 6, 7, 17, 6, 10, 3, 3, 1, 1, 16, 7, 5,
4, 18, 6, 14, 23, 15, 18, 22, 4, 9, 9, 8, 1, 15, 4, 16,
4, 9, 16, 8, 23, 10, 1, 7, 20, 6, 10, 10, 2, 7, 18, 2}

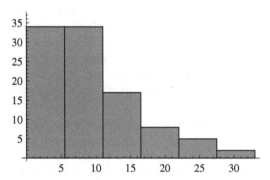

Figure 1.1 - Sample histogram of minimum of five values from {1, 2, ... , 50}

The histogram for the sample minimum that we just saw, together with your experiments, hints at a very deep idea, which is the fifth member of our cast. A random variable observed many times will give a list of values that follows a characteristic pattern, with some random variation. The relative frequencies of occurrence (that is the number of occurrences divided by the number of replications) of each possible value of the random variable, which are the histogram rectangle heights, will stabilize around some theoretical probability. Much as a probability measure can be put on a sample space of simple events, a *probability distribution* can be put on the collection of possible values of a random variable. To cite a very simple example, if 1/3 of the applicants for a job are men and 2/3 are women, if one applicant is selected randomly, and if we define a random variable

$$X = \begin{cases} 1 & \text{if applicant is a man} \\ 0 & \text{if applicant is a woman} \end{cases}$$

then the two possible values of X are 1 and 0, and the probability distribution for this X gives probability 1/3 to value 1 and 2/3 to value 0.

The last member of our cast is *expectation* or *expected value* of a random variable. Its purpose is to give a single number which is an average value that the random variable can take on. But if these possible values of the random variable are not equally likely, it doesn't make sense to compute the average as a simple arithmetical average. For example, suppose the number of 911 emergency calls during an hour is a random variable X whose values and probability distribution are as in the table below.

value of X	2	4	6	8
probability	1/10	2/10	3/10	4/10

What could we mean by the average number of calls per hour? The simple arithmetical average of the possible values of X is (2+4+6+8)/4 = 5, but 6 and 8 are more likely than 2 and 4, so they deserve more influence. In fact, the logical thing to do is to define the expected value of X, denoted by $E[X]$, to be the weighted average of the possible values, using the probabilities as weights:

$$E[X] = \frac{1}{10} \times 2 + \frac{2}{10} \times 4 + \frac{3}{10} \times 6 + \frac{4}{10} \times 8 = \frac{60}{10} = 6$$

We will be looking at all of our characters again in a more careful way later, but I hope that the seeds of intuitive understanding have been planted. You should reread this introductory section from time to time as you progress to see how the intuition fits with the formality.

Exercises 1.1

1. Most state lotteries have a version of a game called Pick 3 in which you win an amount, say $500, if you correctly choose three digits in order. Suppose that the price of a Pick 3 ticket is $1. There are two possible sample spaces that might be used to model the random experiment of observing the outcome of the Pick 3 experiment, according to whether the information of the winning number is recorded, or just the amount you won. Describe these two sample spaces clearly, give examples of outcomes for both, and describe a reasonable probability measure for both.

2. A grand jury is being selected, and there are two positions left to fill, with candidates A, B, C, and D remaining who are eligible to fill them. The prosecutor decides to select a pair of people by making slips of paper with each possible pair written on a slip, and then drawing one slip randomly from a hat. List out all outcomes in the sample space of this random phenomenon, describe a probability measure consistent with the stated assumptions, and find the probability that A will serve on the grand jury.

3. In an effort to predict performance of entering college students in a beginning mathematics course on the basis of their mathematics ACT score, the following data were obtained. The entries in the table give the numbers of students among a group of 179 who had the indicated ACT score and the indicated grade.

		Grade				
		A	B	C	D	F
	23	2	5	16	31	3
ACT	24	6	8	12	15	2
	25	7	10	15	6	1
	26	8	15	13	4	0

(a) A student is selected randomly from the whole group. What is the probability that this student has a 25 ACT and received a B? What is the probability that this student received a C? What is the probability that this student has an ACT score of 23?

(b) A student is selected randomly from the group of students who scored 24 on the ACT. What is the probability that this student received an A?

(c) A student is selected randomly from the group of students who received B's. What is the probability that the student scored 23 on the ACT? Is it the same as your answer to the last question in part (a)?

4. Find the sample space for the following random experiment. Four people labeled A, B, C, and D are to be separated randomly into a group of 2 people and two groups of one person each. Assume that the order in which the groups are listed does not matter, only the contents of the groups.

5. Consider the experiment of dealing two cards from a standard 52 card deck (as in blackjack), one after another without replacing the first card. Describe the form of all outcomes, and give two examples of outcomes. What is the probability of the event "King on 1st card?" What is the probability of the event "Queen on 2nd card?"

6. In the Pick 3 lottery example of Exercise 1, consider the random variable X that gives the net amount you win if you buy one ticket (subtracting the ticket price). What is the probability distribution of X? What is the expected value of X?

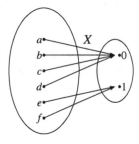

Exercise 7

7. An abstract sample space has outcomes a, b, c, d, e, and f as shown on the left in the figure. Each outcome has probability 1/6. A random variable X maps the sample space to the set $\{0,1\}$ as in the diagram. What is the probability distribution of X? What is the expected value of X?

8. Two fair coins are flipped at random and in succession. Write the sample space explicitly, and describe a reasonable probability measure on it. If X is the random variable that counts the number of heads, find the probability distribution of X and compute the expected value of X.

9. Using the data in Exercise 3, give an empirical estimate of the probability distribution of the random variable X that gives the ACT score of a randomly selected student. What is the expected value of X, and what is its real meaning in this context?

10. (*Mathematica*) Use the DrawIntegerSample command introduced in the section to build a command that draws a desired number of samples, each of a desired size, from the set $\{1, 2, ..., popsize\}$. For each sample, the command appends the maximum number in the sample to a list. This list can be used to create a histogram of the distribution of maxima. Repeat the experiment of taking 50 random samples of size 4 from $\{1, 2, ..., 7\}$ several times. Describe in words the most salient features of the histograms of the maximum. Use one of your replications to give an empirical estimate of the probability distribution of the maximum.

11. (*Mathematica*) The command SimulateMysteryX[m] illustrated below will simulate a desired number m of values of a mystery random variable X. First load the KnoxProb7`Utilities` package, then run the command several times using specific numerical values for m. Try to estimate the probability distribution of X and its expected value.

```
SimulateMysteryX[10]
```

12. (*Mathematica*) Write a command in *Mathematica* which can take as input a list of pairs $\{p_i, x_i\}$ in which a random variable X takes on the value x_i with probability p_i and return the expectation of X. Test your command on the random variable whose probability distribution is in the table below.

value of X	0	1	2	3
probability	1/4	1/2	1/8	1/8

13. Two flights are selected randomly and in sequence from a list of departing flights covering a single day at an airport. There were 100 flights in the list, among which 80 departed on time, and 20 did not. Write a description of the sample space of the experiment of selecting the two flights, and use the multiplication principle to count how many elements the sample space has. Now let X be the number of flights in the sample of two that were late. Find the probability distribution of X.

1.2 Properties of Probability

We will begin this section with examples of the concrete construction of probability measures for sample spaces with finitely many outcomes. With our feet on good intuitive ground, we will then be ready to set up a formal axiom system which permits rigorous proof of some important properties of probability. This is the way that most mathematical research actually proceeds: from concrete to abstract rather than the other way around.

Example 1 If, as we said in Section 1.1, a probability measure gives likelihood to events, then at its most basic level it gives likelihood to simple events, that is, outcomes. As an example, I recently saw a breakdown of advanced placement scores for Illinois 12th graders on the subject test in English Language and Composition. Scores range from 1 to 5, and the numbers of test takers in these categories for the particular year I looked at were: 1: 74, 2: 293, 3: 480, 4: 254, 5: 130, for a total of 1231 students. Consider first the random experiment of drawing a student at random from this group and observing the test score.

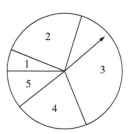

Figure 1.2 - Sampling from a discrete population

There are 5 possible outcomes, namely, the scores 1-5. We can think of the experiment of randomly sampling a student score as spinning the arrow on the pie chart of Figure 2, which has been constructed so that the angles of the circular wedges are proportional to the frequencies in the categories that were listed in the last paragraph.

This example illustrates several important features of the subject of probability. First, there is an underlying physics that drives the motion and final resting position of the arrow. But differences in the starting point, the angular momentum imparted to the arrow, and even slight wind currents and unevenness of the surface of the spinner give rise to a non-deterministic or random phenomenon. We cannot predict with certainty where the arrow will land. It could be philosophically argued that if we knew all of these randomizing factors perfectly, then the laws of physics would perfectly predict the landing point of the arrow, so that the phenomenon is not really random. Perhaps this is true in most, if not all, phenomena that we call random. Nevertheless, it is difficult to know these factors, and the ability to measure likelihoods of events that we model as random is useful to have.

Second, this example shows the two main ways of arriving at an assignment of probabilities. We might make simplifying assumptions which lead to a logical assignment of probability. Or we might repeat the experiment many times and observe how frequently events occur. We were fortunate enough here to have the background data from which our spinner was constructed, which would indicate by simple counting that if the spinner was truly spun "randomly" we should assign probabilities

$$P[1] = \frac{74}{1231}, P[2] = \frac{293}{1231}, P[3] = \frac{480}{1231}, P[4] = \frac{254}{1231}, P[5] = \frac{130}{1231} \tag{1}$$

to the five outcomes. But step back for a moment and consider what we would have to do with our spinner if we weren't blessed with the numbers. We would have no recourse but to perform the experiment repeatedly. If, say, 1000 spins produced 70 1's, then we would estimate P[1] empirically by $70/1000 = .07$, and similarly for the other outcomes.

Third, we can attempt to generalize conclusions. This data set itself can be thought of as a sort of repeated spinning experiment, in which each student represents a selection from a larger universe of students who could take the test. The probabilities in formula (1), which would be exact probabilities for the experiment of selecting one student from this particular batch, become estimates of probabilities for a different experiment; that of selecting 1231 students from the universe of all students, past and present, who could take the test. Whether our particular data set is a "random" sample representing that universe without bias is a very serious statistical question, best left for a statistics course with a good unit on samples and surveys.

Using the probabilities in (1), how should probability be assigned to composite events like "1 or 3?" The logical thing to do is to pool together the probabilities of all outcomes that are contained in the composite event. Then we would have

$$P[1 \text{ or } 3] = P[1] + P[3] = \frac{74}{1231} + \frac{480}{1231} = \frac{554}{1231} \tag{2}$$

So the probability of an event is the total of all probabilities of the outcomes that make up the event.

Two implications follow easily from the last observation: the empty event $\emptyset$ which contains no outcomes has probability zero. Also, the event that the spinner lands somewhere is certain, that is, it has probability 1 (100%). Another way to look at it is that the sample space, denoted by Ω say, is the composite event consisting of all outcomes. In this case we have:

$$
\begin{aligned}
P[\Omega] &= P[1 \text{ or } 2 \text{ or } 3 \text{ or } 4 \text{ or } 5] \\
&= P[1] + P[2] + \cdots + P[5] \\
&= \frac{74}{1231} + \frac{293}{1231} + \cdots + \frac{130}{1231} \\
&= 1
\end{aligned}
\tag{3}
$$

In summary, so far we have learned that outcomes can be given probabilities between 0 and 1 using theoretical assumptions or empirical methods, and probabilities for composite events are the totals of the outcome probabilities for outcomes in the event. Also $P[\emptyset] = 0$ and $P[\Omega] = 1$. Try the following activity, which suggests another property. You will learn more about that property shortly.

Activity 1 For the spinner experiment we looked at P[1 or 3]. Consider now the two events "at least 3" and "between 1 and 3 inclusive." What is the probability of the event {"at least 3" or "between 1 and 3 inclusive"}? How does this probability relate to the individual event probabilities? Can you deduce a general principle regarding the probability of one event or another occurring?

Sometimes we are interested in the set theoretic complement of an event, i.e., the event that the original event does not occur. Does the probability of the complementary event relate to the probability of the event? For the spinner, look at the events "at least 3" and "less than 3," which are complements. We have, using formula (1):

$$P[\text{at least 3}] = P[3 \text{ or } 4 \text{ or } 5] = \frac{480}{1231} + \frac{254}{1231} + \frac{130}{1231} = \frac{864}{1231}$$

$$P[\text{less than 3}] = P[1 \text{ or } 2] = \frac{74}{1231} + \frac{293}{1231} = \frac{367}{1231}$$

(4)

Notice that the two probabilities add to one, that is for these two events E and E^c:

$$P[E] + P[E^c] = 1 \tag{5}$$

This is not a coincidence. Together E and E^c contain all outcomes, and they have no outcomes in common. The total of their probabilities must therefore equal 1, which is the probability of the sample space. ∎

Example 2. Let's look at another random phenomenon. Recall from Section 1.1 the experiment of selecting a sequence of two numbers at random from the set {1, 2, 3, 4, 5} with no repetition of numbers allowed. Such an ordered selection without replacement is known as a *permutation*, in this case a permutation of 2 objects from 5. *Mathematica* can display all possible permutations as follows. The command

KPermutations[list, k]

in the KnoxProb7`Utilities` package returns a list of all ordered sequences of length k from the list given in its first argument, assuming that a list member, once selected, cannot be selected again.

```
Needs["KnoxProb7`Utilities`"]
```

```
KPermutations[{1, 2, 3, 4, 5}, 2]
```

```
{{1, 2}, {2, 1}, {1, 3}, {3, 1}, {1, 4}, {4, 1},
 {1, 5}, {5, 1}, {2, 3}, {3, 2}, {2, 4}, {4, 2}, {2, 5},
 {5, 2}, {3, 4}, {4, 3}, {3, 5}, {5, 3}, {4, 5}, {5, 4}}
```

```
Length[KPermutations[{1, 2, 3, 4, 5}, 2]]
```

20

As the Length output reports, there are 20 such permutations. These 20 are the outcomes that form the sample space of the experiment. The randomness of the drawing makes it quite reasonable to take the theoretical approach to assigning probabilities. Each outcome should have the same likelihood. Since the total sample space has probability 1, the common likelihood of the simple events, say x, satisfies

$$x + x + \cdots + x \,(20 \text{ times}) = 1 \implies 20\,x = 1 \implies x = 1/20 \tag{6}$$

In other words, our probability measure on the sample space is defined on outcomes as:

$$P[\{1, 2\}] = P[\{2, 1\}] = P[\{1, 3\}] = \cdots = P[\{5, 4\}] = 1/20 \tag{7}$$

and for composite events, we can again total the probabilities of outcomes contained in them.

Activity 2 In the example of selecting two positive integers from the first five, find the probability that 1 is in the sample. Explain why your result is intuitively reasonable.

For instance,

P[either 1 or 2 is in the sample]
  = P[{1, 2}, {2, 1}, {1, 3}, {3, 1}, {1, 4}, {4, 1}, {1, 5}, {5, 1},
      {2, 3}, {3, 2}, {2, 4}, {4, 2}, {2, 5}, {5, 2}]
  = 14/20

and

P[2 is not in sample]
  = P[{1, 3}, {3, 1}, {1, 4}, {4, 1}, {1, 5}, {5, 1}, {3, 4}, {4, 3}, {3, 5},
      {5, 3}, {4, 5}, {5, 4}]
  = 12/20

By counting outcomes you can check that P[2 is in sample] = 8/20, which is complementary to P[2 is not in sample] = 12/20 from above. Also notice that

$$P[1 \text{ in sample}] \ = \ \frac{8}{20} = P[2 \text{ in sample}]$$

however,

$$P[1 \text{ in sample or 2 in sample}] \ = \ \frac{14}{20} \neq \frac{8}{20} + \frac{8}{20}$$

We see that the probability of the event "either 1 is in the sample or 2 is in the sample" is not simply the sum of the individual probability that 1 is in the sample plus the probability that 2 is in it. The reason is that in adding $8/20 + 8/20$ we let the outcomes {1, 2} and {2, 1} contribute twice to the sum. This overstates the true probability, but we now realize that it is easy to correct for the double count by subtracting $2/20$ which is the amount of probability contributed by the doubly counted outcomes. (See Theorem 2 below.) ∎

Activity 3 Use *Mathematica* to display the ordered samples of size 3 without replacement from {1, 2, 3, 4, 5, 6}. What is the probability that 1 is in the sample? What is the probability that neither 5 nor 6 is in the sample?

Armed with the experience of our examples, let us now develop a mathematical model for probability, events, and sample spaces that proceeds from a minimal set of axioms to derive several of the most important basic properties of probability. A few other properties are given in the exercises.

Definition 1.   A *sample space* Ω of a random experiment is a set whose elements ω are called *outcomes*.

Definition 2 below is provisional – we will have to be more careful about what an event is when we come to continuous probability.

Definition 2.   An *event A* is a subset of the sample space Ω hence an event is a set of outcomes. The collection of all events is denoted by $\mathcal{H}$

Suppose for example we consider another sampling experiment in which three letters are to be selected at random, without replacement and without regard to order from the set of letters {a, b, c, d}. The Combinatorica` package in *Mathematica* (which is loaded automatically when you load KnoxProb7`Utilities`) has the commands

$$KSubsets[list, k] \qquad and \qquad Subsets[universe]$$

which, respectively, return all subsets of size k taken from the given list of objects, and all subsets of the given universal set. We can use these to find the sample space of the experiment, and the collection of events $\mathcal{H}$

```
Ω = KSubsets[{a, b, c, d}, 3]
```

```
{{a, b, c}, {a, b, d}, {a, c, d}, {b, c, d}}
```

```
H = Subsets[Ω]
```

```
{{}, {{a, b, c}}, {{a, b, d}}, {{a, c, d}},
 {{b, c, d}}, {{a, b, c}, {a, b, d}}, {{a, b, c}, {a, c, d}},
 {{a, b, c}, {b, c, d}}, {{a, b, d}, {a, c, d}},
 {{a, b, d}, {b, c, d}}, {{a, c, d}, {b, c, d}},
 {{a, b, c}, {a, b, d}, {a, c, d}}, {{a, b, c}, {a, b, d}, {b, c, d}},
 {{a, b, c}, {a, c, d}, {b, c, d}}, {{a, b, d}, {a, c, d}, {b, c, d}},
 {{a, b, c}, {a, b, d}, {a, c, d}, {b, c, d}}}
```

Notice that the empty set {} is the first subset of Ω that is displayed; next are the simple events {{a, b, c}}, {{a, b, d}} etc.; next are the two element composite events; and so forth.

In all subsequent work we will continue a practice that we have done so far, which is to blur the distinction between *outcomes* ω, which are members of the sample space, and *simple events* $\{\omega\}$ which are sets consisting of a single outcome. This will enable us to speak of the "probability of an outcome" when in fact probability is a function on events, as seen in the next definition.

Definition 3. A *probability measure* P is a function taking the family of events $\mathcal{H}$ to the real numbers such that

(i) $P[\Omega] = 1$;

(ii) For all $A \in \mathcal{H}$ $P[A] \geq 0$;

(iii) If A_1, A_2, $\cdots$ is a sequence of pairwise disjoint events then
$$P[A_1 \cup A_2 \cup \cdots] = \sum_i P[A_i]$$

Properties (i), (ii), and (iii) are called the *axioms of probability*. Our mathematical model so far corresponds with our intuition: a random experiment has a set of possible outcomes called the sample space, an event is a set of these outcomes, and probability attaches a real number to each event. Axiom (i) requires the entire sample space to have probability 1, axiom (ii) says that all probabilities are non-negative numbers, and axiom (iii) says that when combining *disjoint* events, the probability of the union is the sum of the individual event probabilities. For finite sample spaces this axiom permits us to just define probabilities on outcomes, and then give each event a probability equal to the sum of its (disjoint) outcome probabilities. We will encounter various ways of constructing probability measures along the way, which we must verify to satisfy the axioms. Once having done so, other properties of probability listed in the theorems below are ours for free. They follow from the axioms alone, and not from any particular construction, which of course is the huge advantage of basing a mathematical model on the smallest possible collection of axioms.

Theorem 1. If A is an event with complement A^c, then

$$P[A] + P[A^c] = 1 \tag{8}$$

Therefore, $P[A] = 1 - P[A^c]$ and $P[A^c] = 1 - P[A]$.

Proof. The events A and A^c are pairwise disjoint, by definition of complementation. By the third axiom,

$$P[A \cup A^c] = P[A] + P[A^c]$$

Also, $A \cup A^c = \Omega$ By the first axiom, $P[\Omega] = 1$, therefore

$$1 = P[\Omega] = P[A \cup A^c] = P[A] + P[A^c] \quad \blacksquare$$

Theorem 1 was anticipated earlier. It is generally used in cases where the complement of an event of interest is easier to handle than the original event. If we have the probability of one of the two, then we have the other by subtracting from 1.

The next result is based on the idea that the total probability of a union can be found by adding the individual probabilities and correcting if necessary by subtracting the probability in the double-counted intersection region.

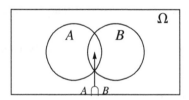

Figure 1.3 - The probability of a union is the total of the event probabilities minus the intersection probability

Theorem 2. If A and B are arbitrary events, then

$$P[A \cup B] = P[A] + P[B] - P[A \cap B] \tag{9}$$

In particular if A and B are disjoint, then

$$P[A \cup B] = P[A] + P[B]$$

Proof. (See Figure 3.) Since A can be expressed as the disjoint union $A = (A \cap B) \cup (A \cap B^c)$, and B can be expressed as the disjoint union $B = (A \cap B) \cup (B \cap A^c)$ we have, by axiom (iii),

$$
\begin{aligned}
P[A] + P[B] &= P[A \cap B] + P[A \cap B^c] + P[A \cap B] + P[B \cap A^c] \\
&= P[A \cap B^c] + P[B \cap A^c] + 2\,P[A \cap B]
\end{aligned}
$$

Therefore,

$$P[A] + P[B] - P[A \cap B] = P[A \cap B^c] + P[B \cap A^c] + P[A \cap B] \qquad (10)$$

But it is easy to see from Figure 3 that the events $(A \cap B^c)$, $(B \cap A^c)$, and $(A \cap B)$ are pairwise disjoint and their union is $A \cup B$. Thus, by axiom (iii) of probability again,

$$P[A \cap B^c] + P[B \cap A^c] + P[A \cap B] = P[A \cup B] \qquad (11)$$

Combining (10) and (11) proves the first assertion. The second assertion follows directly as a specialization of axiom (iii) to two events. Alternatively, if A and B are disjoint, then $A \cap B = \emptyset$ and by Exercise 11, $P[\emptyset] = 0$. ■

Activity 4   Try to devise an analogous thoerem to Theorem 2 for a three set union. (Hint: if you subtract all paired intersection probabilities, are you doing too much?) See Exercise 14 for a statement of the general result.

You may be curious that the axioms only require events to have non-negative probability. It should also be clear that events must have probability less than or equal to 1, because probability 1 represents certainty. In the exercises you are asked to give a rigorous proof of this fact, stated below as Theorem 3.

Theorem 3. For any event A, $P[A] \leq 1$. ■

Another intuitively obvious property is that if an event B has all of the outcomes in it that another event A has, and perhaps more, then B should have higher probability than A. This result is given as Theorem 4, and the proof is Exercise 12.

Theorem 4. If A and B are events such that $A \subset B$, then $P[A] \leq P[B]$. ■

Theorems 3 and 4 help to reassure us that our theoretical model of probability corresponds to our intuition, and they will be useful in certain proofs. But Theorems 1 and 2 on complements and unions are more useful in computations. We would like to finish this section by adding a third very useful computational result, usually called the *Law of Total Probability*. Basically it states that we can "divide and conquer," that is break up the computation of a complicated probability $P[A]$ into a sum of easier probabilities of the form $P[A \cap B]$. The imposition of condition B in addition to A presumably simplifies the computation by limiting or structuring the outcomes in the intersection. (See Figure 4.) All of our computational results will be used regularly in the rest of the text, and the exercise set for this section lets you begin to practice with them.

Figure 1.4 - The Law of Total Probability

Theorem 5. Let A be an arbitrary event, and assume that B_1, B_2, B_3, ... , B_n is a partition of the sample space Ω, that is, the sets B_i are pairwise disjoint and $\bigcup_{i=1}^{n} B_i = \Omega$. Then,

$$P[A] = \sum_{i=1}^{n} P[A \cap B_i] \tag{12}$$

Proof. We will use induction on the number of events B_i beginning with the base case $n = 2$. In the base case, it is clear that $B_2 = B_1^c$, and therefore $A = (A \cap B_1) \cup (A \cap B_1^c)$ and the two sets in this union are disjoint. By Theorem 2,

$$P[A] = P\left[A \cap B_1\right] + P\left[A \cap B_1^c\right] = P\left[A \cap B_1\right] + P\left[A \cap B_2\right]$$

The anchor step of the induction is complete.

Now assume that for all partitions consisting of n events formula (12) is true. Let B_1, B_2, B_3, ... , B_{n+1} be a partition of $n + 1$ events; we must show (12) with the upper index of summation replaced by $n + 1$. Define the events C_i by:

$$C_1 = B_1, \ C_2 = B_2, \ \cdots, \ C_{n-1} = B_{n-1}, \ C_n = B_n \cup B_{n+1}$$

Then the C_i's are also a partition of Ω (see Activity 5 below). By the inductive hypothesis

$$
\begin{aligned}
P[A] &= \sum_{i=1}^{n} P[A \cap C_i] \\
&= \sum_{i=1}^{n-1} P[A \cap B_i] + P[A \cap (B_n \cup B_{n+1})] \\
&= \sum_{i=1}^{n-1} P[A \cap B_i] + P[(A \cap B_n) \cup (A \cap B_{n+1})] \\
&= \sum_{i=1}^{n-1} P[A \cap B_i] + P[A \cap B_n] + P[A \cap B_{n+1}] \\
&= \sum_{i=1}^{n+1} P[A \cap B_i]
\end{aligned}
\tag{13}
$$

This finishes the inductive step of the proof. ∎

Activity 5 Why are the C_i's in the proof of the Law of Total Probability a partition of Ω ? Why is the fourth line of (13) true?

Exercises 1.2

1. Consider the experiment of rolling one red die and two white dice randomly. Observe the number of the face that lands up on the red die and the sum of the two face up numbers on the white dice.

(a) Describe the sample space, giving the format of a typical outcome and two specific examples of outcomes.

(b) Define a probability measure on this sample space. Argue that it satisfies the three axioms of probability.

red die

	1	2	3	4	5	6
2						3 b
3	2 b		1 b			
4				1 b		
5			1 b		2 b	
6				1 b		
7		1 b			HR	
8	1 b			1 b		
9		2 b				1 b
10				1 b		
11			1 b		1 b	
12	2 b					

white sum (labels on the left for rows 2–12)

Exercise 2

2. This is a continuation of Exercise 1. A simulated baseball game is played in such a way that each batter has a card like the one shown in the figure. One red die and two white dice are rolled. The outcome of the batter's plate appearance is decided by finding the column corresponding to the red die and the row corresponding to the sum of the white dice. The entry in that row and that column could be a base hit (1b), a double (2b), a triple (3b), a home run (HR) or a blank, which would mean that the batter is out. For the batter shown, what is the probability that he will get a hit of any kind? If you were designing such a card to simulate a hitter who historically hits about a home run every 20 plate appearances, where would you place HR symbols on the card?

3. On a driver's education test there are three questions that are considered most important and most difficult. The questions are multiple choice, with four alternative answers labeled a, b, c, and d on each question. Exactly one answer is correct for each question. A test taker guesses randomly on these three questions.

(a) Describe the sample space, giving the format of a typical outcome and two specific examples of outcomes.

(b) Define a probability measure on this sample space. Argue that it satisfies the three axioms of probability.

(c) What is the probability that the random guesser gets at least two questions right?

4. (*Mathematica*) Use *Mathematica* to display the sample space of all possible random samples without order and without replacement of three names from the list of names {Al, Bubba, Celine, Delbert, Erma}. How many events are in the collection of all events $\mathcal{H}$? Describe a probability measure for this experiment, and find the probability that either Bubba or Erma is in the sample.

5. Explain why each of the following attempts P_1 and P_2 is not a good definition of a probability measure on the space of outcomes $\Omega = \{a, b, c, d, e, f, g\}$.

	P_1	P_2
a	.12	.16
b	.05	.20
c	.21	.20
d	.01	.12
e	.04	.08
f	.28	.15
g	.13	.12

6. (*Mathematica*) Use *Mathematica* to display the sample space of all possible random samples in succession and without replacement of two numbers from the set {1, 2, ..., 10}. Describe a probability measure, and find the probability that 10 is not in the sample.

7. A 1999 Internal Revenue Service study showed that 44.9 million U.S. taxpayers reported earnings of less than $20,000, 20.4 million earned between $20,000 and $30,000, 16.2 million earned between $30,000 and $40,000, 41.9 million earned between $40,000 and $100,000, and 10.5 million earned more than $100,000.

(a) Write the outcomes for the experiment of sampling one taxpayer at random, and define a probability measure. In such a random sample of one taxpayer, what is the probability that the reported earnings is at least $30,000?

(b) Write the outcomes for the experiment of sampling two taxpayers at random and with replacement, that is, once a taxpayer has been selected that person goes back into the pool and could be selected again. Define a probability measure. (Hint: for a succession of two events which do not depend on one another, it makes sense to compute the probability that they both occur as the product of their probabilities. You can convince yourself of this by considering a very simple experiment like the flip of two coins.) Find the probability that at least one of the two earned more than $100,000.

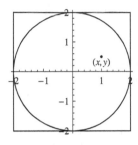

Exercise 8

8. Suppose that a point with random x and y coordinates in $[-2, 2]$ is selected. By this we mean that the probability that the random point lies in a region R inside the square is the proportion of the square's area that R has.

(a) Argue that this means of defining a probability measure on the square $[-2, 2] \times [-2, 2]$ satisfies the axioms of probability.

(b) Find the probability that the random point will be in the third quadrant.

(c) Find the probability that the random point is not in the circle of radius 2 shown in the picture.

(d) Find the probability that the random point is either in the third quadrant or not in the circle of radius 2.

9. If A and B are events such that $P[A \cap B] = .24$, $P[A] = .61$, and $P[B] = .37$, find $P[(A \cup B)^c]$.

10. Prove Theorem 3.

11. Prove that $P[\emptyset] = 0$.

12. Prove Theorem 4.

13. List out all events for the sample space $\Omega = \{\omega_1, \omega_2, \omega_3\}$. Find the probability of each simple event if $P[\{\omega_1, \omega_2\}] = .6$, $P[\{\omega_2, \omega_3\}] = .8$, and $P[\{\omega_1, \omega_3\}] = .6$.

14. Prove that for arbitrary events A, B, and C:

$$P[A \cup B \cup C] = P[A] + P[B] + P[C]$$
$$-(P[A \cap B] + P[A \cap C] + P[B \cap C]) + P[A \cap B \cap C]$$

15. The general result for the probability of a union, which has Theorem 2 and Exercise 14 as special cases, is called the *law of inclusion and exclusion*. Let A_1, A_2, A_3, ..., A_n be a collection of n events. To simplify notation, let A_{ij} stand for a typical intersection of two of them $A_i \cap A_j$, let A_{ijk} stand for a three-fold intersection $A_i \cap A_j \cap A_k$, etc. The law states that

$$P[A_1 \cup A_2 \cup A_3 \cup ... \cup A_n]$$
$$= \sum_i P[A_i] - \sum_{i<j} \sum P[A_{ij}] + \sum_{i<j<k} \sum \sum P[A_{ijk}] - \cdots$$
$$+ (-1)^{n+1} P[A_1 \cap A_2 \cap A_3 \cap ... \cap A_n]$$

In other words, one alternately adds and subtracts all k-fold intersection probabilities as k goes from 1 to n. Use this result in the following classical problem. Four men leave their hats at a hat check station, then on departing take a hat randomly. Find the probability that at least one man takes his own hat.

16. Recall Exercise 5 of Section 1.1 (the blackjack deal). Use the Law of Total Probability, Theorem 5, to find P[queen on 2nd] in a well-justified way.

17. There are two groups of experimental subjects, one with 2 men and 3 women and the other with 3 men and 2 women. A person is selected at random from the first group and placed into the second group. Then a person is randomly selected from the second group. Use the Law of Total Probability to find the probability that the person selected from the second group is a woman.

In this book we will include as appropriate problems taken from the list of sample questions for the Society of Actuaries/Casualty Actuarial Society qualifying exam P (Probability). These problems have been obtained with permission from the societies at the website www.soa.org/files/pdf/P-09-05ques.pdf.

Sample Problems from Actuarial Exam P

18. The probability that a visit to a primary care physician's (PCP) office results in neither lab work nor referral to a specialist is 35%. Of those coming to a PCP's office, 30% are referred to specialists and 40% require lab work. Determine the probability that a visit to a PCP's office results in both lab work and referral to a specialist.

19. An urn contains 10 balls: 4 red and 6 blue. A second urn contains 16 red balls and an unknown number of blue balls. A single ball is drawn from each urn. The probability that both balls are the same color is 0.44. Calculate the number of blue balls in the second urn.

20. An auto insurance company has 10,000 policyholders. Each policyholder is classified as: (i) young or old; (ii) male or female; and (iii) married or single. Of these policyholders, 3000 are young, 4600 are male, and 7000 are married. The policyholders can also be classified as 1320 young males, 3010 married males, and 1400 young married persons. Finally, 600 of the policyholders are young married males. How many of the company's policyholders are young, female, and single?

21. An insurer offers a health plan to the employees of a large company. As part of this plan, the individual employees may choose exactly two of the supplementary coverages A, B, and C, or they may choose no supplementary coverage. The proportions of the company's employees that choose coverages A, B, and C are 1/4, 1/3, and 5/12 respectively. Determine the probability that a randomly chosen employee will choose no supplementary coverage.

1.3 Simulation

To simulate means to reproduce the results of a real phenomenon by artificial means. We are interested here in phenomena that have a random element. What we really simulate are the outcomes of a sample space and/or the states of a random variable in accordance with the probability measure or distribution that we are assuming. Computers give a fast and convenient way of carrying out these sorts of simulations.

There are several reasons to use simulation in problems. Sometimes we do not yet have a good understanding of a phenomenon, and observing it repeatedly may give us intuition about a question related to it. In such cases, closed form analytical results might be derivable, and the simulation mostly serves the purpose of suggesting what to try to prove. But many times the system under study is complex enough that analysis is either very difficult or impossible. (Queueing problems – see Section 6.3 – are good illustrations of both situations.) Then simulation is the only means of getting approximate information, and the tools of probability and statistics also allow us to measure our confidence in our conclusions. Another reason to simulate is somewhat surprising. In some inherently non-probabilistic problems we can use a simulated random target shooting approach to gain information about physical measurements of size and shape. A simple example of such a problem solving method (traditionally called *Monte Carlo simulation* in reference to the European gaming center) is the example on area of a region in Exercise 8.

I have chosen to give the subject emphasis and early treatment in this book for another reason. There is more at stake for you intellectually than gaining intuition about systems by observing simulations. Building an algorithm to solve a problem by simulation helps to teach you about the subtle concepts of the subject of probability, such as probability measures and random variables. Thus, here and elsewhere in this book I will be asking you to carefully study the details of programs I have written, and to write some simulation programs in *Mathematica* yourself. You will invest some time and mental effort, but the payoff will be great.

We have already experienced simulation in the form of selecting ordered samples without replacement of two numbers from the set {1, 2, 3, 4, 5}. In that context we looked at the empirical histogram of the minimum of the two numbers. We will treat simulations of samples again in more detail in the next section. For the time being let us see how to use *Mathematica's* tools for generating streams of random numbers to solve problems.

Most of the kind of simulations that we will do are adaptations of the simple process of choosing a random real number in the interval [0, 1]. The word "random" here could be replaced by the more descriptive word "uniform," because the assumption will be that probability is distributed uniformly through [0, 1], with sets getting large probability only if their size is large, not because they happen to be in some favored location in [0, 1]. More precisely, we suppose that the sample space of the random experiment is $\Omega = [0, 1]$, the collection of events $\mathcal{H}$ consists of all subsets of [0, 1] for which it is possible to define their length, and the probability measure P gives probability to a set equal to its length. Then for instance:

$$P[\,[0,\,1]\,] = 1, \quad P\left[\left(\frac{1}{2},\,\frac{3}{4}\right]\right] = \frac{1}{4}, \quad P\left[\left[0,\,\frac{1}{4}\right] \cup \left[\frac{2}{3},\,1\right]\right] = \frac{1}{4} + \frac{1}{3} = \frac{7}{12},$$

$$P\left[\left\{\frac{5}{6}\right\}\right] = 0$$

Activity 1 Try to describe a uniform distribution of probability on the interval [2,5]. What would be the probability associated to the event [2,3]? [9/4, 13/4] $\cup$ (4,5]? {2.5}?

In *Mathematica's* kernel there are several probability distribution objects, including one for the uniform distribution, and also a means of sampling one or more random numbers from a distribution. Fittingly,

UniformDistribution[{a,b}]

is the syntax for the uniform distribution on the interval [a,b], and

RandomReal[distribution]   and   RandomReal[distribution, n]

return, respectively, one or a list of *n* random numbers from the given distribution. Here are some examples of their use.

```
dist = UniformDistribution[{2, 5}];
RandomReal[dist]
RandomReal[dist]
RandomReal[dist, 5]
```

4.42891

2.00398

{2.14133, 2.06261, 3.20962, 3.47772, 3.93109}

Notice that new random numbers in [2, 5] are returned each time that RandomReal is called. Incidentally, RandomReal may be used with an empty argument when the distribution being sampled from is the uniform distribution on [0, 1], which is what we will usually need.

How do *Mathematica* and other programs obtain such random numbers? There are several means that are possible, but the most common type of random number generator is the *linear congruential generator* which we describe now. The ironic fact is that our supposedly random stream of numbers is only simulated in a deterministic fashion in such a way as to be indistinguishable statistically from a list of truly random numbers. Because of this, the stream of numbers is called *pseudo-random*. The algorithm is as follows. Begin with a single non-negative integer called a *seed*. (Computers can use internal contents such as collections of digits in their clocks as seeds, which change frequently.) To get the first pseudo-random number, compute a new value of the seed by:

$$\text{new value of } seed = (multiplier \times seed + increment) \bmod modulus \qquad (1)$$

where *multiplier*, *increment*, and *modulus* are positive integer constants. Then, using the newly computed *seed*, return the number

$$rand = seed/modulus \qquad (2)$$

Subsequent random numbers are found in the same way, updating *seed* by formula (1) and dividing by *modulus*. Notice that because the initial *seed* is non-negative and all of the constants are positive, *seed* will stay non-negative, which makes $rand \geq 0$ as well. And since the mod operation returns a value less than *modulus*, *seed* will be less than *modulus*, hence *rand* will be less than 1. So each pseudo-random number that we produce will lie in [0, 1).

　　Now the larger the value of *modulus*, the more possible values of *seed* there can be; hence, the more real numbers that can be returned when formula (2) is applied. In fact, there would be exactly *modulus* number of possible *seed* values. But *modulus*, *multiplier*, and *increment* must be carefully chosen to take advantage of this, otherwise, the sequence of pseudo-random numbers could cycle with a very small period. As a simple example, if the constants were as follows:

$$\text{new } seed = (2 \times seed + 2) \bmod 8$$

then possible *seed* values are 0, 1, 2, 3, 4, 5, 6, and 7, but it is easy to check that for these seeds the new seeds are, respectively, 2, 4, 6, 0, 2, 4, 6, 0, so that after the first value of *seed*, subsequent seeds can only take on one of four values. Even worse, should *seed* ever equal 6, then new seeds will continue to be 6 forever, because $(2 \times 6 + 2) \bmod 8$ is again 6.

　　Therefore, the choice of the constants is critical to making a stream of numbers that has the appearance of randomness. There is a lot of number theory involved here which we cannot hope to cover. The book by Rubinstein (see References) gives an introduction. It is reported there that the values

$$multiplier = 2^7 + 1, \quad increment = 1, \quad modulus = 2^{35}$$

have been successful, but these are not the only choices. We can build our own function similar to the RandomReal command in *Mathematica* to generate a list of uniform[0,1] numbers using these values as follows.

```
MyRandomArray[initialseed_, n_] :=
  Module[{seed, rand, thearray, mult, inc, modul, i},
         mult = 2^7 + 1;
         inc = 1;
         modul = 2^35;
         thearray = {};
         seed = initialseed;
         Do[seed = Mod[mult * seed + inc, modul];
            rand = N[ seed / modul];
            AppendTo[thearray, rand], {i, 1, n}];
         thearray]
```

The module begins with the list of local variables used in the subsequent lines, and sets the values above for the multiplier, here called *mult*, the increment *inc*, and the modulus *modul*. After initializing an empty list of random numbers and setting the initial value of *seed* according to the input *initialseed*, the module enters a loop. In each of the *n* passes through the loop, three things are done. The *seed* is updated by the linear congruence relation (1), the new random number *rand* is computed by formula (2), and then *rand* is appended to the array of random numbers. Here is a sample run. You should try running the command again with the same initial seed to see that you get the same sequence of pseudo-random numbers, then try a different seed.

```
MyRandomArray[2^24 + 19, 10]
```

```
{0.0629884, 0.125497, 0.189176, 0.403682, 0.0749859,
 0.673181, 0.840295, 0.398108, 0.355958, 0.918557}
```

Activity 2 Revise the random number generating command above using the following constants, taken from a popular computer science text.

multiplier = 25173   increment = 13849   modulus = 65536

Use the command Histogram[datalist,numrectangles] to plot a histogram of 50 random numbers using 4 rectangles. Do you see what you expected?

One advantage to the pseudo-random approach is that simulation experiments are reproducible. If one starts with the same seed, then the same stream of random numbers will ensue, which will produce the same experimental results. This is advantageous because you may want to compare the performance of similar systems using the same random inputs to ensure fairness. *Mathematica* permits this by using the command

SeedRandom[n]

which sets the initial seed value as the integer *n*. If you use an empty argument to SeedRandom, it resets the seed using the computer's clock time. In the sequence of commands below, the seed is set to 79, two random numbers are generated, then the seed is reset to 79, and the same random number .624058 is returned as was returned the first time. We then reset the seed "randomly" using the computer clock, and the first random number that is returned is something other than .624058.

```
SeedRandom[79];
RandomReal[]
RandomReal[]
```

0.624058

0.782677

```
SeedRandom[79];
RandomReal[]
```

0.624058

```
SeedRandom[];
RandomReal[]
```

0.295358

Activity 3 Perform a similar test with SeedRandom using a seed of 591 and requesting a RandomReal[UniformDistribution[{0,1}], 5]. Reset the seed by computer clock time and call for another random list of five numbers to see that different random numbers are returned.

Example 1 It is time to build a simulation model for an actual system. A *random walk* on the line is a random process in which at each time instant 0, 1, 2, ... an object occupies a point with integer coordinates shown as circled numbers in Figure 5.

Figure 1.5 - Schematic diagram of a random walk

The position on the line cannot be foretold with certainty, but it satisfies the condition that if the position is n at one time instant, then it will either be $n + 1$ with probability p or $n - 1$ with probability $1 - p$ at the next time instant, as indicated by the annotated arrows in the figure. Variations are possible, including allowing the object to wait at its current position with some probability, enforcing boundary behavior such as absorption or reflection to confine the random walk to a bounded interval, and increasing the dimension of the set on which the random walk moves to 2, 3, or higher. Random walks have been used as models for diffusion of heat, the motion of economic quantities such as stock prices, the behavior of populations, and the status of competitive games among many other applications, and their fascinating properties have kept probabilists busy for many decades.

For our purposes suppose we are interested in estimating the probability that a random walk which starts at a point $n \geq 0$ reaches a point $M \geq n$ before it reaches 0. This is a famous problem called the *gambler's ruin problem*, which really asks how likely it is that a gambler's wealth that follows a random walk reaches a target level of M before bankruptcy.

We have an initial position n followed by a sequence of random positions X_1, X_2, X_3, ... The underlying randomness is such that at each time a simple random experiment determines whether the move is to the right (probability p) or the left (probability $1 - p$). So to simulate the random walk we only need to be able to iteratively simulate the simple experiment of adding 1 to the current position with probability p and subtracting 1 with probability $1 - p$. We will stop when position 0 or position M is reached, and make note of which of the two was the final position. By running such a simulation many times and observing the proportion of times that the final state was M, we can estimate the probability that the final state is M.

To implement the plan in the preceding paragraph, we need a way of randomly returning +1 with probability p and −1 with probability $1 - p$. But since we know how to simulate uniform real random numbers on $[0,1]$, and the probability of an interval is equal to its length, let's say that if a simulated uniform random number is less than or equal to p then the result is +1, otherwise −1. Then we have the desired properties of motion, because

$$P[\text{move right}] = P[[0, \ p]] = \text{length}[[0, \ p]] = p$$
$$P[\text{move left}] = P[(p, \ 1]] = \text{length}[(p, \ 1]] = 1 - p$$

Here is a *Mathematica* function that does the job. A uniform random number is selected and compared to p. If the $\leq$ comparison is true, then +1 is returned, else −1 is returned. So that we can simulate random walks with different p values, we make p an argument. One function call with $p = .5$ is displayed; you should try a few others.

```
StepSize[p_] := If[RandomReal[] ≤ p, 1, -1]
```

```
StepSize[.5]
```

−1

To simulate the process once, we let the initial position *n* be given, as well as the right step probability *p* and the boundary point *M*. We repeatedly move by the random StepSize amount until the random walk reaches 0 or *M*. The function below returns the complete list of states that the random walk occupied. Look carefully at the code to see how it works. The variable *statelist* records the current list of positions occupied by the random walk, the variable *done* becomes true when either the boundary 0 or *M* is reached, and variables *currstate* and *nextstate* keep track of the current and next position of the random walk. (The ‖ syntax means logical "or.")

```
SimulateRandomWalk[p_, n_, M0_] :=
    Module[{statelist, done, currstate, nextstate},
        statelist = {n};
        currstate = n;
        done = (currstate ≤ 0) || (currstate ≥ M0);
          While[Not[done],
                nextstate = currstate + StepSize[p];
                AppendTo[statelist, nextstate];
                currstate = nextstate;
        done = (currstate ≤ 0) || (currstate ≥ M0)];
        statelist]
```

Below is an example run of the command. We include a list plot of the sequence of positions to help your understanding of the motion of the random walk. You should rerun this command, changing parameters as desired to get more intuition about the behavior of the random walk. A more systematic investigation is suggested in Exercise 6. We set a random seed first for reproducibility, but if you are in the electronic version of the book and want to see different paths of the random walk, just delete the SeedRandom command.

```
SeedRandom[464758];
thewalk = SimulateRandomWalk[0.5, 4, 8]
ListPlot[thewalk,
  PlotStyle → {PointSize[0.015], Black}]
```

```
{4, 5, 4, 5, 6, 7, 6, 5, 4, 5, 6, 5, 6,
 5, 4, 5, 6, 5, 6, 5, 4, 3, 4, 3, 2, 1, 2, 3,
 4, 5, 6, 5, 4, 3, 4, 3, 4, 3, 2, 3, 2, 1, 0}
```

Figure 1.6 - A random walk on [0, 8]

Activity 4 Roughly how would you describe the dependence of the probability that the random walk reaches M before 0 on the value of p? How does it depend on the value of n? How could you use SimulateRandomWalk to check your intuition about these questions?

Continuing the random walk example, to estimate the probability of hitting M before 0 we must repeat the simulation many times, with new streams of random numbers. We merely take note of the last element of the simulated list of positions, which will be either 0 or M, count the number of times that it is M, and divide by the number of repetitions to estimate the probability. Here is a command that does this. Any input values we like for p, n, and M can be used. We also let the number of replications of the simulation be a fourth argument.

```
AbsorptionPctAtM[p_, n_, M0_, numreps_] :=
    Module[{absatM, empprob, poslist, lastpos, i},
        absatM = 0;
        Do[poslist = SimulateRandomWalk[p, n, M0];
            lastpos = Last[poslist];
    If[lastpos == M0, absatM = absatM + 1, Null],
                {i, 1, numreps}];
            empprob = N[absatM / numreps]]
```

The variable *absatM*, initialized to zero, keeps track of how many times so far the walk has been absorbed at M. The algorithm is simple: we repeat exactly *numreps* times the actions of simulating the list of positions (called *poslist*) of the random walk, then pick out the last element *lastpos*, and increment *absatM* if necessary. For $p = .4$, $n = 3$, and $M = 5$ here are the results of two runs of 100 simulated random walks each.

```
SeedRandom[342 197];
AbsorptionPctAtM[.4, 3, 5, 100]
AbsorptionPctAtM[.4, 3, 5, 100]
```

0.39

0.42

There is good consistency in the results of our runs. About 40% of the time, for these parameters, the random walk reaches $M = 5$ before it hits 0. There will be a way of finding the probability exactly by analytical methods later. (See Exercise 13.) ∎

Exercises 1.3

1. For a uniform distribution on the interval $[a, b]$ of real numbers, find the open interval probability $P[(c, d)]$, the closed interval probability $P[[c, d]]$, and the probability of a single point $P[\{d\}]$, where $a < c < d < b$.

2. (*Mathematica*) Use the MyRandomArray command from the section with initial seeds of 2, 5, 11, and 25 to generate lists of 10 pseudo-random numbers. Look carefully at your lists. (It may help to ListPlot them.) Do the sequences appear random? Where is the source of the problem, and what might be done to correct it?

3. (*Mathematica*) Write a command to simulate an outcome from a finite sample space with outcomes a, b, c, and d with probabilities p_1, p_2, p_3, p_4, respectively.

4. (*Mathematica*) Write a command to simulate a two-dimensional random walk starting at $(0, 0)$. Such a random process moves on the grid of points in the plane with integer coordinates such that wherever it is now, at the next instant of time it moves to one of the immediately adjacent grid points: right with probability r, left with probability l, up with probability u, and down with probability d, where $r + l + u + d = 1$.

5. Consider a linear congruential pseudo-random number generator whose sequence of seed values satisfies the recursive relation

$$X_{n+1} = (aX_n + c) \bmod m$$

Assume that $m = 2^p$ for some positive integer p. The generator is called *full period* if given any initial seed all of the values 0, ... , $m - 1$ will appear as seed values before repetition occurs. Show that if the generator is full period, then c must be an odd number.

6. (*Mathematica*) For the random walk with $p = .3$ and $M = 6$, use the AbsorptionPctAtM command in the section to try to estimate the functional dependence of the probability of absorption at M on the starting state n.

7. (*Mathematica*) Use *Mathematica* to simulate ten samples of size 50 from the uniform distribution on [0, 10]. Study the samples by plotting histograms with 5 rectangles of each of the ten samples. How many observations do you expect in each category? Comment on the consistency of the histograms from one sample to another.

8. (*Mathematica*) While simulation is most productively used with random systems such as the random walk, it is interesting to note that certain numerical approximations of physical quantities like area and volume can be done by Monte Carlo techniques. Develop a *Mathematica* command to approximate the area between the function $f(x) = e^{-x^2}$ and the x-axis on the interval [0,1], which is graphed below. Compare the answers you get from repeated simulations to a numerical integration using *Mathematica*'s NIntegrate command. (Hint: The square [0,1]×[0,1] contains the desired area A, and the area of the square is 1. If you figuratively shoot a random arrow at the square, it should fall into the shaded region a proportion A of the time. Simulate a large number of such random arrows and count the proportion that fall into A.)

Exercise 8   Exercise 10

9. Referring to Exercise 8, why would you not be able to use Monte Carlo simulation to find the area bounded by the x and y-axes, the line $x = 1$, and the curve $y = 1/x$? Could you do it for the curve $y = 1/\sqrt{x}$?

10. (*Mathematica*) A *Galton-Watson board* is a triangular grid of pegs such as the one shown, encased in plastic or glass and stood on end. A large number of small marbles escape from a hole in the top exactly in the middle of the board as shown, bounce from side to side with equal probability as they fall down the board, and eventually land in the bins in the bottom. Build a Galton-Watson simulator for the small five level board shown using arbitrarily many marbles, run it several times, and comment on the shape that is typically formed by the stacks of marbles that gather in the bottom bins.

11. (*Mathematica*) A dam on a 50-foot deep lake is operated to let its vents pass water through at a constant rate so that 1 vertical inch of lake depth is vented per week. The water begins at a level of 5 feet below the top of the dam. A rainy season begins in which each

week a random and uniformly distributed depth of water between 2 and 5 inches is added to the lake. Simulate this process, and study how long it takes for the water to overflow the top of the dam. Do the simulations match your intuition and expectations about the situation?

12. (*Mathematica*) A device uses two components in such a way that the device will no longer function if either of the components wears out. Suppose that each component lifetime is random and uniformly distributed on the interval of real numbers [10, 20]. Write a command to simulate the lifetime of the device many times over and investigate the probability law of the device lifetime empirically. About what is the average lifetime?

13. In the random walk problem it is possible to make progress toward an analytical solution to the computation of $f_M(i) = P[\text{walk reaches } M \text{ before } 0 \text{ beginning at state } i]$ using recursion. First, solve the trivial case $M = 2$, that is, compute intuitively $\{f_2(0), f_2(1), f_2(2)\}$. Then try to extend to the case $M = 3$ by expressing each $f_3(i)$ in terms of f_3 values for states neighboring state i.

14. (*Mathematica*) Two small rental car establishments in a city are such that cars rented at one location may be returned to the other. Suppose that as a simplified model we assume that from one week to another, at each location, a random number of cars move from that location to the other location. The two locations always keep at least one vehicle on hand, so the number of vehicles moving out of each can be no greater than the initial inventory level minus 1. Write a function in *Mathematica* that takes as arguments the initial car inventory levels of each location and the number of time steps to simulate. This program should simulate successive inventory levels of each location, and return the list of inventory levels for location 1. Use it to examine both time trends and the distribution of inventory levels in scenarios in which the initial inventory levels are equal and in which they are drastically unequal. In the latter scenario, do the inventory levels seem to equalize, and if so how many time steps does it take?

15. (*Mathematica*) Consider again the gambler's ruin problem with $p = .4$, $n = 3$, and $M = 5$. Use simulation to estimate the probability distribution of the number of steps it takes the random walk to reach one of the boundary states.

1.4 Random Sampling

In this section we focus on a particular kind of random experiment that we have glimpsed before: sampling randomly from a universal set. In so doing, we are led to the area of mathematics known as *combinatorics*, the study of counting. The reason why combinatorics and random sampling are so important is that one of a statistician's best tools for obtaining information about a population inexpensively and reliably is by taking a random sample from it, as opposed to a comprehensive census.

There are several main types of sampling situations, depending on whether the objects being sampled are taken in sequence or just in a batch, and whether the objects are replaced as sampling proceeds, and so become eligible to be sampled again, or the sampling is done without replacement.

Sampling in Sequence without Replacement

If the order in which sampled items are taken matters and sampling is done without replacing previously sampled items, then the sample is a *permutation*. In the following discussion let there be *n* objects in the universe and suppose the sample is to be of size *k*. We can utilize the command KPermutations[list, k] (see Section 1.2) in the KnoxProb7`Utilities` package to display all of the permutations. Here for example are the permutations of 2 objects chosen from the set {*a*, *b*, *c*, *d*, *e*}.

```
Needs["KnoxProb7`Utilities`"];
KPermutations[{a, b, c, d, e}, 2]
```

```
{{a, b}, {b, a}, {a, c}, {c, a}, {a, d}, {d, a},
 {a, e}, {e, a}, {b, c}, {c, b}, {b, d}, {d, b}, {b, e},
 {e, b}, {c, d}, {d, c}, {c, e}, {e, c}, {d, e}, {e, d}}
```

The same package also contains the command

RandomKPermutation[list, k]

which randomly selects a permutation of *k* elements from the given list, as shown below.

```
RandomKPermutation[{a, b, c, d, e}, 3]
```

```
{e, b, c}
```

Do the following activity to begin the problem of finding the number of permutations of *k* objects from *n*.

Activity 1 Without utilizing *Mathematica*, how many permutations of 1 object from *n* are there? Now use KPermutations to count the permutations of 2 objects from 2, 2 objects from 3, 2 from 4, 2 from 5, and 2 from 6. Can you detect a pattern and speculate about the number of permutations of 2 objects from *n*?

It is not hard to find the number of permutations of 2 objects from *n* mathematically, and the line of reasoning carries over to permutations of *k* objects from *n*. Let $P_{n,k}$ denote this number of *k*-permutations. To determine a unique 2-permutation, you must pick the first sampled item, which can be done in n ways, and then the second, which can be done in $n - 1$ ways. The result is therefore $n(n - 1)$ by the multiplication principle. Here is the general theorem, which can be proved similarly (see Exercise 4).

Theorem 1. The number of ways of selecting k objects in sequence and without replacement from a population of n objects is

$$P_{n,k} = n(n - 1)(n - 2)\cdots(n - k + 1) = \frac{n!}{(n - k)!} \tag{1}$$

where $n! = n(n - 1)(n - 2)\cdot 2\cdot 1$. ∎

Example 1 In the example above of selecting 2 objects randomly in sequence and without replacement from the set $\{a, b, c, d, e\}$, what is the probability that the sample will not include d nor e?

The size of the sample space is $n(\Omega) = 5\cdot 4 = 20$ by Theorem 1. To count the number of outcomes in the event that the sample includes neither d nor e, note that in this case the sample must be drawn from the subset $\{a, b, c\}$. There are $3\cdot 2 = 6$ ways of drawing such a sample, again by Theorem 1. Therefore, the probability of this event is $6/20 = .3$. ∎

Example 2 A hand of blackjack is dealt in sequence and without replacement so that the first card is face down and the second is face up. The hand is referred to as a blackjack hand if it contains an ace and a face card, that is, a king, queen, or jack. Let us find the probability of being dealt a blackjack.

The sample space is $\Omega = \{(x, y) : x, y \text{ are cards and } x \neq y\}$. Since there are 52 cards in an ordinary deck, the size of the sample space is

$$n(\Omega) = 52\cdot 51$$

by Theorem 1. There are two disjoint cases making up the event $A = \{\text{hand is a blackjack}\}$, namely, the cards are dealt in the order ace, face card or they are in the order face card, ace. The number of outcomes in the event A is therefore the total

$$n(A) = 4\cdot 12 + 12\cdot 4$$

since there are 4 aces and 12 face cards in the deck. Hence the probability of a blackjack is about .036, as below. ∎

$$N\left[\frac{4 * 12 + 12 * 4}{52 * 51}\right]$$

0.0361991

Example 3 Recall the command called DrawIntegerSample[a, b, k] in the KnoxProb7`Utili-ties` package which presumes that the universe being sampled from is {a, a + 1, ..., b}. By default it will select a random permutation from this space, but you can also include the option Ordered->False to sample without regard to order (see the next subsection) and/or the option Replacement->True to draw a sample with replacement. We can use this command to study aspects of drawing random samples more deeply, in the way that we looked at the minimum of the sampled values in Section 1.1. Exercise 7 asks you to explore the average of the sample members. Here is a similar experiment on the *range* of the data, that is, the largest value minus the smallest.

    For instance, if 10 numbers are randomly selected from the set {1, 2, ..., 50} and the sample elements turn out to be 42, 30, 2, 45, 27, 6, 15, 31, 29, and 40, then the range is 45 − 2 = 43. The command below initializes an empty list of ranges, then for a desired number of replications, it draws a new sample, computes the range, and appends this new range to the list of ranges. The final list of ranges is returned.

```
SimulateRange[numreps_, a_, b_, sampsize_] :=
    Module[{thesample, rangelist, i, newrange},
        rangelist = {};
        Do[
        thesample = DrawIntegerSample[a, b, sampsize];
        newrange = Max[thesample] - Min[thesample];
        AppendTo[rangelist, newrange],
     {i, 1, numreps}];
            rangelist]
```

Below is one run of the command for 100 replications of a sample of 10 integers from the first 50. You should try sampling again a few times. You will see that the histogram is rather consistent in its shape: it is asymmetrical with a longer left tail than right tail, it has a tall peak in the mid 40's, which is probably about where the average range is located, and a very high percentage of the time the range is at least 30. ■

```
SeedRandom[124];
ranges = SimulateRange[100, 1, 50, 10]
```

{45, 41, 44, 48, 48, 44, 46, 43, 48, 39, 46, 39, 43, 44, 30,
 36, 42, 45, 40, 41, 46, 38, 45, 43, 45, 43, 46, 45, 40, 45,
 47, 31, 43, 41, 48, 33, 45, 38, 32, 39, 38, 39, 46, 37,
 46, 39, 31, 39, 46, 47, 36, 41, 42, 45, 44, 31, 45, 45,
 36, 40, 47, 43, 41, 42, 41, 48, 47, 47, 41, 45, 42, 47,
 38, 45, 44, 38, 43, 38, 47, 48, 36, 33, 31, 37, 45, 41,
 39, 40, 39, 43, 35, 46, 33, 32, 31, 43, 46, 47, 42, 44}

Histogram[ranges, 10, ChartStyle → Gray]

Figure 1.7 - Histogram of ranges

Example 4 In a random sample of given size, with what likelihood does each individual appear in the sample? Recall we asked this question in Section 1.1 about a sample of two numbers from the set {1, 2, 3, 4, 5}. If the universe being sampled from is $U = \{1, 2, 3, ..., n\}$, and a sample of k individuals is drawn in sequence and without replacement, then the sample space is

$$\Omega = \left\{(x_1, x_2, \cdots, x_k) : x_i \in U \text{ for all } i, \text{ and } x_i \neq x_j \text{ for all } i \neq j\right\}$$

and by Theorem 1, the cardinality of Ω is

$$n(\Omega) = n(n-1)(n-2) \cdots (n-k+1)$$

We suppose that the randomness assumption means that all outcomes are equally likely and hence the probability of an event E is its cardinality $n(E)$ divided by this $n(\Omega)$.

Consider the event A that individual 1 is in the sample. That individual could have been sampled first, or second, or third, etc., so that A can be broken apart into k disjoint subsets

$$A = B_1 \cup B_2 \cup \cdots \cup B_k$$

where $B_i = $ "1 occurs on i^{th} draw." Therefore, by the third axiom of probability, $\text{P}[A] = \sum_{i=1}^{k} \text{P}[B_i]$. But also, for each i,

$$P[B_i] = (n - 1)(n - 2) \cdots ((n - 1) - (k - 1) + 1)/n(\Omega)$$

since once individual 1 is known to be in position i, the rest of the positions are filled by selecting a permutation of $k - 1$ individuals from the remaining $n - 1$. Therefore,

$$P[A] = \frac{k(n - 1)(n - 2) \cdots (n - k + 1)}{n(n - 1)(n - 2) \cdots (n - k + 1)} = \frac{k}{n}$$

The same reasoning holds for every other individual in the population. So we have shown the rather intuitive result that if a sample in sequence and without replacement of k objects from n is taken, every individual in the universe has an equal chance of k/n of appearing in the sample. ∎

Activity 2 Try to carry through similar reasoning to that of Example 4 if the sample is drawn in sequence with replacement. Does the argument break down? If it does, try another approach (such as complementation).

Sampling in a Batch without Replacement

Now let us look at sampling without replacement when the order in which sample values are drawn is of no consequence, i.e., the sample may as well have been drawn all at once in a batch. Most card games follow this prescription, but there are some other very important situations of this type too. A sample without order or replacement of k objects from n is called a *combination* of k chosen from n.

Recall the *Mathematica* command KSubsets[list, k] in the package Combinatorica` which returns all subsets of size $k$ from the given list. (This package is loaded automatically when you load KnoxProb7`Utilities`.) It therefore returns the sample space of all combinations of k chosen from n. Here are the three-element combinations taken from $\{a, b, c, d, e\}$.

```
KSubsets[{a, b, c, d, e}, 3]
```

{{a, b, c}, {a, b, d}, {a, b, e}, {a, c, d}, {a, c, e},
  {a, d, e}, {b, c, d}, {b, c, e}, {b, d, e}, {c, d, e}}

Notice that there are 10 combinations of 3 elements chosen from 5. The same package has a command

RandomKSubset[list, k]

which returns one such combination selected at random. Below are a couple of instances, and as we would anticipate, we get a different randomly selected subset the second time than the first time.

```
SeedRandom[67 653];
RandomKSubset[{a, b, c, d, e}, 3]
RandomKSubset[{a, b, c, d, e}, 3]
```

{b, d, e}

{a, b, e}

Let us try to find a formula for

$C_{n,k}$ = number of combinations of k objects chosen from n

If at first k objects are selected in a batch from n, then it is possible to think of subsequently putting them into a sequence. By doing so we would create a unique permutation of k objects from n. The number of possible permutations is known from Theorem 1. And it is fairly easy to see that there are $k! = k(k-1)(k-2) \cdot 2 \cdot 1$ ways of sequencing the k selected objects, by the multiplication principle. Therefore, by the multiplication principle again,

$$\sharp \text{ permutations } = \sharp \text{ combinations } \cdot \sharp \text{ ways of sequencing the combination}$$
$$\implies P_{n,k} = C_{n,k} \cdot k!$$

If we now bring in the formula for $P_{n,k}$ from Theorem 1, we have the following result.

Theorem 2. The number of combinations of k objects selected in a batch without replacement from n is

$$C_{n,k} = \frac{n!}{k! \cdot (n-k)!} \quad \blacksquare \tag{2}$$

You may recognize $C_{n,k}$ as the same quantity as the binomial coefficient $\binom{n}{k}$ from the binomial theorem and other contexts. This coefficient is usually read as "n choose k," and we now know why. Incidentally, *Mathematica* can compute this quantity using the Binomial[n, k] function, which is in the kernel.

```
C6choose3 = Binomial[6, 3]
```

20

Activity 3 Check to see whether empirically the probability that a particular element of the universe is sampled is dependent on the element. For a sample of k from n in a batch without replacement, what does this probability seem to be? (See also Exercise 10.) You may use the command below, which inputs the population size n (assuming the population is coded as $\{1, 2, ..., n\}$), the sample size k, the population member m between 1 and n to be checked, and the number of replications *numreps*. It outputs the proportion of the replications for which m was in the sample.

```
ProportionMInSample[n_, k_, m_, numreps_] :=
  Module[{mcount, simsample, i},
     mcount = 0;
    Do[simsample =
       DrawIntegerSample[1, n, k, Ordered -> False];
       If[MemberQ[simsample, m],
       mcount = mcount + 1, Null],
              {i, 1, numreps}];
       mcount / numreps]
```

Example 5 Theorem 2 is extremely useful in answering probability questions. For example let us investigate whether it is reasonable to suppose that a population of 1000 tax returns has no more than 20% with errors on them, if a random sample of 100 returns from this population yielded 40 with errors. The proportion in the sample is twice that which is assumed for the population. Is it likely that such an anomaly could happen by chance?

The 20% population figure, and not 19%, 18%, or anything less than 20%, is the most likely to have caused a large number of erroneous returns in the sample (see Exercise 13), so we will recast the question in the following way. If there are exactly 200 among the 1000 returns that have errors, how likely is it that a random sample of 100 has 40 erroneous returns or more? We say 40 *or more* because any large number of errors, not just identically 40, gives evidence against the 20% hypothesis. If the probability of 40 or more errors is quite small, then there is strong evidence in the observed data against the 20% hypothesis.

Theorem 2 allows us to write the probability of exactly k erroneous returns in the sample, hence $100 - k$ good returns, as

$$\frac{\binom{200}{k}\binom{800}{100-k}}{\binom{1000}{100}}, \; k = 0, 1, 2, ..., 100$$

This is because there are $\binom{1000}{100}$ equally likely samples of size 100 in the sample space, and

there are $\binom{200}{k}$ ways of selecting k erroneous returns from the group of 200 of them, and

$\binom{800}{100-k}$ ways of selecting the remaining correct returns from the 800 correct returns in the population.

We can calculate the probability of 40 or more erroneous returns by summing these numbers from 40 to 100, or equivalently by complementation, summing from 0 to 39 and subtracting from 1. We use the latter approach to reduce computation.

```
N[1 - Sum[ Binomial[200, k] * Binomial[800, 100 - k] / Binomial[1000, 100] ,

   {k, 0, 39}]]
```

8.85353×10^{-7}

Since the probability is only on the order of 10^{-6}, the hypothesized 20% error rate in the population of tax returns is in very serious doubt. ∎

Example 6 As indicated earlier, Theorem 2 has implications of importance for card players. In poker, a hand of 2-pair means that there are two cards of one rank (such as 4's or 7's), two cards of a different rank, and a single card that doesn't match either of the first two ranks. A hand of 3-of-a-kind consists of three cards of one rank (such as 6's), and two unmatching singletons (such as one king and one ace). Which hand is rarer, and hence more valuable?

For the experiment of drawing a batch of 5 cards without replacement from a standard deck of 52 cards we want to compare the probabilities of the 2-pair event and the 3-of-a-kind event. The sample space has $C_{52,5}$ elements. For the 2-pair event, to determine a unique hand we must first select in no particular order two ranks for the pairs. This can be done in $C_{13,2}$ ways, since there are 13 possible ranks 2, 3, ... , Ace. For each rank we must select two of the suits (hearts, diamonds, clubs, or spades) from among the four possible suits to determine the exact content of the pairs. Then we must select a final card to fill the hand from among the $52 - 4 - 4 = 44$ cards left in the deck which do not match our selected ranks. Thus, the following is the number of 2-pair hands:

```
num2pairhands =
  Binomial[13, 2] * Binomial[4, 2] * Binomial[4, 2] * 44
```

123 552

For 3-of-a-kind, we must pick a rank from among the 13 for our 3-of-a-kind card, then we select 3 from among the 4 suits for that rank, and then two other different ranks from the 12 remaining ranks, and a suit among the four possible for each of those unmatched cards. Therefore, this expression gives the number of 3-of-a-kind hands:

```
num3ofakindhands =
  Binomial[13, 1] * Binomial[4, 3] * Binomial[12, 2] * 4 * 4
```

54 912

Below are the two probabilities. Since the 2-pair hand is over twice as likely to be dealt as the 3-of-a-kind hand, the 3-of-a-kind hand is more valuable. ■

```
{Ptwopair, P3ofkind} =
  {num2pairhands / Binomial[52, 5],
  num3ofakindhands / Binomial[52, 5]}
```

$$\left\{ \frac{198}{4165}, \frac{88}{4165} \right\}$$

Other Sampling Situations

There are four possible sampling scenarios that allow for the cases of sequential or batch sampling and replacement or no replacement. The two scenarios in which there is no replacement have been covered. To sample with replacement but in a batch is an unusual scenario that we will not consider here. But the final possibility of sampling in sequence and with replacement is an important one that deserves a few words.

Once again let there be n objects in the population and let the sample be of size k. Since we can now reuse previously sampled objects, there are n possibilities for each of the k positions in the sample. Directly from the multiplication principle there are

$$n \times n \times \cdots \times n \ (k \text{ times}) = n^k \tag{3}$$

possible such random samples. For example, if $\{1, 2, ..., n\}$ is the population, the probability that the first sample object is a 3 is

$$\frac{1 \cdot n^{k-1}}{n^k} = \frac{1}{n}$$

since the restriction that 3 is in position 1 means that there is just one way of filling that position, and still there are n ways of filling each of the other $k - 1$ positions. There is nothing special about object 3, nor is there anything special about position 1. Each population member has a probability of $1/n$ of appearing in each position.

The special case $n = 2$ comes up frequently, and will in fact be the subject of a section in Chapter 2. If there are two possibilities, call them 0 and 1, for each of the k sample positions, then there are 2^k possible sequences of length k of these 0's and 1's. Try this activity, which will be very useful later.

> **Activity 4** How many sequences of 0's and 1's are there such that there are exactly m 1's in the sample, for m between 0 and k? First try writing down all sequences with $k = 3$ and counting how many of them have m 1's for $m = 0, 1, 2, 3$. Next move to $k = 4$, and then try to generalize. (Hint: Think what must be done to uniquely determine a sequence of k terms with m 1's in it.)

Another sampling situation arises in statistical experimentation in which a researcher wants to split a population of subjects at random into several groups, each of which will receive a different treatment. This is called *partitioning* the test group. The random assignment of subjects to groups is done so as to reduce the possibilty of systematic *confounding*, in which members within one or more of the groups share some common feature apart from the fact that they receive the same treatment. For example, if all of the members of one group were over 60 years of age, and all members of another were under 30 years of age, then the researcher would have a hard time distinguishing whether a difference in response to treatment between groups was due to the different treatments they received or due to the confounding factor of age.

Example 7 A medical researcher wants to divide 15 healthy adult male cigarette smokers numbered 1, 2, 3, ... , 15 randomly into three equally sized groups. Group 1 will be given a placebo and told that it is a drug to reduce the impulse to smoke, group 2 will be given a nicotine patch, and group 3 will be given nicotine gum to chew. The intent is to let some time pass and then measure the effectiveness of the smoke reduction methods, as well as other physical indicators of nicotine related effects. In how many ways can the smokers be partitioned? If after partitioning, the placebo group contains subjects 1, 2, 3, 4, and 5, is there cause to doubt the efficacy of the randomization procedure?

The sample space of all partitions of 15 subjects into three groups has the form below, which is a triple of sets of size 5 drawn from the set $\{1, 2, ..., 15\}$.

$$\Omega = \left\{ (\{x_1, x_2, x_3, x_4, x_5\}, \{x_6, x_7, x_8, x_9, x_{10}\}, \{x_{11}, x_{12}, x_{13}, x_{14}, x_{15}\}) : \right.$$
$$\left. x_i \in \{1, 2, ..., 15\} \text{ and } x_i \neq x_j \text{ for all } i \neq j \right\}$$

This suggests a way of counting the elements in Ω. First we must pick a set of 5 subjects from 15 in a batch and without replacement to form the placebo group. Then from the remaining 10 subjects we sample 5 more for the nicotine patch group, and finally we are forced to use the remaining 5 subjects for the nicotine gum group. By the multiplication principle there are

$$\binom{15}{5}\binom{10}{5}\binom{5}{5} = \frac{15!}{5!\ 10!}\ \frac{10!}{5!\ 5!}\ \frac{5!}{5!\ 0!} \cdot 5! = \frac{15!}{5!\ 5!\ 5!}$$

elements of Ω. (Notice that the rightmost factor in the product of binomial coefficients is just equal to 1.) *Mathematica* tells us how many this is:

```
Binomial[15, 5] * Binomial[10, 5]
```

756 756

If a partition is selected at random, then each outcome receives probability 1/756,756 and events have probability equal to their cardinality divided by 756,756. Now the number of partitions in the set of all partitions for which the placebo group is fixed at {1, 2, 3, 4, 5} can be found similarly: we must sample 5 from 10 for the patch group, and we are left with a unique set of 5 for the gum group. Thus, the probability that the placebo group is {1, 2, 3, 4, 5} is

$$\frac{\binom{10}{5}}{\binom{15}{5}\binom{10}{5}} = \frac{1}{\binom{15}{5}}$$

which is the following number

```
1 / Binomial[15, 5]
```

$$\frac{1}{3003}$$

There is less than a one in 3000 chance that this placebo group assignment could have arisen randomly, but we should not be too hasty. Some placebo group subset must be observed, and we could say the same thing about the apparently random subset {3, 5, 8, 12, 13} as we did about {1, 2, 3, 4, 5}: it is only one in 3000 likely to have come up. Should we then doubt any partition we see? The answer is not a mathematical one. Since {1, 2, 3, 4, 5} did come up, we should look at the sampling process itself. For instance if slips of paper were put into a hat in order and the hat wasn't sufficiently shaken, the low numbers could have been on top and therefore the first group would tend to contain more low numbers. If numbers are sampled using a reputable random number generator on a computer, one would tend to be less suspicious. ■

For a population of size n and k groups of sizes n_1, n_2, ..., n_k, you should check that the reasoning of Example 7 extends easily to yield the number of partitions as

$$\frac{n!}{n_1! \, n_2! \cdots n_k!} = \begin{pmatrix} n \\ x_1 \ x_2 \ \dots x_k \end{pmatrix}$$

This expression is called the *multinomial coefficient*, and *Mathematica* can compute it using the function:

$$\text{Multinomial}[x1, x2, ..., xk]$$

The computation below gives the same result as in Example 7.

Multinomial[5, 5, 5]

---

756 756

In Exercise 18 you are asked to try to code the algorithm for a function RandPartition[poplist, numgroups, sizelist], which returns a random partition of the given population, with the given number of groups, whose group sizes are as in the list *sizelist*. Again you will find the reasoning of Example 7 to be helpful.

Exercises 1.4

1. A police lineup consists of five individuals who stand in a row in random order. Assume that all lineup members have different heights.
   (a) How many possible lineups are there?
   (b) Find and justify carefully the probability that the tallest lineup member is in the middle position.
   (c) Find the probability that at least one lineup member is taller than the person in the first position.

2. A bank machine PIN number consists of four digits. What is the probability that a randomly chosen PIN will have at least one digit that is 6 or more? (Assume that digits can be used more than once.)

3. A pushbutton combination lock entry to a vehicle consists of a five digit number, such as 94470 or 53782. To gain entry to the vehicle, one pushes a sequence of five buttons on the door in which two digits are assigned to each button, specifically the first button is 0-1, the second is 2-3, etc. How many distinct lock combinations are there? (Assume that digits can be used more than once in a combination.)

4. Use mathematical induction to prove Theorem 1.

5. (*Mathematica*) (This is a famous problem called the *birthday problem*.) There are n people at a party, and to liven up the evening they decide to find out if anyone there has the same birthday as anyone else. Find the probability that this happens, assuming that there are 365 equally likely days of the year on which each person at the party could be born. Report your answer in the form of a table of such probabilities for values of n from 15 to 35.

6. (*Mathematica*) Use DrawIntegerSample to study the distribution of the maximum sample member in a batch random sample without replacement of five integers from {1, 2, ..., 40}. Discuss the most important features of the distribution.

7. (*Mathematica*) The *mean* of a random sample X_1, X_2, ..., X_n is the arithmetical average $(X_1 + X_2 + \cdots + X_n)/n$ of the sample values. *Mathematica* can compute the sample mean with the command Mean[list]. Use DrawIntegerSample to compare the spread of the distribution of sample means of a sequenced random sample with replacement to that of sample means of a sequenced random sample without replacement. (By "spread" we mean not only the distance between the largest and smallest observations, but also the degree to which observed values tend to differ from the center of the distribution.) Is there a difference, and if so, is there a reasonable explanation for it?

8. How many total subsets of a set of size n are there? Devise a counting approach that shows that this number is a power of 2.

9. If $C(n, k)$ denotes the binomial coefficient $\binom{n}{k} = \frac{n!}{k!\,(n-k)!}$, then prove that:

$$C(2n + 1, 0) + C(2n + 1, 1) + C(2n + 1, 2) + \cdots + C(2n + 1, n) = 2^{2n}.$$

10. For a random sample of k subjects chosen from n in a batch without replacement, show that the probability that a particular subject is in the sample is k/n.

11. Which of these two poker hands is more valuable: a full house (3 cards of one rank and 2 cards of another) or a flush (5 cards of the same suit which are not simultaneously a straight flush, that is, consecutive ranks)?

12. An MP3 player user has 10 albums stored in the memory of the player, consisting of respectively 9, 9, 8, 10, 8, 9, 10, 9, 10, and 10 songs. The user plans to make a new song playlist by choosing two of the albums with 9 songs, two of the albums with 10 songs, and one of the albums with 8 songs, and randomly selected two songs each from the chosen albums. Assuming that the order in which the songs are played matters, how many playlists can be created?

13. (*Mathematica*) In Example 5, argue that the sample of 100 returns with 40 errors is more likely to happen if the true proportion of erroneous returns in the population is 20% than if it is some $p < 20\%$. You may need *Mathematica* to exhaustively check all of the possible values of p.

14. A major cookie manufacturer has an inspection plan that determines whether each carton it prepares will be shipped out for sale under its brand name or will be rejected and repackaged and sold under a generic name. A carton contains 50 packages of cookies, and a random sample of 10 packages will be removed from the carton and carefully inspected. If one or more sample packages are found that have significant cookie breakage, the whole carton will be sold as generic.

(a) If a particular carton has 3 bad packages, what is the probability that it will be sold as generic?

(b) (*Mathematica*) Under the current inspection plan, how many bad packages would be necessary in the carton in order that the probability of rejecting the carton should be at least 90%?

(c) (*Mathematica*) For a carton with 3 bad packages, what should the sample size be so that the probability of rejecting the carton is at least 75%?

15. A fair die is rolled 10 times. Find the probability that there will be at least 2 6's among the 10 rolls.

16. For a partition of n population subjects into k groups of sizes n_1, n_2, ..., n_k, find and carefully justify the probability that subject number 1 is assigned to the first group.

17. In how many distinguishable ways can the letters of the word Illinois be arranged?

18. (*Mathematica*) Write the *Mathematica* command RandPartition described at the end of the section.

19. We looked at pseudo-random number generation in Section 1.3, in which a sequence of numbers x_1, x_2, x_3, x_4, ... is generated which simulates the properties of a totally random sequence. One test for randomness that is sensitive to upward and downward trends is to consider each successive pair, (x_1, x_2), (x_2, x_3), (x_3, x_4), ... and generate a sequence of symbols U (for up) or D (for down) according to whether the second member of the pair is larger or smaller than the first member of the pair. A sequence such as UUUUUUUUD-DDDDDD with a long upward trend followed by a long downward trend would be cause for suspicion of the number generation algorithm. This example sequence has one so-called *run* of consecutive U's followed by one *run* of D's for a total of only 2 runs. We could diagnose potential problems with our random number generator by saying that we will look at a sequence such as this and reject the hypothesis of randomness if there are too few runs. For sequences that are given to have 8 U symbols and 7 D symbols, if our random number generator is working well, what is the probability that there will be 3 or fewer runs?

20. In large sampling problems, such as polling registered voters in the United States, it is not only impractical to take an exhaustive census of all population members, it is even impractical to itemize all of them in an ordered list $\{1, 2, ... , n\}$ so as to take a simple random sample from the list. In such cases a popular approach is to use *stratified sampling*, in which population members are classified into several main strata (such as their state of residence), which may not be equally sized, and then further into one or more substrata (such as their county, township, district, etc.) that are nested inside the main strata. A stratified sample is taken by sampling at random a few strata, then a few substrata within it, etc. until the smallest strata level is reached at which point a random sample of individuals is taken from that (cont.).

Here suppose there are three main strata, each with three substrata. Each of the nine substrata will also have the same number of individuals, namely 20. Two main strata will be sampled at random in a batch without replacement, then in each of the sampled strata two substrata will be sampled in a batch without replacement, and then four individuals from each substratum will be sampled in a batch without replacement to form the stratified sample. Find the probability that the first individual in the first substrata of the first stratum will be in the stratified sample. Is the probability of being in the sample the same for all individuals in this case?

1.5 Conditional Probability

In many problems, especially in random phenomena that take place over time, we are interested in computing probabilities of events using partial knowledge that some other event has already occurred. These kinds of probabilities are called *conditional probabilities*, and we will define them and study their properties and applications in this section.

A simple and intuitive way to begin is with the following example. For quite a few years the state of Illinois has been engaged in a sometimes bitter battle over a proposal for a third airport in the Chicago metropolitan area. The goal is to help relieve the burden on O'Hare and Midway airports. South suburban Peotone is often mentioned as a possible site for the new airport. Suppose that a survey drew responses to the proposal from four regions: north and northwest suburban Chicago, the city itself, south and southwest suburban Chicago, and "downstate" Illinois (a strange usage of the term intended to mean anything outside of the Chicago area). The numbers of people jointly in each region and opinion category are in the table below, and totals are computed for each row and column.

	in favor	opposed	don't care	total
north suburbs	85	34	20	139
Chicago	126	28	64	218
south suburbs	12	49	5	66
downstate	43	72	25	140
total	266	183	114	563

An individual is sampled at random from the group of 563 survey respondents. Because there are 183 who oppose the Peotone airport proposal, the probability that the sampled individual opposes the new airport is 183/563 ≈ .33. But suppose that it is known that the sampled individual is from the south suburbs. Then only the 66 people in line 3 of the table are eligible, among whom 49 oppose the proposal. Therefore, the conditional probability that the sampled individual opposes the Peotone airport given that he or she lives in the south suburbs is 49/66 ≈ .74, which is much higher than the unconditional probability. In Exercise 1 you are asked to compute several other conditional probabilities similarly.

The main idea is that if an event A is known to have happened, the sample space reduces to those outcomes in A, and every other event B reduces to its part on A, that is $A \cap B$. This motivates the following definition.

Definition 1. If A is an event with $P[A] > 0$, and B is an event, then the *conditional probability* of B given A is

$$P[B \mid A] = \frac{P[A \cap B]}{P[A]} \tag{1}$$

In the Peotone airport example, B is the event that the sampled individual opposes the Peotone airport, A is the event that he or she is from the south suburbs, and $A \cap B$ is the event that the individual both opposes the proposal and is from the south suburbs. Thus, $P[A \cap B] = 49/563$, $P[A] = 66/563$, and the conditional probability is

$$\frac{49/563}{66/563} = \frac{49}{66}$$

as if the group of survey respondents in the south suburbs were the only ones in a reduced sample space for the experiment. Be aware though that there is nothing about the definition expressed by formula (1) that requires outcomes to be equally likely.

Example 1 Among fifty gas stations in a city, there are twenty that raised their prices yesterday. A small sample of ten of the fifty stations is taken in sequence and without replacement. Find the conditional probability that there were at least three stations in the sample that increased their price given: (a) the first two in the sample increased; and (b) at least two in the sample increased. Do you expect the two answers to be the same?

The universe being sampled from is the entire collection of 50 gas stations, 20 of which raised the price and 30 did not. For part (a), given that the first two stations increased their price, the full sample space reduces to the samples of size 8 from the remaining gas stations other than the two that were picked. In that sample space, the population being sampled from includes 18 stations who raised the price and 30 who did not. The conditional probability of the event that there are at least three stations in the whole sample that increased is therefore the probability in the reduced sample space that at least one more station increased. By complementation, this is:

$$P[\text{at least 3 increases} \mid 1\text{st 2 increased}] = 1 - P[\text{no increases in final 8 samples}$$

$$= 1 - \frac{30 \cdot 29 \cdot 28 \cdot 27 \cdot 26 \cdot 25 \cdot 24 \cdot 23}{48 \cdot 47 \cdot 46 \cdot 45 \cdot 44 \cdot 43 \cdot 42 \cdot 41}$$

```
N[ 1 -  30 * 29 * 28 * 27 * 26 * 25 * 24 * 23  ]
        ───────────────────────────────────────
        48 * 47 * 46 * 45 * 44 * 43 * 42 * 41
```

0.984489

For part (b) we want to compute the probability of at least 3 increases in the sample given that at least 2 stations in the sample increased. By the definition of conditional probability, this is

$$
\begin{aligned}
\text{P[at least 3 increases} \mid \text{at least 2 increases]} \;&=\; \frac{\text{P[at least 3 increases} \cap \text{at least 2 increases]}}{\text{P[at least 2 increases]}} \\[2mm]
&=\; \frac{\text{P[at least 3 increases]}}{\text{P[at least 2 increases]}} \\[2mm]
&=\; \frac{1-\text{P[0,1, or 2 increases]}}{1-\text{P[0 or 1 increases]}}
\end{aligned}
$$

There are no increases in the sample of size 10 if and only if all stations in the sample belonged to the set of 30 non-increasing stations. This occurs with probability

$$
\text{P[0 increases]} = \frac{30\cdot 29\cdot 28\cdot 27\cdot 26\cdot 25\cdot 24\cdot 23\cdot 22\cdot 21}{50\cdot 49\cdot 48\cdot 47\cdot 46\cdot 45\cdot 44\cdot 43\cdot 42\cdot 41}
$$

There is exactly 1 increase in 10 equally likely cases in which the station that increased is in position i in the sample, and all other positions are filled by stations that did not increase prices. Note that the station that increased can be selected in 20 different ways and the rest of the sample is a permutation of 9 stations selected from 30 that didn't increase, hence this event has probability

$$
\text{P[1 increase]} = 10\cdot \frac{20\cdot 30\cdot 29\cdot 28\cdot 27\cdot 26\cdot 25\cdot 24\cdot 23\cdot 22}{50\cdot 49\cdot 48\cdot 47\cdot 46\cdot 45\cdot 44\cdot 43\cdot 42\cdot 41}
$$

We also need the probability that there are exactly 2 stations in the sample that increased prices. First, there are $\binom{10}{2} = 45$ pairs of positions in the ordered sample where the increasing stations could appear, then there are $20\,'19$ pairs of increasing stations that could be in those positions, and lastly the rest of the sample is a permutation of 8 stations selected from the 30 that didn't increase. The probability of 2 increases is therefore

$$
\text{P[2 increases]} = 45\cdot \frac{(20\cdot 19)\cdot 30\cdot 29\cdot 28\cdot 27\cdot 26\cdot 25\cdot 24\cdot 23}{50\cdot 49\cdot 48\cdot 47\cdot 46\cdot 45\cdot 44\cdot 43\cdot 42\cdot 41}
$$

Using *Mathematica*, and defining a utility function Perm[n,k] to be the number of permuations of k objects taken from n, the desired conditional probability P[at least 3 increases | at least 2 increases] comes out to be the following

```
Perm[n_, k_] := Binomial[n, k] * Factorial[k];
N[(1 - (Perm[30, 10] + 10 * 20 * Perm[30, 9] +
        45 * 20 * 19 * Perm[30, 8]) / Perm[50, 10]) / (1 -
        (Perm[30, 10] + 10 * 20 * Perm[30, 9]) / Perm[50, 10])]
```

0.888304

Comparing part (a) to part (b), the event that the first two are increasing stations is more specific than the event that at least two in the sample are increasing, and it is more favorable to the event of at least 3 increases. There is no reason to suspect that the conditional probabilities would come out equal. In the latter case, there are still many more remaining outcomes in the reduced sample space, which serves to lower the conditional probability. ∎

Activity 1 Referring to Example 1, recompute the conditional probability in part (b) in the case of sampling in a batch.

Multiplication Rules

The definition of conditional probability can be rewritten as

$$P[A \cap B] = P[A] \cdot P[B \mid A] \tag{2}$$

which can be used to compute intersection probabilities. One thinks of a chain of events in which A happens first; the chance that B also happens is the probability of A times the probability of B given A. For instance, if we draw two cards in sequence and without replacement from a standard deck, the probability that both the first and second are aces is $\frac{4}{52} \cdot \frac{3}{51}$, that is, the probability that the first is an ace times the conditional probability that the second is an ace given that the first is an ace. For three aces in a row, the probability would become $\frac{4}{52} \cdot \frac{3}{51} \cdot \frac{2}{50}$, in which the third factor is the conditional probability that the third card is an ace given that both of the first two cards are aces.

The example of the three aces suggests that formula (2) can be generalized as in the following theorem. You are asked to prove it in Exercise 7. Notice that from stage to stage one must condition not only on the immediately preceding event but all past events.

Theorem 1. Let B_1, B_2, ..., B_n be events such that each of the following conditional probabilities is defined. Then

$$P[B_1 \cap B_2 \cap \cdots \cap B_n] = $$
$$P[B_1] \cdot P[B_2 \mid B_1] \cdot P[B_3 \mid B_1 \cap B_2] \cdots P[B_n \mid B_1 \cap \cdots \cap B_{n-1}] \tag{3}$$

Example 2 Exercise 5 in Section 1.4 described the following famous problem. Suppose you ask a room of n people for their birthdays, one by one. What is the probability that there will be at least one set of matching birthdays?

In the exercise you were expected to use combinatorics to solve the problem. Here we take a conditional probability approach using formula (3). It is convenient to study the complementary event of no matching birthdays. Let B_1 be the event that there is no match after the first person is asked, which has probability 1. Let B_2 be the event that there is no match after the first two people, which has conditional probability 364/365 given B_1 since one birthday is disallowed. Similarly, if B_3 is the event that there is no match after the first three people, then B_3 has conditional probability 363/365 given the event $B_1 \cap B_2$ of no previous matches. For n people we have

$$
\begin{aligned}
P[\text{at least 1 match}] &= 1 - P[\text{no matches after } n \text{ people}] \\
&= 1 - P[B_1 \cap B_2 \cap \cdots \cap B_n] \\
&= 1 - \frac{364}{365} \cdot \frac{363}{365} \cdots \cdot \frac{365-n+1}{365}
\end{aligned}
$$

Here is a *Mathematica* function to compute that probability.

```
f[n_] := 1 - Product[ (365 - i) / 365 , {i, 0, n - 1}]
```

For example for, respectively, 20 and 50 people,

```
{N[f[20]], N[f[50]]}
```

```
{0.411438, 0.970374}
```

If you have not done the exercise, you might be surprised at how large these probabilities are. Here is a plot that shows how the probability depends on the number of people n. The halfway mark is reached at about 23 people. You should try finding the smallest n for which there is a 90% probability of a match. ∎

```
birthdays = Table[N[{n, f[n]}], {n, 5, 50}];
ListPlot[birthdays,
  PlotStyle → {Black, PointSize[0.02`]},
  AxesLabel → {n, "prob. at least 1 match"}]
```

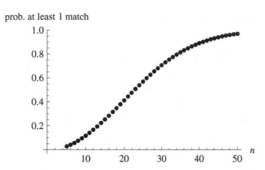

prob. at least 1 match

Figure 1.9 - The probability of matching birthdays as a function of n

---

Activity 2 Try this similar problem. Suppose the n people are each asked for the hour and minute of the day when they were born. How many people would have to be in the room to have at least a 50% chance of matching birth times?

The next proposition, though quite easy to prove, is very powerful. It goes by the name of the *Law of Total Probability*. We saw an earlier version of it in Section 2 which only dealt with intersection probabilities. The proof simply takes that result and adds on the multiplication rule (2).

Theorem 2. If A_1, A_2, ..., A_n is a partition of the sample space, that is, the A's are pairwise disjoint and their union is Ω, then

$$P[B] = \sum_{i=1}^{n} P[B \mid A_i] P[A_i] \tag{4}$$

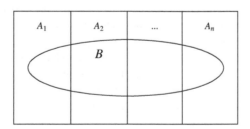

Figure 1.10 - The Law of Total Probability

Proof. The events $A_i \cap B$ are pairwise disjoint, and their union is B (see Figure 10). Thus, by the Law of Total Probability for intersections and the multiplication rule (2),

$$P[B] = \sum_{i=1}^{n} P[A_i \cap B] = \sum_{i=1}^{n} P[B \mid A_i] P[A_i] \tag{5}$$

(Note: It is easy to argue that the conclusion still holds when the A's are a partition of the event B but not a partition of all of Ω Try it.) ■

The Law of Total Probability is very useful in staged experiments where B is a second stage event and it is difficult to directly find its probability; however, conditional probabilities of B given first stage events are easier. Formula (4) lets us condition B on a collection of possible first stage events A_i and then "un-condition" by multiplying by $P[A_i]$ and adding over all of these possible first stage cases. The next example illustrates the idea.

Example 3 Each day a Xerox machine can be in one of four states of deterioration labeled 1, 2, 3, and 4 from the best condition to the worst. The conditional probabilities that the machine will change from each possible condition on one day to each possible condition on the next day are shown in the table below.

tomorrow

		1	2	3	4
	1	3/4	1/8	1/8	0
today	2	0	3/4	1/8	1/8
	3	0	0	3/4	1/4
	4	0	0	0	1

For instance, P[machine in state 2 tomorrow | machine in state 1 today] = 1/8. Given that the machine is in state 1 on Monday, let us find the probability distribution of its state on Thursday.

A helpful device to clarify the use of the law of total probability is to display the possible outcomes of a stage of a random phenomenon as branches on a tree. The next stage emanates from the tips of the branches of the current stage. The branches can be annotated with the conditional probabilities of making those transitions. For the Monday to Tuesday transition, the leftmost level of the tree in Figure 11 illustrates the possibilities. The leftmost 1 indicates the known state on Monday. From state 1 we can go to state 1 on Tuesday with probability 3/4, and the other two states each with probability 1/8. From each possible machine state 1, 2, and 3 on Tuesday we can draw the possible states on Wednesday at the next level. Then from Wednesday, the Thursday states can be drawn to complete the tree. (We omit labeling the transition probabilities on the Wednesday to Thursday branches to reduce clutter in the picture; they can be found from the table.)

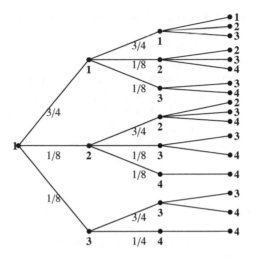

Figure 1.11 - Monday to Thursday transition probabilities

To see how the law of total probability works in this example, consider first the simpler problem of finding the probability that the state on Wednesday is 2. We condition and uncondition on the Tuesday state, which can be 1, 2, or 3 (however, if it is 3 then Wednesday's state cannot be 2).

$$
\begin{aligned}
P[\text{Wed} = 2] \; = \; & P[\text{Wed} = 2 \mid \text{Tues} = 1]\,P[\text{Tues} = 1] \\
& + P[\text{Wed} = 2 \mid \text{Tues} = 2]\,P[\text{Tues} = 2] \\
= \; & \frac{1}{8}\cdot\frac{3}{4} + \frac{3}{4}\cdot\frac{1}{8} = \frac{3}{16}
\end{aligned}
$$

Notice that this is the sum of path probabilities in Figure 11 for paths ending in 2 on Wednesday. A path probability is the product of the branch probabilities on that path. Applying the same reasoning for Thursday,

$$
\text{ThursdayProb1} = \frac{3}{4} * \frac{3}{4} * \frac{3}{4}
$$

$$
\text{ThursdayProb2} = \frac{3}{4} * \frac{3}{4} * \frac{1}{8} + \frac{3}{4} * \frac{1}{8} * \frac{3}{4} + \frac{1}{8} * \frac{3}{4} * \frac{3}{4}
$$

$$
\frac{27}{64}
$$

$$
\frac{27}{128}
$$

Similarly you can check that the probability that the state on Thursday is 3 equals 63/256, and the probability that it is 4 equals 31/256. Notice that these four numbers sum to 1 as they ought to do. ∎

Activity 3 For the previous example, finish the computation of the Wednesday probabilities. Look closely at the sum of products in each case, and at the table of transition probabilities given at the start of the problem. Try to discover a way of using the table directly to efficiently calculate the Wednesday probabilities, and check to see if the idea extends to the Thursday probabilities. For more information, see Section 6.1 on Markov chains, of which the machine state process is an example.

Bayes' Formula

The law of total probability can also be used easily to derive an important formula called *Bayes' Formula*. The setting is as above, in which a collection of events A_1, A_2, ..., A_n partitions Ω. Then $P[B]$ can be expressed in terms of the conditional probabilities $P[B \mid A_i]$, which we suppose are known. The question we want to consider now is: is it possible to reverse these conditional probabilities to find $P[A_i \mid B]$ for any $i = 1, 2, ..., n$? The importance of being able to do so is illustrated by the next example. Roughly the idea is that diagnostic tools such as lie detectors, medical procedures, etc. can be pre-tested on subjects who are known to have a condition, so that the probability that they give a correct diagnosis can be estimated. But we really want to use the diagnostic test later on unknown individuals and conclude something about whether the individual has the condition given a positive test result. In the pre-testing phase the event being conditioned on is that the individual has the condition, but in the usage phase we know whether or not the test is positive, so the order of conditioning is being reversed.

All we need to derive *Bayes' formula*, which is (6) below, is the definition of the conditional probability $P[A_i \mid B]$ and the law of total probability:

$$P[A_i \mid B] = \frac{P[A_i \cap B]}{P[B]} = \frac{P[B \mid A_i] P[A_i]}{\sum_{j=1}^{n} P[B \mid A_j] P[A_j]} \tag{6}$$

We close the section with a nice application.

Example 4 As of 2006, the Centers for Disease Control estimate of the number of people in the U.S. who were infected with the HIV virus was around 1.1 million among a total population of about 300 million, which is a proportion of about $p = .0037$. In considering policy issues like the institution of universal HIV screening, one must be aware of the possibility of needlessly frightening uninfected people with a positive screening result. Also, a person with a positive screening result could then be subjected unnecessarily to further costly, time-consuming, but more accurate diagnostic procedures to confirm the condition. Suppose for our example that a screening procedure is accurate at a level q (near 1) for people who do have HIV, and it gives an incorrect positive test with probability r (near 0), for people who do not have HIV. If a randomly selected individual has a positive screening result, what is the probability that this person actually has HIV? If the screening could be redesigned to improve performance, which would be the most important quantity to improve: q or r?

We introduce the following notation, suggested by the problem statement:

$$p = P[\text{randomly selected person has HIV}]$$
$$q = P[\text{test positive} \mid \text{person has HIV}]$$
$$r = P[\text{test positive} \mid \text{person does not have HIV}]$$

By Bayes' formula,

$$P[\text{person has HIV} \mid \text{test positive}] =$$
$$\frac{P[\text{person has HIV and test positive}]}{P[\text{test positive}]} = \frac{q \cdot p}{q \cdot p + r(1-p)}$$

(Make sure you understand the last formula.) We would like to see how best to increase this probability: by improving q (making it closer to 1) or r (making it closer to 0). We should also investigate whether the decision is different for high p vs. low p. Let us study some graphs of the function above for values of p near 0 (due to the relative rarity of the disease), q near 1, and r near 0.

$$\mathtt{f[p\_, q\_, r\_] := \frac{p*q}{p*q+r\,(1-p)}}$$

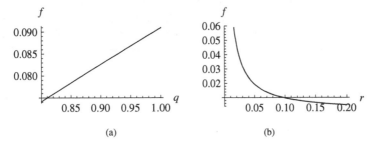

(a)   (b)

Figure 1.12 - (a) Probability of HIV given positive test, as function of q; (b) Probability of HIV given positive test, as function of r ($p = .001$ in both cases)

The results are very striking for a small p value of .001, and would be more so for even smaller p's: increasing q toward 1 produces only roughly linear increases in our objective f, to a level even less than .1. In other words, even if the test was made extremely accurate on persons with HIV, it is still unlikely that someone with a positive HIV screening actually is infected. This may be due to the very small p in the numerator, which is overcome by the term $r(1 - p)$ in the denominator. By contrast, the rate of increase of f as r decreases toward 0 is much more rapid, at least once the screening reaches an error rate of less than .05. When the probability of an incorrect positive test on a person without HIV is sufficiently reduced, the probability that truly HIV is present when the screening test reads positive dramatically increases toward 1. A 3D plot shows this effectively. (The code to produce it is in the closed cell above Figure 13.) Change in f due to increasing q is almost imperceptible in comparison to reducing r, at least when r is small enough.

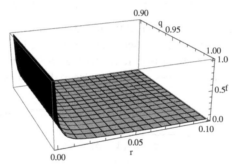

Figure 1.13 - Probability of HIV given positive test, as function of q and r ($p = .001$)

When p is not quite so small, as for instance in the case where the disease under study is a cold or flu passing quickly through a small population instead of HIV, the preference for decreasing r to increasing q is not quite as dramatic. Try repeating these graphs for some other parameter values, such as $p = .2$ to continue the investigation. (See also Exercise 14.) ■

Activity 4 Does our function f shown in Figure 12(b) reach a finite limit as $r \longrightarrow 0$? If so, what is it?

Exercises 1.5

1. In the example on the Peotone airport proposal at the start of the section, find

 (a) P[in favor | Chicago]
 (b) P[downstate | don't care]
 (c) P[north suburbs or Chicago | in favor or don't care]

2. Under what condition on $A \cap B$ will $P[A \mid B] = P[A]$? Try to interpret the meaning of this equation.

3. In the Peotone airport example, a sample of size 4 in sequence and without replacement is taken from the group of 563 survey respondents. Find the conditional probability that all four sample members oppose the proposal given that at least three of them are from the south suburbs.

4. Find $P[A^c \cap B^c]$ if $P[A] = .5$, $P[B] = .4$, and $P[A \mid B] = .6$.

5. In a small group of voters participating in a panel discussion, four are Republicans and three are Democrats. Three different people will be chosen in sequence to speak. Find the probability that
(a) all are Republicans
(b) at least two are Republicans given that at least one is a Republican

6. In a family with 3 children, what is the probability of at least 2 boys given that at least 1 child is a boy?

7. Prove Theorem 1.

8. Show that if A is a fixed event of positive probability, then the function $Q[B] = P[B \mid A]$ taking events B into $\mathbb{R}$ satisfies the three defining axioms of probability.

9. Blackjack is a card game in which each of several players is dealt one card face down which only that player can look at, and one card face up which all of the players can see. Players can draw as many cards as they wish in an effort to get the highest point total until their total exceeds 21, at which point they lose automatically. (An ace may count 1 or 11 points at the discretion of the player, all face cards count 10 points, and other cards of ranks 2-10 have point values equal to their rank.)

Suppose that you are in a game of blackjack with one other person. Your opponent's face up card is a 7, and you are holding an 8 and a 5. If you decide to take exactly one more card and your opponent takes no cards, what is the probability that you will beat your opponent's total?

Exercise 10

10. A two-dimensional random walk moves on the integer grid shown in the figure. It begins at (0,0) and on each move it has equal likelihood of going up, down, right, or left to points adjacent to its current position. What is the probability that after three moves it is at the point (0,1)?

11. (*Mathematica*) Write a conditional probability simulator for the following situation. The sample space Ω has six outcomes a, b, c, d, e, and f, with probabilities 1/6, 1/12, 1/12, 1/6, 1/3, and 1/6, respectively. Define event A as $\{a, b, c\}$ and event B as $\{b, c, d, e\}$. For a given number of trials n, your simulator should sample an outcome from Ω according to the given probabilities, and it should return the proportion of times, among those for which the outcome was in B, that the outcome was also in A. What should this proportion approach as the number of trials becomes very large?

12. (*Mathematica*) Recall the random walk example from Section 1.3 in which we used the command AbsorptionPctAtM to empirically estimate the probability that the random walk, starting at state 3, would be absorbed at state 5 for a right step probability of .4. Let $f(n)$ be the probabilty of absorption at 5 given that the random walk starts at n, for $n = 0, 1, 2, 3, 4$, 5. Use the law of total probability to get a system of equations for $f(n)$, and solve that system in *Mathematica*. Check that the solution is consistent with the simulation results.

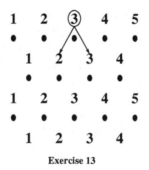

Exercise 13

13. A coin drops down a pegboard as shown in the figure, bouncing right or left with equal probability. If it starts in postion 3 in row 1, find the probability that it lands in each of the slots 1, 2, 3, and 4 in the fourth row.

14. (*Mathematica*) In the Bayes' Theorem investigation of HIV testing at the end of this section, suppose that instead our goal is to minimize the probability that a person who tests negative actually has the disease. Now what is the most important quantity to improve: q or r?

15. I am responsible for administering our college mathematics placement exam for first-year students. Suppose that students from some past years broke down into precalculus grade and placement exam score categories as follows:

grade

		A	B	C	D	F
placement	below 10	2	3	14	8	5
	10 – 20	7	20	41	5	1
	above 20	9	15	18	4	2

Assuming that the next incoming class behaves similarly, use the table directly, then use Bayes' formula to estimate the probability that someone who scores between 10 and 20 gets at least a C. Do you get the same result using both methods?

16. A credit card company studies past information on its cardholders. It determines that an index of riskiness on a scale from 1 to 5, based on factors such as number of charge accounts, total amount owed, family income, and others, might help identify new card applicants who will later default on their debts. They find that among their past defaulters, the distribution of index values was: 1:2%, 2: 15%, 3: 22%, 4: 28%, 5: 33%. Among non-defaulters the distribution of index values was: 1:28%, 2: 25%, 3: 18%, 4: 16%, 5: 13%. It is known that about 5% of all card holders default. What is the probability that a new card applicant with an index of 5 will default? Repeat the computation for the other index values 4 through 1.

Sample Problems from Actuarial Exam P

17. A public health researcher examines the medical records of a group of 937 men who died in 1999 and discovers that 210 of the men died from causes related to heart disease. Moreover, 312 of the 937 men had at least one parent who suffered from heart disease, and of these 312 men, 102 died from causes related to heart disease. Determine the probability that a man randomly selected from this group died of causes related to heart disease, given that neither of his parents suffered from heart disease.

18. A doctor is studying the relationship between blood pressure and heartbeat abnormalities in her patients. She tests a random sample of her patients and notes their blood pressures (high, low, or normal) and their heartbeats (regular or irregular). She finds that: (i) 14% have high blood pressure; (ii) 22% have low blood pressure; (iii) 15% have an irregular heartbeat; (iv) of those with an irregular heartbeat, one-third have high blood pressure; (v) of those with normal blood pressure, one-eighth have an irregular heartbeat. What portion of the patients selected have a regular heartbeat and low blood pressure?

19. An insurance company issues life insurance policies in three separate categories: standard, preferred, and ultra-preferred. Of the company's policyholders, 50% are standard, 40% are preferred, and 10% are ultra-preferred. Each standard policyholder has probability .010 of dying in the next year, each preferred policyholder has probability .005 of dying in the next year, and each ultra-preferred policyholder has probability .001 of dying in the next year. A policyholder dies in the next year. What is the probability that the deceased policyholder was ultra-preferred?

20. The probability that a randomly chosen male has a circulation problem is .25. Males who have a circulation problem are twice as likely to be smokers as those who do not have a circulation problem. What is the conditional probability that a male has a circulation problem, given that he is a smoker?

1.6 Independence

The approach to probability that we have been taking so far draws heavily on random sampling and simulation. These concepts are also very helpful in understanding the subject of this section: *independence* of events in random experiments. For example, what is the crucial difference between sampling five integers in sequence from $\{1, 2, ..., 10\}$ without replacement as compared to sampling them with replacement? In the no replacement scenario, once a number, say 4, is sampled, it cannot appear again, and other numbers are a little more likely to appear later in the sequence because 4 is no longer eligible. When sampling with replacement, 4 is just as likely to appear again later as it was the first time, and other numbers also have the same likelihood of appearing, regardless of whether 4 was sampled. In this case, the occurrence of an event like: "4 on 1st" has no effect on the probability of an event like: "7 on 2nd." This is the nature of independence, and sampling with replacement permits it to happen while sampling without replacement does not.

Another way to look at independence involves the computation of intersection probabilities. Take the example above and use combinatorics to compute the probability that the first two sampled numbers are 4 and 7. Since we still must account for the other three members of the sample, the probability of this event, assuming no replacement, is

$$P[4 \text{ on } 1\text{ st}, \ 7 \text{ on } 2\text{ nd}] = \frac{1 \cdot 1 \cdot 8 \cdot 7 \cdot 6}{10 \cdot 9 \cdot 8 \cdot 7 \cdot 6} = \frac{1}{10} \cdot \frac{1}{9} = \frac{1}{90}$$

The probability of the same event, assuming replacement, is

$$P[4 \text{ on } 1\text{ st}, \ 7 \text{ on } 2\text{ nd}] = \frac{1 \cdot 1 \cdot 10 \cdot 10 \cdot 10}{10 \cdot 10 \cdot 10 \cdot 10 \cdot 10} = \frac{1}{10} \cdot \frac{1}{10} = \frac{1}{100}$$

which differs from the answer obtained under the assumption of no replacement. Moreover, assuming again that the sample is drawn with replacement,

$$P[4 \text{ on } 1\text{ st}] = \frac{1 \cdot 10 \cdot 10 \cdot 10 \cdot 10}{10 \cdot 10 \cdot 10 \cdot 10 \cdot 10} = \frac{1}{10}, \ P[7 \text{ on } 2\text{ nd}] = \frac{10 \cdot 1 \cdot 10 \cdot 10 \cdot 10}{10 \cdot 10 \cdot 10 \cdot 10 \cdot 10} = \frac{1}{10}$$

hence

$$P[4 \text{ on } 1\text{ st}, \ 7 \text{ on } 2\text{ nd}] = \frac{1}{10} \cdot \frac{1}{10} = P[4 \text{ on } 1 \text{ st}] \cdot P[7 \text{ on } 2 \text{ nd}]$$

You should check that for sampling without replacement, the joint probability does not factor into the product of individual event probabilities.

Factorization can occur when three or more events are intersected as well. For instance when the sample is taken with replacement,

$$P[4 \text{ on 1 st}, \ 7 \text{ on 2 nd}, \ 4 \text{ on 3 rd}] = \frac{1 \cdot 1 \cdot 1 \cdot 10 \cdot 10}{10 \cdot 10 \cdot 10 \cdot 10 \cdot 10} = \frac{1}{10} \cdot \frac{1}{10} \cdot \frac{1}{10}$$

$$P[4 \text{ on 1 st}] = \frac{1 \cdot 10 \cdot 10 \cdot 10 \cdot 10}{10 \cdot 10 \cdot 10 \cdot 10 \cdot 10} = \frac{1}{10} = P[7 \text{ on 2 nd}] = P[4 \text{ on 3 rd}]$$

Therefore,

$$P[4 \text{ on 1 st}, \ 7 \text{ on 2 nd}, \ 4 \text{ on 3 rd}] = P[4 \text{ on 1 st}] \cdot P[7 \text{ on 2 nd}] \cdot P[4 \text{ on 3 rd}]$$

These three events are independent, and in fact it is easy to show that any subcollection of two of the events at a time satisfies this factorization condition also.

Having laid the groundwork for the concept of independence in the realm of sampling, let us go back to the general situation for our definition.

Definition 1. Events $B_1, B_2, \ldots B_n$ are called *mutually independent* if for any subcollection of them $B_{i_1}, B_{i_2}, \ldots B_{i_k}, 2 \leq k \leq n$,

$$P\left[B_{i_1} \cap B_{i_2} \cap \ldots \cap B_{i_k}\right] = P\left[B_{i_1}\right] P\left[B_{i_2}\right] \cdots P\left[B_{i_k}\right] \tag{1}$$

In particular for two independent events A and B, $P[A \cap B] = P[A] \cdot P[B]$, which means that

$$P[B \mid A] = \frac{P[A \cap B]}{P[A]} = \frac{P[A] \cdot P[B]}{P[A]} = P[B] \tag{2}$$

In words, if A and B are independent, then the probability of B does not depend on whether A is known to have occurred.

Activity 1 For a roll of two fair dice in sequence, check that the events $B_1 =$ "1 on 1st" and $B_2 =$ "6 on 2nd" are independent by writing out the sample space and finding $P[B_1 \cap B_2]$, $P[B_1]$, and $P[B_2]$. Compare $P[B_2 \mid B_1]$ to $P[B_2]$.

Example 1 Four customers arrive in sequence to a small antique store. If they make their decisions whether or not to buy something independently of one another, and if they have individual likelihoods of 1/4, 1/3, 1/2, and 1/2 of buying something, what is the probability that at least three of them will buy?

The problem solving strategy here is of the *divide and conquer* type; divide the large problem into subproblems, and conquer the subproblems using independence. First, the event that at least three customers buy can be expressed as the disjoint union of five subevents, from which we can write:

$$P[\text{at least 3 buy}] = P[\text{customers 1, 2, 3 buy and not 4}] + P[1, 2, 4 \text{ buy and not}$$
$$+P[1, 3, 4 \text{ buy and not 2}] + P[2, 3, 4 \text{ buy and not 1}]$$
$$+P[1, 2, 3, 4 \text{ buy}]$$

Each subevent is an intersection of independent events, so by formula (1) and the given individual purchase probabilities,

$$P[\text{at least 3 buy}] =$$

$$\left(\frac{1}{4}\right)\left(\frac{1}{3}\right)\left(\frac{1}{2}\right)\left(\frac{1}{2}\right) + \left(\frac{1}{4}\right)\left(\frac{1}{3}\right)\left(\frac{1}{2}\right)\left(\frac{1}{2}\right) + \left(\frac{1}{4}\right)\left(\frac{1}{2}\right)\left(\frac{1}{2}\right)\left(\frac{2}{3}\right)$$

$$+\left(\frac{1}{3}\right)\left(\frac{1}{2}\right)\left(\frac{1}{2}\right)\left(\frac{3}{4}\right) + \left(\frac{1}{4}\right)\left(\frac{1}{3}\right)\left(\frac{1}{2}\right)\left(\frac{1}{2}\right) = \frac{1}{6} \quad \blacksquare$$

Example 2 One frequently sees the notion of independence come up in the study of two-way tables called *contingency tables*, in which individuals in a population are classified according to each of two characteristics. Such a table appeared in Section 1.5 in reference to the Peotone airport proposal. The first characteristic was the region in which the individual lived (4 possible regions), and the second characteristic was the individual's opinion about the airport issue (3 possible opinions). If one person is sampled at random from the 563 in this group, you can ask whether such events as "the sampled person is from the north suburbs" and "the sampled person is opposed to the new airport" are independent. Using the counts in the table, we can check that $P[\text{north and opposed}] \approx .0604$, $P[\text{north}] \approx .2469$, $P[\text{opposed}] \approx .3250$, and $P[\text{north}] \cdot P[\text{opposed}] \approx .0803$. So the joint probability $P[\text{north and opposed}]$ is not equal to the product of the individual probabilities $P[\text{north}] \cdot P[\text{opposed}]$. Hence we conclude that these two events are not independent. There may be something about being from the north suburbs which changes the likelihood of being opposed to the airport. (You can conjecture that north suburban people might be more inclined to favor the airport, since it is not in their own backyard.)

However, this is only a small sample from a much larger population of Illinois residents. If you want to extend the conclusion about independence to the whole population of Illinois, you have the problem that probabilities like .0604, .2469, etc. are only estimates of the true probabilities of "north and opposed," "north," etc. for the experiment of sampling an individual from the broader universe. Sampling variability alone could account for some departures from equality in the defining condition for independence $P[A \cap B] = P[A]\,P[B]$. We pursue this idea further in the next example. $\blacksquare$

Example 3 Suppose that each individual in a random sample of size 80 from a population can be classified according to two characteristics A and B. There are two possible values or types for each characteristic. Assume that there are 30 people who are of type 1 for characteristic A, 50 of type 2 for A, and 40 each of types 1 and 2 for characteristic B. Draw one individual at random from the sample of 80. Find the unique frequencies x, y, w, and z in the table below which make events of the form "individual is type i for A", $i = 1,2$, independent of events of the form "individual is type j for B", $j = 1,2$.

		char B		
		1	2	total
char A	1	x	y	30
	2	w	z	50
	total	40	40	80

The interesting thing in this example is that the answer <u>is</u> unique; the sample of 80 can only come out in one way in order to make characteristic A events independent of characteristic B events. Without the independence constraint, x could take many values, from which y, w, and z are uniquely determined. If $x = 2$ for instance, y must be 28, w must be 38, and z must be 12 in order to produce the marginal totals in the table. In general, letting x be the free variable, the table entries must be:

		char B		
		1	2	total
char A	1	x	$30 - x$	30
	2	$40 - x$	$10 + x$	50
	total	40	40	80

For independence, we want in addition,

$$P[A = 1, B = 1] = P[A = 1]P[B = 1]$$
$$\implies \frac{x}{80} = \frac{30}{80} \cdot \frac{40}{80} \implies x = 15$$

The unique set of table entries is then $x = 15$, $y = 15$, $w = 25$, and $z = 25$. You can check (see also Exercise 3) that $\{A = 1\}$ is independent of $\{B = 2\}$ for these table entries, and also $\{A = 2\}$ is independent of $\{B = 1\}$ and of $\{B = 2\}$.

But it seems a lot to expect for the data in a sample to line up so perfectly if the real question of interest is whether in the whole population characteristic A is independent of characteristic B. There is a well-known test in statistics for independence of characteristics in contingency tables like ours, but here let us just use simulation to give us an idea of the degree of variability we can expect in the table.

Suppose our sample of 80 is taken from a universe of 200 individuals, and the universe is set up so that the characteristics are independent. One such configuration of the universe is given by the table on the left below. Let us shorten the notation a bit by using the code $A1$ to stand for the event $A = 1$, $A1B1$ for the event $A = 1$ and $B = 1$, etc. Each individual in the sample can be of type $A1B1$ with probability $60/200 = .3$, and similarly can be of type $A1B2$ with probability .3, type $A2B1$ with probability .2 and type $A2B2$ with probability .2. A sample of size 80 would be expected to have these proportions of the 80 sample members in the four categories. The expected counts in the sample are in the table on the right below; for example, the expected number of $A1B1$ individuals in the sample is $80\dot{~}(.3)$ = 24.

		char B					char B		
		1	2	total			1	2	total
char A	1	60	60	120	char A	1	24	24	48
	2	40	40	80		2	16	16	32
	total	100	100	200		total	40	40	80

universe   **sample**

The next *Mathematica* function simulates a sample of a desired size (such as 80), in sequence and with replacement, from the population categorized as in the universe table, tallies up the frequencies in each category for the sample, and presents the results in tabular format. After initializing the counter variables to zero, it selects a random number between 0 and 1, and based on its value, increments one of the counters. The cutoffs used in the Which function are chosen so that the category probabilities of .3, .3, .2, and .2 are modeled. The marginal totals are then computed for the tabular display, and the output is done.

```
SimContingencyTable[sampsize_] :=
 Module[{A1, A2, B1, B2, A1B1, A1B2, A2B1, A2B2,
   nextsample, i, outtable}, A1B1 = 0; A1B2 = 0;
  A2B1 = 0; A2B2 = 0; Do[nextsample = RandomReal[];
  Which[nextsample < 0.3`, A1B1 = A1B1 + 1,
    0.3` ≤ nextsample < 0.6`, A1B2 = A1B2 + 1,
    0.6` ≤ nextsample < 0.8`, A2B1 = A2B1 + 1,
    0.8` ≤ nextsample ≤ 1, A2B2 = A2B2 + 1],
   {i, 1, sampsize}]; A1 = A1B1 + A1B2;
  A2 = A2B1 + A2B2; B1 = A1B1 + A2B1; B2 = A1B2 + A2B2;
  outtable = {{"A/B", 1, 2, "Total"},
    {1, A1B1, A1B2, A1}, {2, A2B1, A2B2, A2},
    {"Total", B1, B2, sampsize}}; TableForm[outtable]]
```

Here are three runs of the command. For replicability, we set the seed of the random number generator.

```
SeedRandom[34 567];
SimContingencyTable[80]
SimContingencyTable[80]
SimContingencyTable[80]
```

```
A/B     1   2   Total
1       25  27  52
2       14  14  28
Total   39  41  80

A/B     1   2   Total
1       21  30  51
2       14  15  29
Total   35  45  80

A/B     1   2   Total
1       27  25  52
2       13  15  28
Total   40  40  80
```

The expected numbers under the independence assumption (24's in the first row and 16's in the second) are not always very close to what is observed. Try running the commands again, and you will regularly find discrepancies in the frequencies of at least 3 or 4 individuals. But the more important issue is this. The characteristics are independent in the population, but one must be rather lucky to have perfect independence in the sample. The statistical test for independence of categories that we referred to above must acknowledge that contingency tables from random samples have sampling variability, and should reject independence only when the sample discrepancies from independence are "large enough." ■

Activity 2   What configuration of the universe in the above example would lead to independence of characteristics A and B and totals of 100 in each of A1, A2, B1, and B2? Revise the SimContingencyTable command accordingly, and do a few runs using sample sizes of 100 to observe the degree of variability of the table entries.

Example 4   Electrical switches in a circuit act like gates which allow current to flow or not according to whether they are closed or open, respectively. Consider a circuit as displayed in Figure 14 with three switches arranged in parallel.

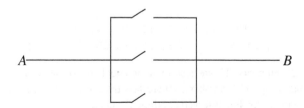

Figure 1.14 - A parallel circuit

Current coming from point A may reach point B if and only if at least one of the switches is closed. If the switches act independently of one another, and they have probabilities p_1, p_2, and p_3 of being closed, what is the probability that current will flow?

This is an easy one. The event that current will flow is the complement of the event that it will not flow. The latter happens if and only if all three switches are open. Switch i is open with probability $1 - p_i$. Thus, by independence,

$$P[\text{current flows}] = 1 - P[\text{current doesn't}] = 1 - (1 - p_1)(1 - p_2)(1 - p_3) \quad \blacksquare$$

Example 5 Our last example of this section again deals with simulation and contingency tables. A good pseudo-random number generator as described in Section 3 ought to have the property that each pseudo-random number that is generated appears to be independent of its predecessors (even though it is of course a deterministic function of its immediate predecessor). Checking for validity of the generator therefore requires a check for independence of each member from each other in a stream of numbers. There are several ways to do this, and here is one that is sensitive to unwanted upward or downward trends in the stream of numbers.

We can be alerted to a possible violation of independence if some property that independence implies is contradicted by the data. For example, in a stream of 40 numbers we can classify each number according to two characteristics: whether it is in the first group of 20 or the second, and whether it is above some fixed cutoff or below it. If the numbers are truly random and independent of one another, the group that a number is in should have no effect on whether that number is above or below the cutoff. If the two-by-two contingency table that we generate in this way shows serious departures from the tallies that we would expect under independence, then the assumption of independence is in doubt.

Here is a stream of 40 random integers between 1 and 20, simulated using DrawIntegerSample.

```
Needs["KnoxProb7`Utilities`"];
SeedRandom[11753];
datalist =
  DrawIntegerSample[1, 20, 40, Replacement -> True]
```

```
{16, 9, 7, 3, 3, 3, 15, 17, 1, 6, 3, 6, 19,
 18, 1, 6, 2, 12, 12, 18, 7, 13, 2, 14, 10, 20,
 1, 16, 14, 14, 9, 13, 4, 8, 15, 4, 15, 9, 1, 14}
```

Among the first 20 numbers, 12 are less than or equal to 10 and 8 are greater than 10. Among the second group of 20 numbers, 10 are less than or equal to 10 and 10 are greater. We therefore have the following observed contingency table.

		size		
		≤ 10	> 10	total
group	1	12	8	20
	2	10	10	20
	total	22	18	40

Equal numbers of 10 observations are to be expected in each category. Interestingly, this particular table is coming very close to independence, because the fact that the row totals are 20 means that the 22 and 18 observations in the two columns should split about evenly between the two rows, which they do. There is very little evidence against independence in this run of the experiment. ■

Activity 3 Repeat the drawing of a sample of 40 from {1, 2, ..., 20} a few more times. Do you observe many tables that are far from expectations? Try doubling the sample size to 80, and using four groups of 20 instead of two. What category counts are expected? Do you see striking departures from them in your simulations?

Exercises 1.6

1. A student guesses randomly on ten multiple choice quiz questions with five alternative answers per question. Assume that there is only one correct response to each question, and the student chooses answers independently of other questions. What is the probability that the student gets no more than one question right?

2. Consider the experiment of randomly sampling a single number from {1, 2, ..., 10}. Is the event that the number is odd independent of the event that it is greater than 4?

3. If A and B are independent events, prove that each of the following pairs of events are independent of one another: (a) A and B^c; (b) A^c and B; (c) A^c and B^c.

4. If events A and B have positive probability and are disjoint, is it possible for them to be independent?

5. Prove that if events A, B, C, and D are independent of one another, then so are the groups of events (a) A, B, and C; (b) A and B.

6. Prove that both Ω and $\emptyset$ are independent of every other event $A \subset \Omega$.

7. (*Mathematica*) Referring to Example 3, run the SimContingencyTable command a few times each for sample sizes of 40, 80, and 120. Compare the variabilities for the three sample sizes.

8. A sample space has six outcomes: a, b, c, d, e, and f. Define the event A as {a, b, c, d} and the event B as {c, d, e, f}. If each of a, b, e, and f has probability 1/8, find probabilities on c and d that make A and B independent events, or explain why it is impossible to do so.

9. Three coins are flipped in succession.

(a) If outcomes are equally likely, show that "H on 1st" is independent of "T on 2nd" and show that "H on 2nd" is independent of "T on 3rd."

(b) Put probabilities on the sample space consistent with the assumptions of independence of different flips and an unfair weighting such that tails is 51% likely to occur on a single flip.

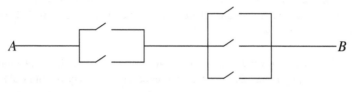

Exercise 10

10. In the figure is a part of an electrical circuit as in Example 4, except that two parallel groups of switches are connected in series. Current must pass through both groups of switches to flow from A to B. If each switch in the first group has probability p of being closed, each switch in the second group has probability q of being closed, and the switches operate independently, find the probability that current will flow from A to B.

11. (*Mathematica*) A system has three independent components which work for a random, uniformly distributed amount of time in [10, 20] and then fail. Each component can take on the workload of the other, so the system fails when the final component dies. Simulate 1000 such systems, produce a histogram of the system failure times, and estimate the probability that the system fails by time 15. What is the exact value of that probability?

12. In Example 4 we were able to find the probability that current flows from A to B easily by complementation. Compute the same probability again without complementation.

13. (*Mathematica*) In Example 5 we set up the integer sampling process so that replacement occurs. If the pseudo-random number generator behaves properly we would not expect any evidence against randomness. But the question arises: is this test sensitive enough to detect real independence problems? Try drawing integer samples again, this time of 40 integers from {1, 2, ..., 80} without replacement. Using characteristics of group (1st 20, 2nd 20) and size (40 or below, more than 40), simulate some contingency tables to see whether you spot any clear departures from independence. Would you expect any?

14. If events B_1, B_2, B_3, and B_4 are independent, show that

(a) $P[B_1 \mid B_3 \cap B_4] = P[B_1]$

(b) $P[B_1 \cap B_2 \mid B_3 \cap B_4] = P[B_1 \cap B_2]$

(c) $P[B_1 \cup B_2 \mid B_3] = P[B_1 \cup B_2]$

15. A stock goes up or down from one day to another independently. Each day the probability that it goes up is .4. Find the probability that in a week of five trading days the stock goes up at least four times.

16. Two finite or countable families of events $\mathbf{A} = \{A_1, A_2, \ldots\}$ and $\mathbf{B} = \{B_1, B_2, \ldots\}$ are called *independent* if every event $A_i \in \mathbf{A}$ is independent of every event $B_j \in \mathbf{B}$. Suppose that two dice are rolled one after the other. Let $\mathbf{A}$ consist of basic events of the form "1st die $= n$" for $n = 1, 2, \ldots, 6$, together with all unions of such basic events. Similarly let $\mathbf{B}$ be the events that pertain to the second die. Show that each basic event in $\mathbf{A}$ is independent of each basic event in $\mathbf{B}$, and use this to show that the families $\mathbf{A}$ and $\mathbf{B}$ are independent.

Sample Problem from Actuarial Exam P

17. An actuary studying the insurance preferences of automobile owners makes the following conclusions: (i) an automobile owner is twice as likely to purchase collision coverage as disability coverage; (ii) the event that an automobile owner purchases collision coverage is independent of the event that he or she purchases disability coverage; (iii) the probability that an automobile owner purchases both collision and disability coverages is .15. What is the probability that an automobile owner purchases neither collision nor disability coverage?

CHAPTER 2

DISCRETE DISTRIBUTIONS

2.1 Discrete Random Variables, Distributions, and Expectations

We begin Chapter 2 by consolidating many of the ideas that you encountered in Chapter 1, and applying them in the context of discrete random variables. Here are the two main concepts.

Definition 1. A *discrete random variable* X is a function taking the sample space Ω of a random phenomenon to a discrete (i.e., finite or countable) set E called its *state space*. For each outcome $\omega \in \Omega$, the point $x = X(\omega)$ is called a *state* of the random variable.

Definition 2. A discrete random variable X is said to have *probability mass function* (abbr. p.m.f.) $f(x)$ if for each state x,

$$f(x) = P[X = x] = P[\{\omega \in \Omega : X(\omega) = x\}]$$

So a random variable X maps outcomes to states, which are usually numerical valued. The p.m.f. of the random variable gives the likelihoods that X takes on each of its possible values. Notice that by Axiom 2 of probability, $f(x) \geq 0$ for all states $x \in E$, and also by Axiom 1, $\sum_{x \in E} f(x) = 1$. These are the conditions for a function f to be a valid probability mass function.

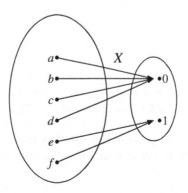

Figure 2.1 - A discrete random variable X

73

Example 1 Exercise 7 in Section 1.1 illustrates the ideas well. We repeat the diagram of that exercise here for convenience as Figure 1. Assuming that all outcomes $\{a, b, c, d, e, f\}$ in Ω are equally likely, and X maps outcomes to states $\{0, 1\}$ as shown, then the probability mass function of X is

$$f(0) = P[X = 0] = P[\{a, b, c, d\}] = \frac{4}{6} = \frac{2}{3}$$
$$f(1) = P[X = 1] = P[\{e, f\}] = \frac{2}{6} = \frac{1}{3} \qquad \blacksquare$$

Example 2 Suppose that a list of grades for a class is in the table below, and the experiment is to randomly draw a single student from the class.

grade	A	B	C	D	F	total
number of students	4	8	6	2	1	21

The sample space Ω is the set of all students, and the random variable X operates on Ω by returning the grade of the student selected. Then the state space of X is $E = \{A, B, C, D, F\}$ and the probability mass function is

$$f(A) = \frac{4}{21}; \; f(B) = \frac{8}{21}; \; f(C) = \frac{6}{21}; \; f(D) = \frac{2}{21}; \; f(F) = \frac{1}{21} \qquad (1)$$

The probability that the randomly selected student has earned at least a B for instance is

$$P[X = A \text{ or } X = B] = \; P[X = A] + P[X = B] = \; f(A) + f(B) = \frac{12}{21} \quad \blacksquare$$

The last computation in Example 2 illustrates a general rule. If X has p.m.f. f, and S is a set of states, then by the third axiom of probability,

$$P[X \in S] = \sum_{x \in S} P[X = x] = \sum_{x \in S} f(x) \qquad (2)$$

The p.m.f. therefore completely characterizes how probability distributes among the states, so that the probability of any interesting event can be found. We often use the language that f characterizes (or is) the *probability distribution of X* for this reason.

Activity 1 Encode the letter grades in Example 2 by the usual integers 4, 3, 2, 1, and 0 for *A-F*, respectively. Write a description of, and sketch a graph of, the function $F(x) = P[X \le x]$ for $x \in \mathbb{R}$.

Example 3 If you did Exercise 3 of Section 1.3 you discovered how to simulate values of a random variable with a known p.m.f. on a finite state space. Let us apply the ideas here to write a command to repeatedly simulate from the grade distribution in Example 2.

This can be done by encoding simulated uniform[0,1] observations in accordance with the grade distribution in formula (2). If, for example, such a uniform number takes a value in [0, 4/21) we can call the result an *A*. Since the probability that a simulated uniform[0,1] observation falls into an interval is equal to the length of the interval, state *A* will have probability 4/21 as desired. State *B* must have probability 8/21, so we can call the result a *B* if the uniform number falls into [4/21, 12/21), which has length 8/21. Proceeding similarly for states *C*, *D*, and *F*, we have the following command. (Encode *A* as 4, *B* as 3, etc. as in Activity 1.)

```
SimulateGrades[numreps_] :=
 Module[{therand, i, gradelist, nextgrade},
  gradelist = {};
  Do[therand = RandomReal[];

   nextgrade = Which[therand < 4/21, 4,
    4/21 ≤ therand < 12/21, 3, 12/21 ≤ therand < 18/21,
    2, 18/21 ≤ therand < 20/21, 1, therand ≥ 20/21, 0];
   AppendTo[gradelist, nextgrade], {i, 1, numreps}];
  gradelist]
```

```
Needs["KnoxProb7`Utilities`"];
SeedRandom[98 921];
Histogram[SimulateGrades[100], 5, ChartStyle → Gray]
```

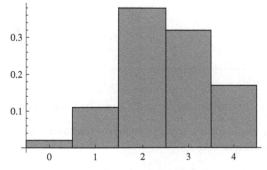

Figure 2.2 - Simulated grade distribution

Since the probabilities of states 0-4 are .05, .09, .29, .38, and .19, this particular empirical histogram fits the theoretical distribution fairly well. You might have noticed that although we are interested in simulating the selection of a random student, what we actually did in the command was to simulate a random variable Y taking [0, 1] to {4, 3, 2, 1, 0} as in Figure 3, with the same distribution as X. To produce the picture we have used *Mathematica*'s Which function to define the step function, and a utility contained in the KnoxProb7`Utilities` package called

PlotStepFunction[function, domain, jumplist]

which plots a right continuous step function on an interval with the given domain, with jumps at the given list of x values. It has a DotSize option as shown to control the size of the dots, and also accepts the AxesOrigin and PlotRange options, as does the Plot command. ∎

$$
\begin{aligned}
&\texttt{F[x\_] := Which}\left[x < \frac{4}{21},\ 4,\ \frac{4}{21} \leq x < \frac{12}{21},\ 3,\right.\\
&\quad \frac{12}{21} \leq x < \frac{18}{21},\ 2,\ \frac{18}{21} \leq x < \frac{20}{21},\ 1,\ x \geq \frac{20}{21},\ 0\bigg];\\
&\texttt{PlotStepFunction}\left[\texttt{F[x]},\ \{x,\ 0,\ 1\},\ \left\{\frac{4}{21},\ \frac{12}{21},\ \frac{18}{21},\ \frac{20}{21}\right\},\right.\\
&\quad \texttt{AxesOrigin} \rightarrow \{0,\ -0.1\},\ \texttt{PlotRange} \rightarrow \{-0.1,\ 4.1\},\\
&\quad \texttt{DotSize} \rightarrow 0.02,\ \texttt{AspectRatio} \rightarrow .5\bigg]
\end{aligned}
$$

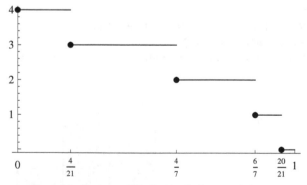

Figure 2.3 - Random variable Y taking [0,1] to numerical grades

There is an alternative characterization of the distribution of a random variable as given in the definition below.

Definition 3. A discrete random variable X with state space $E \subset \mathbb{R}$ is said to have *cumulative distribution function* (abbr. c.d.f.) $F(x)$ if for all $x \in \mathbb{R}$,

$$F(x) = P[X \le x] = P[\{\omega \in \Omega : X(\omega) \le x\}] \tag{3}$$

The c.d.f. accumulates total probability lying to the left of and including point x. It also characterizes the complete distribution of probability, because one can move back and forth from it to the probability mass function. For instance, suppose that a random variable has state space $E = \{.2, .6, 1.1, 2.4\}$ with probability masses 1/8, 1/8, 1/2, and 1/4, respectively. Then for the four states,

$$F(.2) \quad = \quad P[X \le .2] \quad = \quad f(.2) = \frac{1}{8}$$

$$F(.6) \quad = \quad P[X \le .6] \quad = \quad f(.2) + f(.6) = \frac{1}{8} + \frac{1}{8} = \frac{1}{4}$$

$$F(1.1) \quad = \quad P[X \le 1.1] \quad = \quad f(.2) + f(.6) + f(1.1) = \frac{1}{8} + \frac{1}{8} + \frac{1}{2} = \frac{3}{4}$$

$$F(2.4) \quad = \quad P[X \le 2.4] \quad = \quad f(.2) + f(.6) + f(1.1) + f(2.4) = \frac{1}{8} + \frac{1}{8} + \frac{1}{2} + \frac{1}{4} = 1$$

Between the states there is no probability, so that a full definition of F is

$$F(x) = \begin{cases} 0 & \text{if} \quad x < .2 \\ \frac{1}{8} & \text{if} .2 \le x < .6 \\ \frac{1}{4} & \text{if} .6 \le x < 1.1 \\ \frac{3}{4} & \text{if} 1.1 \le x < 2.4 \\ 1 & \text{if} \quad x \ge 2.4 \end{cases} \tag{4}$$

```
F[x_] := Which[x < 0.2, 0, 0.2 ≤ x < 0.6, 1/8,
      0.6 ≤ x < 1.1, 1/4, 1.1 ≤ x < 2.4, 3/4, x ≥ 2.4, 1];
PlotStepFunction[F[x], {x, 0, 2.6},
   {0.2, 0.6, 1.1, 2.4}, DotSize → 0.02,
   AxesOrigin → {0, -0.01}, PlotRange → {-0.01, 1.01}]
```

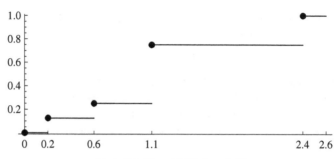

Figure 2.4 - The c.d.f. F in formula (4)

It is clear that if we know f we can then construct F, but notice also that the probability masses on the states are the amounts by which F jumps; for example $f(.6) = 1/8 = 1/4 - 1/8 = F(.6) - F(.6^-)$, where $F(x^-)$ denotes the left-hand limit of F at the point x. Hence, if we know F we can construct f. In general, the relationships between the p.m.f. $f(x)$ and the c.d.f. $F(x)$ of a discrete random variable are

$$F(x) = \sum_{t \le x} f(t) \text{ for all } x \in \mathbb{R}; \quad f(x) = F(x) - F(x^-) \text{ for all states } x. \tag{5}$$

Figure 4 displays the c.d.f. F for this example. Notice that it begins at functional value 0, jumps by amount $f(x)$ at each state x, and reaches 1 at the largest state.

Activity 2 Consider a p.m.f. on a countably infinite state space $E = \{0, 1, 2, \ldots\}$ of the form $f(x) = \frac{1}{2^{x+1}}$. Argue that f is a valid mass function and find an expression for its associated c.d.f. F. What is the limit of $F(x)$ as $x \to \infty$?

Example 4 The c.d.f. is sometimes a better device for getting a handle on the distribution of a random variable than the p.m.f. For instance, remember that in Chapter 1 we simulated random samples, for each of which we calculated a maximum or minimum sample value. We then used a list of such extrema to give a histogram of their empirical distributions. Here we will find the exact theoretical distribution of a maximum sample value.

Suppose that the random experiment is to draw five numbers randomly in sequence and with replacement from $\{1, 2, \ldots, 20\}$. Let X be the maximum of the sample values, e.g., if $\omega = \{5, 2, 10, 4, 15\}$ then $X(\omega) = 15$. Though it is not clear immediately what is the p.m.f. of X, we can compute the c.d.f. easily. In order for the largest observation X to be less than or equal to a number x, all five sample values (call them X_1, X_2, X_3, X_4, X_5) must be less than or equal to x. Since sampling is done with replacement, we may assume that the five sample values are mutually independent; hence, for $x \in \{1, 2, \ldots, 20\}$,

$$F(x) \;=\; P[X \le x] \;=\; P[X_1 \le x, \, X_2 \le x, \, X_3 \le x, \, X_4 \le x, \, X_5 \le x]$$
$$= \; \prod_{i=1}^{5} P[X_i \le x]$$
$$= \; \left(\tfrac{x}{20}\right)^5$$

The last line follows because for each sampled item, the chance that it is less than or equal to x is the number of such numbers (x) divided by the total number of numbers (20) that could have been sampled. The p.m.f. of X is therefore

$$f(x) \;=\; P[X = x] \;=\; P[X \le x] - P[X \le x - 1]$$
$$= \; \left(\tfrac{x}{20}\right)^5 - \left(\tfrac{x-1}{20}\right)^5 \tag{6}$$

for $x \in \{1, 2, ..., 20\}$. ∎

Activity 3 Use formula (6) above to make a list plot in *Mathematica* of the values of f with the plot points joined. Superimpose it on a histogram of 500 simulated values of X using the DrawIntegerSample command with option Replacement->True in the Knox-Prob7`Utilities` package to observe the fit of the sample to the theoretical distribution.

Mean, Variance, and Other Moments

In Section 1.1 we briefly introduced the notion of expected value. Now we shall give a good definition in the context of discrete random variables, and explore certain special expectations of powers of the random variable called *moments* of the probability distribution. Remember that expected value is the average of the states, with states being weighted by their probabilities.

Definition 4. If X is a discrete random variable with p.m.f. $f(x)$, then the expected value of X is

$$E[X] \;=\; \sum_{x \in E} x \cdot f(x) \tag{7}$$

where the sum is taken over all states. Similarly, the expected value of a function g of X is

$$E[g(X)] = \sum_{x \in E} g(x) \cdot f(x) \tag{8}$$

In the case of the expectation of a function $g(X)$ note that the values of the function $g(x)$ are again weighted by the state probabilities $f(x)$ and summed, so that $E[g(X)]$ is a weighted average of the possible values $g(x)$. Often $E[X]$ is called the *mean of the distribution of X* (or just the *mean of X*), and is symbolized by μ.

Example 5 Suppose that on a particular day, a common stock price will change by one of the increments: $-1/4$, $-1/8$, 0, 1/8, 1/4, or 3/8, with probabilities .06, .12, .29, .25, .18, and .10, respectively. Then the expected price change is

$$\mu = (-1/4)\ (.06) + (-1/8)\ (.12) + (0)\ (.29) + \\ (1/8)\ (.25) + (1/4)\ (.18) + (3/8)\ (.10)$$

0.08375

Let us also find the expected absolute deviation of the price change from the mean. Symbolically, this is $E[\,|X - \mu\,|]$, and it is of importance because it is one way of measuring the spread of the probability distribution of X. It is calculated as the weighted average $\sum_{x \in E} |x - \mu| \cdot f(x)$. In particular for the given numbers in this example,

$$\text{Abs}\,[(-1/4) - \mu]\ *\ (.06) + \text{Abs}\,[(-1/8) - \mu]\ *\ (.12) + \\ \text{Abs}\,[0 - \mu]\ *\ (.29) + \text{Abs}\,[1/8 - \mu]\ *\ (.25) + \\ \text{Abs}\,[1/4 - \mu]\ *\ (.18) + \text{Abs}\,[3/8 - \mu]\ *\ (.10)$$

0.138725

∎

The so-called *moments* of X play a key role in describing its distribution. These are expectations of powers of X. We have already met the mean $\mu = E[X]$, which is the *first moment about 0*. In general, the *rth moment about a point a* is $E[(X - a)^r]$. Next to the mean, the most important moment of a distribution is the second moment about μ, i.e.,

$$\sigma^2 = \text{Var}(X) = E[(X - \mu)^2] \tag{9}$$

which is called the *variance of the distribution of X* (or just the *variance of X*). The variance gives a weighted average of squared distance between states and the average state μ; therefore, it is a commonly used measure of spread of the distribution. Its square root $\sigma = \sqrt{\sigma^2}$ is referred to as the *standard deviation* of the distribution.

For the probability distribution of Example 5, where we computed that $\mu = .08375$, the variance and standard deviation are

$$\sigma\text{squared} = \\ (-1/4 - \mu)^2\ (.06) + (-1/8 - \mu)^2\ (.12) + (0 - \mu)^2\ (.29) + \\ (1/8 - \mu)^2\ (.25) + (1/4 - \mu)^2\ (.18) + (3/8 - \mu)^2\ (.10) \\ \sigma = \sqrt{\sigma\text{squared}}$$

0.0278297

0.166822

Besides the mean and the variance, the moment of most importance is the third moment about the mean $E[(X - \mu)^3]$. It is similar to the variance, but because of the cube it measures an average signed distance from the mean, in which states to the right of μ contribute positively, and states to the left contribute negatively. So for example, if there is a collection of states of non-negligible probability which are a good deal greater than μ, which are not offset by states on the left side of μ, then we would expect a positive value for the third moment. On the left side of Figure 5 is such a probability distribution. Its states are {1, 2, 10, 20}, with probabilities {.25, .50, .125, .125}. Because of the long right tail we say that this distribution is *skewed to the right*. To produce the plot we have used the command

ProbabilityHistogram[statelist, problist]

in KnoxProb7`Utilities`, which requires the list of states and the corresponding list of probabilities for the states. It accepts the options of Graphics together with its own BarColor option to control the look of the diagram.

```
g1 = ProbabilityHistogram[{1, 2, 10, 20},
        {1/4, 1/2, 1/8, 1/8}, BarColor → Gray];
g2 = ProbabilityHistogram[{1, 2, 15, 20},
        {1/8, 1/8, 1/2, 1/4}, BarColor → Gray];
GraphicsRow[{g1, g2}]
```

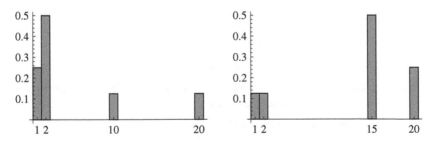

Figure 2.5 - Two skewed probability mass functions

We find the mean and third moment about the mean to be

$$\mu = (.25)\,(1) + (.50)\,(2) + (.125)\,(10) + (.125)\,(20)$$
$$\textbf{thirdmoment} = (.25)\,(1 - \mu)^3 +$$
$$(.50)\,(2 - \mu)^3 + (.125)\,(10 - \mu)^3 + (.125)\,(20 - \mu)^3$$

5.

408.

The third moment about the mean is positive, as expected. Do the next activity to convince yourself that when the distribution has a long left tail, the third moment about the mean will be negative, in which case we call the distribution *skewed to the left*.

Activity 4 Compute the mean and the third moment about the mean of the distribution shown on the right of Figure 5. (The states are {1, 2, 15, 20} and their probabilities are {1/8, 1/8, 1/2, 1/4}.)

Below are some important results about expectation and variance. The first two parts will require some attention to joint probability distributions of two random variables before we can prove them. We postpone these until later in this chapter, but a preview is given in Exercise 19. We will supply proofs of the other parts.

Theorem 1. Let X and Y be random variables (with finite means and variances wherever they are referred to). Then

(a) $E[X + Y] = E[X] + E[Y]$

(b) For any constants a and b, $E[a\,X + b\,Y] = a\,E[X] + b\,E[Y]$

(c) If a is a constant then $E[a] = a$

(d) $\mathrm{Var}(X) = E[X^2] - \mu^2$, where $\mu = E[X]$

(e) If c and d are constants, then $\mathrm{Var}(c\,X + d) = c^2\mathrm{Var}(X)$

Proof. (c) The constant a can be viewed as a random variable X with one state $x = a$ and p.m.f. $f(x) = 1$ if $x = a$ and 0 otherwise. Thus,

$$E[a] = a\cdot P[X = a] = a\cdot 1 = a$$

(d) Since μ and μ^2 are constants and $E[X] = \mu$,

$$\begin{aligned}\mathrm{Var}(X) &= E\big[(X - \mu)^2\big]\\ &= E\big[X^2 - 2\mu X + \mu^2\big]\\ &= E\big[X^2 - 2\mu E[X] + \mu^2\big]\\ &= E\big[X^2\big] - \mu^2\end{aligned}$$

(e) First, the mean of the random variable $c\,X + d$ is $c\,\mu + d$, by parts (b) and (c) of this theorem. Therefore,

$$
\begin{aligned}
\mathrm{Var}(c\,X + d) &= \mathrm{E}\big[\{(c\,X + d) - (c\,\mu + d)\}^2\big] \\
&= \mathrm{E}\big[\{c\,X - c\,\mu\}^2\big] \quad\blacksquare \\
&= c^2\,\mathrm{E}\big[\{X - \mu\}^2\big] = c^2\,\mathrm{Var}(X)
\end{aligned}
$$

This theorem is a very important result which deserves some commentary. Properties (a) and (b) of Theorem 1 make expectation pleasant to work with: it is a linear operator on random variables, meaning that the expected value of a sum (or difference) of random variables is the sum (or difference) of the expected values, and constant coefficients factor out. The intuitive interpretation of part (c), as indicated in the proof, is that a random variable that is constant must also have that constant as its average value. Part (d) is a convenient computational formula for the variance that expresses it in terms of the second moment about 0. For example if the p.m.f. of a random variable X is $f(1) = 1/8$, $f(2) = 1/3$, $f(3) = 1/4$, $f(4) = 7/24$, then

$$
\mu = \mathrm{E}[X] = \frac{1}{8}\cdot 1 + \frac{1}{3}\cdot 2 + \frac{1}{4}\cdot 3 + \frac{7}{24}\cdot 4 = \frac{65}{24}
$$

$$
\mathrm{E}\big[X^2\big] = \frac{1}{8}\cdot 1^2 + \frac{1}{3}\cdot 2^2 + \frac{1}{4}\cdot 3^2 + \frac{7}{24}\cdot 4^2 = \frac{201}{24}
$$

$$
\mathrm{Var}(X) = \frac{201}{24} - \left(\frac{65}{24}\right)^2 = \frac{599}{576}
$$

The last computation is a bit easier than computing the variance directly from the definition $\mathrm{Var}(X) = \mathrm{E}[(X - 65/24)^2]$, though it is not a major issue in light of the availability of technology. The formula in part (d) is actually more useful on some occasions when for various reasons we know the variance and the mean and want to find the second moment. Finally with regard to the theorem, part (e) shows that the variance is not linear as the mean is: constant coefficients factor out as squares, which happens because variance is defined as an expected square. Also, the variance measures spread rather than central tendency, so that the addition of a constant like d in part (e) of the theorem does not affect the variance.

Example 6 A discrete probability distribution on the non-negative integers that we will study later is the *Poisson distribution*. It has a parameter called μ, and is referred to in *Mathematica* as PoissonDistribution[mu]. The probability mass function for this distribution is

$$
f(k) = \frac{e^{-\mu}\,\mu^k}{k!}, \quad k = 0, 1, 2, \dots
$$

We do not know the mean and variance of this distribution yet, but let us simulate large random samples from the Poisson distribution and use the empirical distribution to estimate them. Can we develop a conjecture about how the mean and variance relate to the parameter μ?

As we will be seeing later, the mean μ and variance σ^2 of a distribution can be estimated by the *sample mean* $\overline{X}$ and *sample variance* S^2 calculated as below from a random sample X_1, X_2, ... , X_n from that distribution.

$$\overline{X} = \frac{1}{n} \sum_{i=1}^{n} X_i \; ; \; S^2 = \frac{1}{n-1} \sum_{i=1}^{n} \left(X_i - \overline{X} \right)^2 \tag{10}$$

These are sensible estimators of the distributional mean and variance, since $\overline{X}$ is a simple arithmetical mean of the data values, and S^2 is a version of an average squared distance of the data values from the sample mean. It is a major result in probability theory that these sample estimators will converge to μ and σ^2 as the sample size goes to infinity.

To simulate random samples, note that there is a discrete version of the RandomReal command called RandomInteger that applies to any named discrete distribution object in *Mathematica*. Also, the commands Mean and Variance, when applied to a list, will return the sample mean and variance, respectively. Here is one simulated random sample of size 100 from the Poisson(2) distribution, a histogram, and a report of the sample mean and variance. Results will differ with the simulation, but it appears as if both the mean and variance are close to the parameter $\mu = 2$.

```
P2 = PoissonDistribution[2];
SeedRandom[65 476];
datalist2 = RandomInteger[P2, 100]
Histogram[datalist2, 6, ChartStyle → Gray]
{N[Mean[datalist2]], N[Variance[datalist2]]}
```

```
{3, 3, 5, 2, 4, 0, 3, 0, 4, 2, 5, 2, 2, 3, 3, 0, 4, 0, 2, 3, 1,
 3, 2, 2, 2, 0, 4, 1, 2, 4, 1, 4, 2, 3, 0, 1, 1, 0, 1, 3, 2,
 3, 1, 0, 1, 1, 3, 1, 1, 3, 2, 1, 6, 0, 1, 1, 3, 2, 1, 5, 0,
 1, 4, 2, 2, 2, 3, 1, 1, 4, 2, 2, 1, 6, 1, 4, 3, 1, 1, 1, 2,
 2, 0, 4, 0, 1, 2, 0, 1, 3, 2, 2, 1, 0, 1, 3, 3, 2, 1, 3}
```

{2., 2.0404}

To check to see whether this happens again for another value of μ, observe the simulation below for $\mu = 5$. We suppress the actual sample values and the histogram this time.

```
P5 = PoissonDistribution[5];
SeedRandom[10 237];
datalist5 = RandomInteger[P5, 100];
{N[Mean[datalist5]], N[Variance[datalist5]]}
```

{4.81, 5.18576}

In the electronic version of the text, you should delete the SeedRandom command and try rerunning these simulations several times. Try also increasing the sample size. You should find convincing empirical evidence that the Poisson distribution with parameter μ has mean equal to μ and also has variance equal to μ. ∎

Two Special Distributions

The rest of this chapter is primarily a catalog of the most common discrete distributions, their moments, and their applications. We will begin this catalog here by looking at two simple distributions that arise easily from work we have already done on random sampling: the discrete uniform distribution and the hypergeometric distribution.

Example 7   When one item is sampled at random from a set of consecutive integers $\{a, a + 1, ..., b\}$, since there are $(b - a + 1)$ equally likely states, the probability mass function of the selected item is clearly

$$f(x) = \frac{1}{b - a + 1} \text{ if } x \in \{a, a + 1, ..., b\} \tag{11}$$

which is called the *discrete uniform distribution* on $\{a, a + 1, ..., b\}$.

Many examples of this kind of random experiment spring to mind, from draft lotteries to dice rolling and other games of chance. Because states are equally likely, the probability that the sampled item X falls into a set B is simply the number of states in B divided by the number of states in the whole space.

Usually this distribution is used as a building block in other kinds of problems however. There is a famous story, perhaps apocryphal, about German tanks in World War II, which apparently were numbered sequentially from 1 to the highest number. Allied intelligence had observed a sample of tank numbers, and was interested in using the sample to estimate how many tanks the Germans had.

To formulate a mathematical model, suppose that tank numbers $X_1, X_2, ... , X_n$ are observed in the sample, suppose also that events involving different X_i's are mutually independent (we will have a good definition of independent random variables later), and suppose that each X_i has the discrete uniform distribution on $\{1, 2, ..., \theta\}$, where θ is the total number of tanks in existence. Note that the c.d.f. $F(x) = P[X \leq x]$ of this distribution would have the value (x/θ) at each state x (why?). We can then proceed just as in Example 4 to obtain the c.d.f. and mass function of $Y = \max\{X_1, X_2, ..., X_n\}$, the largest tank number observed in the sample. (You should fill in the details.)

$$G(y) = P[Y \leq y] = \left(\frac{y}{\theta}\right)^n$$

$$g(y) = P[Y = y] = \left(\frac{y}{\theta}\right)^n - \left(\frac{y-1}{\theta}\right)^n, \quad y \in \{1, 2, ..., \theta\}$$

One sensible way of proceeding from here is to have *Mathematica* compute the expected value of this distribution for a fixed sample size n and various values of θ, and choose as our estimate of θ the one for which the expected value of Y comes closest to the highest tank number actually observed in the sample. Suppose for the sake of the example that 40 tanks are observed and the highest tank number in the sample is 125. The first function below is the p.m.f. of the maximum tank number in the sample, as a function of the sample size n and the highest tank number in the population of tanks θ. The second function gives the expected value of the highest tank number in the sample, as a function of n and θ.

```
g[y_, n_, θ_] := (y / θ)ⁿ - ((y - 1) / θ)ⁿ
```

```
ExpectedHighTank[n_, θ_] :=
  NSum[y * g[y, n, θ], {y, 1, θ}]
```

Figure 6 shows the graph of the expected value in terms of θ for a sample size of $n = 40$.

```
ListPlot[
  Table[{θ, ExpectedHighTank[40, θ]}, {θ, 125, 130}],
  AxesLabel → {"θ", "Expected High Tank"},
  PlotStyle → {Black, PointSize[0.02`]}]
```

Figure 2.6 - Expected value of highest tank number as a function of θ

The two values of θ of 127 and 128 have the closest expected high tank number to 125. By evaluating ExpectedHighTank at each of these choices, you will find that $\theta = 128$ is the closest estimate. ∎

Activity 5 Show that the expected value of the discrete uniform distribution on $\{1, 2, ..., n\}$ is $(n + 1)/2$, and the variance is $\left(n^2 - 1\right)/12$.

Example 8 Suppose that each member of a finite population can be classified as either having a characteristic or not. There are numerous circumstances of this kind: in a batch of manufactured objects each of them is either good or defective, in a deck of cards each card is either a face card or not, etc. The particular application that we will look at in this example is the classical capture-recapture model in ecology, in which a group of animals is captured, tagged, and then returned to mix with others of their species. Thus, each member in the population is either tagged or not.

Suppose that there are N animals in total, M of which are tagged. A new random sample of size n is taken in a batch without replacement. Let the random variable X be the number of tagged animals in the new sample. Reasoning as in Section 1.4, the probability that there are x tagged animals in the sample, and hence $n - x$ untagged animals, is

$$P[X = x] = \frac{\binom{M}{x}\binom{N-M}{n-x}}{\binom{N}{n}},$$

(12)

for integers x such that $0 \le x \le M$ and $0 \le n - x \le N - M$

The p.m.f. in formula (12) is called the *hypergeometric* p.m.f. with parameters N, M, and n. The restrictions on the state x guarantee that we must sample a non-negative number of tagged and untagged animals, and we cannot sample more than are available of each type.

For instance, if $N = 20$ animals, $M = 10$ of which were tagged, the probability of at least 4 tagged animals in a sample of size 5 is

$$\frac{\texttt{Binomial[10, 4] * Binomial[10, 1]}}{\texttt{Binomial[20, 5]}} + \frac{\texttt{Binomial[10, 5] * Binomial[10, 0]}}{\texttt{Binomial[20, 5]}}$$

$$\frac{49}{323}$$

Let us try to compute the expected value of the hypergeometric distribution with general parameters N, M, and n, and then apply the result to the animal tagging context. First we need to check that the assumption that the sample was taken in a batch can be changed to the assumption that the sample was taken in sequence, without changing the distribution of the number of tagged animals. By basic combinatorics, if the sample was taken in sequence,

$$P[X = x] = \frac{\text{\# sequences with exactly } x \text{ tagged}}{\text{\# sequences possible}}$$

$$= \frac{(\text{\# orderings of a batch of } n)\cdot(\text{\# batches with } x \text{ tagged and } n-x \text{ untagged})}{\text{\# sequences possible}}$$

$$= \frac{n!\cdot\binom{M}{x}\binom{N-M}{n-x}}{N\cdot(N-1)\cdots(N-n+1)} = \frac{\binom{M}{x}\binom{N-M}{n-x}}{\binom{N}{n}}$$

Since the last quotient is the same as formula (12), we may dispense with the assumption of batch sampling. This allows us to write X, the number of tagged animals in the sample, as

$$X = X_1 + X_2 + \cdots + X_n$$

where

$$X_i = \begin{cases} 1 & \text{if the } i^{\text{th}} \text{ animal in the sample is tagged} \\ 0 & \text{otherwise} \end{cases}$$

By linearity of expectation, $E[X] = \sum_{i=1}^{n} E[X_i]$. But each X_i has the same probability distribution, which gives weight $P[X_i = 1] = M/N$ to state 1 and weight $1 - M/N$ to state 0. Thus, the expected value of the hypergeometric random variable X is

$$E[X] = \sum_{i=1}^{n} \left(1 \cdot \frac{M}{N} + 0 \cdot \left(1 - \frac{M}{N} \right) \right) = \frac{nM}{N} \qquad (13)$$

The mean number of tagged animals in our sample of 5 is therefore 5(10)/20 = 2.5. We will have to wait to compute the variance of the hypergeometric distribution until we know more about dependence between random variables. ■

   Mathematica's kernel contains objects that represent the distributions that we will meet in this chapter, in addition to which are ways of using these objects to find mass function values and cumulative distribution function values, and to produce simulated observations of random variables. The two named distributions that we have worked with in this subsection are written in *Mathematica* as:

   DiscreteUniformDistribution[{a,b}]   (i.e., states $\{a, a + 1, \ldots, b\}$)
   HypergeometricDistribution[n, M, N]   (sample size n, M with characteristic, N total)

and we introduced the PoissonDistribution object in Example 6. When you want a value of the p.m.f. of a named distribution (including its parameter arguments) at a state x, you issue the command

PDF[distribution, x]

(The "D" in "PDF" stands for "density," a word that will be motivated in Chapter 3.) For example, the earlier computation of the probability of at least 4 tagged animals could also have been done as follows. To save writing we give a temporary name to the distribution, and then we call for the sum of the p.m.f. values:

```
dist1 = HypergeometricDistribution[5, 10, 20];
PDF[dist1, 4] + PDF[dist1, 5]
```

$$\frac{49}{323}$$

For cumulative probabilities the function

CDF[distribution, x]

gives the value of the c.d.f. $F(x) = P[X \leq x]$. For example, we can get a list of c.d.f. values at the states of the uniform distribution on $\{1, 2, \ldots, 10\}$ as follows:

```
dist2 = DiscreteUniformDistribution[{1, 10}];
Table[N[CDF[dist2, i]], {i, 1, 10}]
```

{0.1, 0.2, 0.3, 0.4, 0.5, 0.6, 0.7, 0.8, 0.9, 1.}

Notice that the c.d.f. increments by exactly 1/10 at each state.

Here we display a histogram of the hypergeometric distribution with parameters $n = 5$, $M = 10$, $N = 20$ from the animal tagging example. We give a shorthand name to the distribution, write out the list of states longhand in a *Mathematica* list, create the list of probability masses by using Table and the PDF command, and finally call on ProbabilityHistogram.

```
dist1 = HypergeometricDistribution[5, 10, 20];
statelist = {0, 1, 2, 3, 4, 5};
problist = Table[PDF[dist1, x], {x, 0, 5}];
ProbabilityHistogram[
  statelist, problist, BarColor → Gray]
```

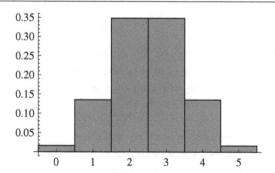

Figure 2.7 - Histogram of the hypergeometric(5,10,20) distribution

Finally, here are 10 simulated observations from the hypergeometric distribution with the same parameters as above.

```
RandomInteger[dist1, 10]
```

{2, 2, 2, 3, 2, 2, 4, 3, 3, 1}

We will see more distributions as we go through this chapter. You may also be interested in looking at other utility functions in the kernel, such as Quantile, which is essentially the inverse of the c.d.f., and Mean which returns the mean of a given distribution.

Exercises 2.1

1. Below are three possible probability mass functions for a random variable whose state space is {1, 2, 3, 4, 5}. Which, if any, is a valid p.m.f.? For any that are valid, give the associated c.d.f.

state	1	2	3	4	5
$f_1(x)$	.01	.42	.23	.15	.10
$f_2(x)$	.20	.30	.20	.15	.20
$f_3(x)$	.16	.31	.18	.10	.25

2. Two fair dice are rolled in succession. Find the p.m.f. of (a) the sum of the two; (b) the maximum of the two.

3. (*Mathematica*) Simulate 1000 rolls of 2 dice, produce a histogram of the sum of the two up faces, and compare the empirical distribution to the theoretical distribution in part(a) of Exercise 2. Do the same for the maximum of the two dice and compare to the theoretical distribution in Exercise 2(b).

4. Devise an example of a random variable that has the following c.d.f.

$$F(x) = \begin{cases} 0 & \text{if } x < 0 \\ 1/8 & \text{if } 0 \le x < 1 \\ 4/8 & \text{if } 1 \le x < 2 \\ 7/8 & \text{if } 2 \le x < 3 \\ 1 & \text{if } x \ge 3 \end{cases}$$

5. Find the c.d.f., and then the p.m.f., of the random variable X that is the minimum of a sample of five integers taken in sequence and with replacement from {1, 2, ..., 20}.

6. Consider a probability distribution on the set of integers $\mathbb{Z}$

$$f(x) = \begin{cases} \frac{2}{9} \cdot \left(\frac{1}{3}\right)^{|x|-1} & \text{if } x = \pm 1, \pm 2, \pm 3, \ldots \\ \frac{1}{3} & \text{if } x = 0 \end{cases}$$

Verify that this is a valid p.m.f., and show that its c.d.f. F satisfies the conditions

$$\lim_{x \to -\infty} F(x) = 0 \; ; \; \lim_{x \to +\infty} F(x) = 1$$

7. (*Mathematica*) Find the p.m.f. of the random variable X that is the largest item in a random sample of size 5 from {1, 2, ..., 20} taken in a batch and without replacement. Verify using *Mathematica* that your function is a valid p.m.f.

8. Why is $E[X - \mu] = 0$ for every random variable X with finite mean μ? Why is the variance of any random variable non-negative? Is there any situation in which the variance can be zero?

9. Find the mean, variance, and third moment about the mean for the probability distribution $f(1.5) = .3$, $f(2) = .2$, $f(2.5) = .2$, $f(3.5) = .3$.

10.   Compare the variances of the two distributions with probability mass functions $f(0) = .25$, $f(1) = .5$, $f(2) = .25$ and $g(0) = .125$, $g(1) = .75$, $g(2) = .125$. Draw pictures of the two p.m.f.'s that account for the differing variances.

11.   Compare the third moments about the mean $E[(X - \mu)^3]$ for the two distributions with probability histograms below.

```
g1 = ProbabilityHistogram[{0, 1, 2},
      {0.25, 0.5, 0.25}, BarColor → Gray];
g2 = ProbabilityHistogram[{0, 2, 4},
      {0.5, 0.25, 0.25}, BarColor → Gray];
GraphicsRow[{g1, g2}]
```

Exercise 11

12. For a random variable with state space {0, 1, 2, ...} and c.d.f. F, show that

$$E[X] = \sum_{n=0}^{\infty} (1 - F(n))$$

(Hint: Start with the right side and write $1 - F(n)$ as a sum.)

13. Derive a computational formula for $E[(X - \mu)^3]$ in terms of moments about 0.

14. Construct an example of a random variable whose mean is $\mu = 1$ and whose variance is $\sigma^2 = 1$

15. Some authors object to presenting the formula $E[g(X)] = \Sigma\ g(x) \cdot f(x)$ as a definition of the expected value of a function of X. Rather, they prefer to say that $Y = g(X)$ defines a new random variable Y, which has its own p.m.f. $f_y(y)$, and expectation $E[Y] = \Sigma\ y \cdot f_y(y)$. Give an argument that the two versions of expectation are the same in the simplest case where g is a 1-1 function.

16. Find the mean and variance of the discrete uniform distribution on the set of states $\{-n, -n + 1, \ldots, 0, 1, \ldots, n - 1, n\}$.

17. If a discrete random variable X has mean 6.5 and second moment about 0 equal to 50, find the mean and variance of $Y = 2X - 6$.

18. Two games are available. Game 1 returns $5 with probability 1/2, and -$1 with probability 1/2. Game 2 returns $15 with probability 1/4, and -$2 with probability 3/4.

(a) Compute the mean and variance of return from the two games. Is one clearly better than the other?

(b) Now suppose that the gambler values a monetary amount of x by the utility function $U(x) = \sqrt{x + 2}$. Which of the two games has the higher expected utility?

19. Here is a sneak preview of the proof of linearity of expectation for two discrete random variables X and Y. First, X and Y have *joint probability mass function* $f(x, y)$ if for all state pairs (x, y),

$$P[X = x, Y = y] = f(x, y)$$

The expectation of a function of X and Y is defined analogously to the one variable case:

$$E[g(X, Y)] = \sum_x \sum_y g(x, y) \cdot f(x, y)$$

Use these definitions to show that $E[aX + bY] = a E[X] + b E[Y]$.

20. (*Mathematica*) In the context of the animal capture-recapture model of Example 7, suppose that $M = 10$ animals are tagged, a sample of size 8 gave 3 tagged animals, and the total population size N is unknown. What is the most likely value for N? (Compute the appropriate probabilities in *Mathematica* for various possible N.)

21. (*Mathematica*) In a batch of 100 lightbulbs, 7 have defective filaments and will last only a very short time. A sample of size 10 is taken in a batch and without replacement. Find the mean number of defective bulbs in the sample exactly, and use *Mathematica* to numerically calculate the variance of the number of defective bulbs.

22. (*Mathematica*) Draw probability histograms of the hypergeometric distribution with $N = 100$ for the cases: (a) $M = 50$ and $n = 10, 20, 30$; (b) $n = 20$ and $M = 40, 50, 60$. Comment on the centers of the histograms.

23. (*Mathematica*) Simulate (at least) 4 random samples of size 24, in sequence and with replacement, from the discrete uniform distribution on $\{1, 2, 3, 4\}$ and plot histograms of the samples. Comment on the variability of the samples. Increase the sample size to 100, and comment.

Sample Problems from Actuarial Exam P

24. An insurance policy pays $100 per day for up to 3 days of hospitalization and $50 per day for each day of hospitalization thereafter. The number of days of hospitalization X is a discrete random variable with probability mass function

$$P[X = k] = \begin{cases} \frac{6-k}{15} & \text{if } k = 1, 2, 3, 4, 5 \\ 0 & \text{otherwise} \end{cases}$$

Determine the expected payment for hospitalization under this policy.

25. A probability distribution of the claim sizes for an auto insurance policy is given in the table below. What percentage of the claims are within one standard deviation of the mean claim size?

claim size	20	30	40	50	60	70	80
probability	.15	.10	.05	.20	.10	.10	.30

2.2 Bernoulli and Binomial Random Variables

One of the simplest discrete distributions is called the *Bernoulli distribution*. This arises in dichotomous situations, for example a flipped coin is either heads or tails, a computer bit is either set to 1 or 0, a sampled item from a population either has a certain characteristic or not, etc. With no perjorative intent we use the conventional language "success" and "failure" for the two possible experimental outcomes. A probability p is attached to the success outcome and $q = 1 - p$ is attached to the failure outcome. Using the computer bit model as our template, a random variable X is said to have the *Bernoulli distribution* if its state space is {0, 1}, and its p.m.f. is as follows:

$$f(1) = P[X = 1] = p$$
$$f(0) = P[X = 0] = q = 1 - p \tag{1}$$

The mean and variance of the Bernoulli distribution are easy to find.

$$\mu = E[X] = 1\,(p) + 0\,(1 - p) = p \tag{2}$$

$$\sigma^2 = E\big[(X - p)^2\big] = (1 - p)^2\,p + (0 - p)^2\,(1 - p) = p(1 - p) \tag{3}$$

Activity 1 Recompute the variance of the Bernoulli distribution using the computational formula for variance $E[X^2] - \mu^2$.

Example 1 In the game of roulette, the roulette wheel has 38 slots into which a ball can fall when the wheel is spun. The slots are numbered 00, 0, 1, 2, ... , 36, and the numbers 0 and 00 are colored green while half of the rest are colored black and the other half red. A gambler can bet $1 on a color such as red, and will win $1 if that color comes up and lose the dollar that was bet otherwise. Consider a $1 bet on red. The gambler's winnings can then be described as a random variable $W = 2X - 1$, where X is 1 if the ball lands in a red slot and $X = 0$ if not. Clearly X has the Bernoulli distribution. The probability of a success, that is red, is $p = 18/38$ if all slots are equally likely. By linearity of expectation and formula (2), the expected winnings are:

$$E[W] = E[2X - 1] = 2E[X] - 1 = 2\left(\frac{18}{38}\right) - 1 = -\frac{2}{38}$$

Since the expected value is negative, this is not a good game for the gambler. (What do you think are the expected winnings in 500 such bets?) The winnings are variable though, specifically by the properties of variance discussed in Section 2.1 together with formula (3),

$$\mathrm{Var}(W) = \mathrm{Var}(2X - 1) = 4\,\mathrm{Var}(X) = 4\left(\frac{18}{38}\right)\left(\frac{20}{38}\right) = \frac{360}{361}$$

Since the variance and standard deviation are high relative to the mean, for the single $1 bet there is a fair likelihood of coming out ahead (exactly 18/38 in fact). We will see later that this is not the case for 500 bets. ∎

Mathematica's syntax for the Bernoulli distribution is

BernoulliDistribution[p]

The functions PDF, CDF, and RandomInteger can be applied to it in the way described at the end of the last section. Here for example is a short program to generate a desired number of replications of the experiment of spinning the roulette wheel and observing whether the color is red (R) or some other color (O). We just generate Bernoulli 0-1 observations with success parameter $p = 18/38$, and encode 0's as O and 1's as R.

```
SimulateRoulette[numreps_] :=
Table[If[
    RandomInteger[BernoulliDistribution[18 / 38]] == 1,
    "R", "O"], {numreps}]
```

```
SimulateRoulette[20]
```

{R, O, O, R, O, O, O, R, O, R, O, R, O, R, O, O, O, O, O, R}

Activity 2 Write a *Mathematica* command to simulate 500 \$1 bets on red in roulette, and return the net winnings. Run the command 20 times. How often are the net winnings positive?

The Bernoulli distribution is most useful as a building block for the more important *binomial distribution*, which we now describe. Instead of a single performance of a dichotomous success-failure experiment, consider n independent repetitions (usually called *trials*) of the experiment. Let the random variable X be defined as the total number of successes in the n trials. Then X has possible states $\{0, 1, 2, ..., n\}$. To find the probability mass function of X, consider a particular outcome for which the number of successes is exactly k. For concreteness, suppose that there are $n = 7$ trials, and we are trying to find the probability that the number of successes is $k = 4$. One such outcome is SSFFSFS, where the four successes occur on trials 1, 2, 5, and 7. By the independence of the trials and the fact that success occurs with probability p on any one trial, the probability of this particular outcome is

$$p \cdot p \cdot (1 - p) \cdot (1 - p) \cdot p \cdot (1 - p) \cdot p = p^4 (1 - p)^3$$

By commutativity of multiplication, it is easy to see that any such outcome with four successes and 3 failures has the same probability. In general, the probability of any particular n trial sequence with exactly k successes is $p^k(1 - p)^{n-k}$. How many different n trial sequences are there that have exactly k successes? A sequence is determined uniquely by sampling k positions from among n in which the S symbols are located. By combinatorics, there are $\binom{n}{k}$ such sequences. Thus the overall probability of exactly k successes in n Bernoulli trials is the following, which defines the *binomial probability mass function*.

$$f(k) = P[X = k] = \binom{n}{k} p^k (1 - p)^{n-k} , \quad k = 0, 1, 2, ..., n \tag{4}$$

As a shorthand to indicate that a random variable X has a particular distribution, we use the symbol " $\sim$ ", and we also use abbreviations to stand for some of the most common distributions and their parameters. Here, to say that X has the binomial distribution with n trials and success probability p, we write

$$X \sim b(n, p)$$

The Bernoulli distribution is the special case of the binomial distribution in which there is only $n = 1$ trial. Notice in (4) that in this case the state space is $\{0, 1\}$, and the p.m.f. formula reduces to

$$f(k) = \binom{1}{k} p^k (1 - p)^k = p^k(1 - p)^{1-k}, \quad k = 0, 1 \tag{5}$$

which is an equivalent form to (1).

Mathematica's binomial distribution object is

BinomialDistribution[n, p]

where the meanings of the arguments are the same as described above. We will show the use of this distribution object in the example below.

Example 2 In the roulette problem, let us try to compute the probability that the gambler who bets \$1 on red 500 times comes out ahead. This event happens if and only if there are strictly more wins than losses among the 500 bets. So we want to compute $P[X \geq 251]$, or what is the same thing, $1 - P[X \leq 250]$, where X is the number of reds that come up. Since X has the binomial distribution with $n = 500$ and $p = 18/38$, we can compute:

```
N[1 - CDF[BinomialDistribution[500, 18 / 38], 250]]
```

0.110664

So our gambler is only around 11% likely to profit on his venture. Another way to have made this computation is to add the terms of the binomial p.m.f. from 251 to 500:

```
NSum[PDF[BinomialDistribution[500, 18 / 38], k],
  {k, 251, 500}]
```

0.110664

(If you execute this command you will find that it takes a good deal longer than the one above using the CDF, because *Mathematica* has some special algorithms to approximate sums of special forms like the binomial CDF.) ∎

To recap a bit, you can identify a binomial experiment from the following properties: it consists of n independent trials, and on each trial either a success (probability p) or a failure (probability $q = 1 - p$) can occur. The binomial p.m.f. in formula (4) applies to the random variable that counts the total number of successes. Shifting the focus over to sampling for a moment, the trials could be repeated samples of one object from a population, replacing the sampled object as sampling proceeds, and observing whether the object has a certain special characteristic (a success) or not (a failure). The success probability would be the proportion of objects in the population that have the characteristic. One of the exercises at the end of this section invites you to compare the probability that k sample members have the characteristic under the model just described to the probability of the same event assuming sampling in a batch without replacement.

Activity 3 Use *Mathematica* to compute the probability that the gambler who bets 500 times on red in roulette comes out at least \$10 ahead at the end. If the gambler plays 1000 times, what is the probability that he will not show a positive profit?

In Figure 8(a) is a histogram of 500 simulated values from the binomial distribution with $n = 10$ trials and $p = 1/2$. Figure 8(b) is a histogram of the probability mass function itself, which shows a similar symmetry about state 5, similar spread, and similar shape.

```
Needs["KnoxProb7`Utilities`"]
datalist =
  RandomInteger[BinomialDistribution[10, 0.5], 500];
pmf[k_] := PDF[BinomialDistribution[10, 0.5], k];
states = {0, 1, 2, 3, 4, 5, 6, 7, 8, 9, 10};
probs = Table[pmf[k], {k, 0, 10}];
GraphicsRow[{Histogram[datalist, 11, "Probability",
    ChartStyle → Gray], ProbabilityHistogram[
    states, probs, BarColor → Gray]}]
```

(a)   (b)

Figure 2.8 - (a) simulated histogram; (b) probability histogram of b(10,.5)

The binomial mass function is always symmetric about its center when $p = 1/2$ (see Exercise 11). In the cases $p > 1/2$ and $p < 1/2$, respectively, the distribution is skewed left and right. (Convince yourself of these facts analytically by looking at the formula for the p.m.f., and intuitively by thinking about the binomial experiment itself.) The command below allows you to manipulate the value of p between a low value of $p = .1$ and a high value of $p = .9$ for the case $n = 10$ and see the effect on the histogram of the binomial mass function. Try to find the symmetric case. You should try adapting it to other values of n in the electronic version of the text.

```
pmf[k_, n_, p_] := PDF[BinomialDistribution[n, p], k];
states = Table[i, {i, 0, 10}];
binprobs[p_] := Table[pmf[k, 10, p], {k, 0, 10}];
Manipulate[ProbabilityHistogram[states,
   binprobs[p], BarColore → Gray], {p, .1, .9}]
```

We now turn to the moments of the binomial distribution. Think about these questions: If you flip 4 fair coins, how many heads would you expect to get? What if you flipped 6 fair coins? Analyze how you are coming up with your guesses, that is, of what simpler things is your final answer composed? What does this mean about the expected value of the number of heads in an odd number of flips, say 7?

After the last paragraph you probably can guess the formula for the mean of the binomial distribution. Compare your hypothesis with the following brute force calculations of means in *Mathematica*, done by summing the probability of a state, times the state, over all states.

```
BinomialPMF[x_, n_, p_] :=
  PDF[BinomialDistribution[n, p], x]
```

```
(* case n = 20, p = .2 *)
Sum[BinomialPMF[k, 20, .2] * k, {k, 0, 20}]
```

4.

```
(* case n = 30, p = .8 *)
Sum[BinomialPMF[k, 30, .8] * k, {k, 0, 30}]
```

24.

The binomial mean is of course $n \cdot p$, which we prove in the next theorem. We also include the formula for the variance of the binomial distribution.

Theorem 1. If $X \sim b(n, p)$, then (a) $E[X] = n \cdot p$; (b) $\text{Var}(X) = n \cdot p \cdot (1 - p)$

Proof. (a) We may factor n and p out of the sum that defines $E[X]$, delete the $k = 0$ term (which equals 0), and cancel the common factors of k to get

$$E[X] \;=\; \Sigma_{k=0}^{n} \, k \cdot \binom{n}{k} \, p^k \, (1 - p)^{n-k}$$

$$\;=\; n \cdot p \cdot \Sigma_{k=1}^{n} \, \frac{(n-1)!}{(k-1)! \, (n-k)!} \, p^{k-1} \, (1 - p)^{n-k}$$

Then we can change variables in the sum to $l = k - 1$, noting that in the exponent of $1 - p$ we can rewrite $n - k$ as $n - 1 - (k - 1)$, which reduces the sum to that of all terms of the $b(n - 1, \; p)$ probability mass function.

$$E[X] \;=\; n \cdot p \cdot \Sigma_{l=0}^{n-1} \, \frac{(n-1)!}{l! \, (n-k)!} \, p^l \, (1 - p)^{n-1-l}$$

$$\;=\; n \cdot p \cdot 1$$

Since this sum equals 1, the desired formula $n \cdot p$ for the mean results.

(b) We use the indirect strategy of computing $E[X \, (X - 1)]$ first, because this sum can be simplified very similarly to the sum above for $E[X]$.

$$E[X \, (X - 1)] \;=\; \Sigma_{k=0}^{n} \, k \cdot (k - 1) \cdot \binom{n}{k} \, p^k \, (1 - p)^{n-k}$$

$$\;=\; n \cdot (n - 1) \cdot p^2 \cdot \Sigma_{k=2}^{n} \, \frac{(n-2)!}{(k-2)! \, (n-k)!} \, p^{k-2} \, (1 - p)^{n-k}$$

The sum on the right adds all terms of the $b(n - 2, \; p)$ p.m.f., hence it equals 1. (Use the change of index $l = k - 2$ to see this.) Therefore, $E[X(X - 1)] = n \cdot (n - 1) \cdot p^2$, and we can write the left side as $E[X^2] - E[X]$. Finally,

$$\begin{aligned}
\mathrm{Var}\,(X) \;&=\; E[X^2] - (E[X])^2 \\
&=\; E[X^2] - E[X] + E[X] - (E[X])^2 \\
&=\; n \cdot (n - 1) \cdot p^2 + np - (np)^2
\end{aligned}$$

which simplifies to the desired formula. ∎

Example 3 An overnight package delivery service guarantees successful overnight delivery. If the true probability that a delivery is successfully made on time is .98, in a group of 1000 packages find the mean and standard deviation of the number of successful deliveries, and compute the probability that at least 975 of the packages will be delivered overnight.

By the formulas for the mean and variance in the last theorem, we compute the mean and standard deviation of the number X of successfully delivered packages as:

```
μ = N[1000 * .98]
σ = N[√1000 * .98 * .02]
```

```
980.
```

```
4.42719
```

For the second question, we have that

$$P[\text{at least 975 successes}] = 1 - P[\text{no more than 974 successes}],$$

which is:

```
1 - N[CDF[BinomialDistribution[1000, .98], 974]]
```

```
0.890067
```

■

Activity 4 We will see later that a very large proportion of the probability weight for any distribution falls within 3 standard deviations (which comes out to about 13 in Example 3) of the mean of the distribution. So the delivery company can be fairly certain that between 967 and 993 of the packages will be delivered on time overnight. Use the CDF command to find the probability of this event.

Example 4 Many attempts have been made to estimate the number of planets in the universe on which there is some form of life, based on some very rough estimates of proportions of stars that have planetary systems, average numbers of planets in those systems, the probability that a planet is inhabitable, etc. These estimates use random variables that are assumed to be modeled by binomial distributions. At the time of this writing, scientists are estimating that 2% of stars have planetary systems. Suppose that only 1% of those systems have a planet that is inhabitable (assume conservatively that there is at most one such planet per system), and that only 1% of inhabitable planets have life. For a galaxy of, say 100 billion $= 10^{11}$ stars, let us find the expected number of stars with a planet that bears life.

Consider each star as a trial, and define a success as a star with a system that includes a planet that has life. Then there are $n = 10^{11}$ trials, and by the multiplication rule for conditional probability, the success probability per star is

$$p = P[\text{life}] = P[\text{has system and has inhabitable planet and life}]$$
$$= (.02)(.01)(.01) = 2 \times 10^{-6}$$

Therefore, the expected value of the number of planets with life is $n\,p = 2 \times 10^5 = 200,000$. Incidentally, the standard deviation of the number of planets with life would be

$$\sigma = \text{N}\left[\sqrt{10^{11} \times \left(2 \times 10^{-6}\right) \left(1 - 2 \times 10^{-6}\right)}\,\right]$$

447.213

As mentioned in the last activity, most of the probability weight for any distribution falls within 3 standard deviations of the mean. Since 3σ is around 1300, which is a great deal less than the mean of 200,000, this means that it is very likely that the number of planets with life is in the vicinity of 200,000 under the assumptions in this example. ∎

Multinomial Distribution

A generalization of the binomial experiment consists of n independent trials, each of which can result in one of k possible categories or values. Think of these values as k different types of success. Category i has probability p_i, and we must have that the sum of all p_i for $i = 1$ to k equals 1. This so-called *multinomial experiment* therefore involves k counting random variables $X_1, X_2, \ldots, X_k$, where X_i is the number of trials among the n that resulted in value i. Since all of the n trials must result in exactly one value, the total of all of these X_i's must come out to n. For example, consider the roll of a fair die $n = 12$ times in succession. On each roll, there are $k = 6$ possible faces that could come up. Random variables associated with this experiment are $X_1 = \#\,1$'s among the 12 rolls, $X_2 = \#\,2$'s among the 12 rolls, etc. In addition to finding how these random variables behave individually, we might be interested in probabilities of events involving intersections, such as the event that there are at least 3 1's rolled and at most 2 6's.

　　Let us consider a somewhat smaller case with unequal outcome probabilities to see the underlying idea. For a multinomial experiment with 10 trials and 4 possible categories on each trial whose probabilities are 1/4, 1/2, 1/8, and 1/8, let us find an expression for the *joint probability mass function* of the category frequencies X_1, X_2, X_3, and X_4 defined by

$$f(x_1, x_2, x_3, x_4) = P[X_1 = x_1 \cap X_2 = x_2 \cap X_3 = x_3 \cap X_4 = x_4]$$

(It is common practice to make such intersections more readable by replacing the $\cap$ symbol by a comma.) Reasoning as in the development of the binomial p.m.f., any sequence of 10 trials that contains x_1 1's, x_2 2's, x_3 3's, and x_4 4's has probability

$$\left(\frac{1}{4}\right)^{x_1} \left(\frac{1}{2}\right)^{x_2} \left(\frac{1}{8}\right)^{x_3} \left(\frac{1}{8}\right)^{x_4}$$

The number of such sequences is the number of ways of partitioning 10 trials into x_1 positions for 1's, x_2 positions for 2's, x_3 positions for 3's, and x_4 positions for 4's. We observed in Section 1.4 that such a partition can be done in $\binom{10}{x_1\ x_2\ x_3\ x_4}$ ways. Hence,

$$P[X_1 = x_1 , X_2 = x_2 , X_3 = x_3 , X_4 = x_4] = \binom{10}{x_1 \ x_2 \ x_3 \ x_4}\left(\frac{1}{4}\right)^{x_1}\left(\frac{1}{2}\right)^{x_2}\left(\frac{1}{8}\right)^{x_3}\left(\frac{1}{8}\right)^{x_4}$$

After the next example we will write the general formula for the joint probability mass function of k multinomial random variables with success probabilities $p_1, p_2, \ldots, p_k$ and n trials.

Example 5 In a study of the on-time performance of a bus service at a particular stop, each day the categories might be: early, on time, and late, say with probabilities .12, .56, and .32, respectively. What is the probability that in an eight day period the bus is early once, on time 4 times, and late 3 times? What is the probability that the bus is late at least 6 times more than it is early?

Denoting the three daily categories as E, O, and L, two of the many configurations that are consistent with the requirements are:

$$\text{E O O O O L L L}$$
$$\text{O E O L O L L O}$$

The probability of each such configuration is clearly $(.12)^1(.56)^4(.32)^3$, by the independence of the trials.   There are $\binom{8}{1 \ 4 \ 3} = \frac{8!}{1! \times 4! \times 3!}$ such configurations. Therefore the overall probability of exactly 1 early arrival, 4 on time arrivals, and 3 late arrivals is

$$P[X_1 = 1, \ X_2 = 4, \ X_3 = 3] = \binom{8}{1 \ 4 \ 3}(.12)^1 (.56)^4 (.32)^3$$

Remember that *Mathematica* can compute the multinomial coefficient as a function of the category frequencies as Multinomial[x1, x2, ..., xk]. Hence, the desired probability is

```
Multinomial[1, 4, 3] * .12 * (.56)⁴ * (.32)³
```

0.108278

For the second question, we must itemize disjoint cases that make up the event that the bus is late at least 6 times more than it is early, and then add their probabilities. Since there are only eight days in the study, the possible cases are $\{X_1 = 0, X_2 = 2, X_3 = 6\}$, $\{X_1 = 0, X_2 = 1, X_3 = 7\}$, $\{X_1 = 1, X_2 = 0, X_3 = 7\}$, and $\{X_1 = 0, X_2 = 0, X_3 = 8\}$. The total probability of the desired event is then:

```
Multinomial[0, 2, 6] * .12⁰ * (.56)² * (.32)⁶ +
  Multinomial[0, 1, 7] * .12⁰ * (.56)¹ * (.32)⁷ +
  Multinomial[1, 0, 7] * .12¹ * (.56)⁰ * (.32)⁷ +
  Multinomial[0, 0, 8] * .12⁰ * (.56)⁰ * (.32)⁸
```

0.0114074

■

In general, the *multinomial distribution* is the joint distribution of the numbers X_1, X_2, ..., X_k of items among n trials that belong to the categories 1, 2, ..., k, respectively. If the category probabilities are p_1, p_2, ... , p_k, where $\Sigma_i p_i = 1$, then the joint probability mass function is

$$f(x_1, x_2, ..., x_k) = P[X_1 = x_1, X_2 = x_2, ..., X_k = x_k]$$
$$= \binom{n}{x_1 \; x_2 \; ... \; x_k} p_1^{x_1} p_2^{x_2} \cdots p_k^{x_k}, \qquad (6)$$
$$\Sigma_i x_i = n, \; x_i \geq 0 \text{ for all } i$$

The individual distributions of random variables that have the joint multinomial distribution are easy to find if we go back to first principles. Consider X_1 for example. Each of the n trials can either result in a category 1 value, or some other. Lump together all of the other categories into one category thought of as a failure, and combine their probabilities into one failure probability. Thus, X_1 counts the number of successes in a binomial experiment with success probability p_1. This reasoning shows that if X_1, ... , X_k have the multinomial distribution with n trials and category probabilities p_1, ..., p_k, then each X_i has the binomial distribution with parameters n and p_i. In a roll of 12 fair dice in succession for example, the number of 6's has the $b(12, 1/6)$ distribution. The final activity of this section asks you to generalize this reasoning.

Activity 5 If X_1, X_2, X_3, X_4, and X_5 have the multinomial distribution with $n = 20$ trials and category probabilities .1, .2, .3, .2, and .2, respectively, find the joint p.m.f. of X_1, X_2, and X_3.

Exercises 2.2

1. Devise two examples other than the ones given at the start of the section of random phenomena that might be modeled using Bernoulli random variables

2. Explain how a random variable X with the b(n, p) distribution can be thought of as a sum of Bernoulli(p) random variables.

3. Find the third moment about the mean for the Bernoulli distribution with parameter p.

4. The Bernoulli distribution can be generalized slightly by removing the restriction that the state space is $\{0, 1\}$. Consider Bernoulli distributions on state spaces $\{a, b\}$, where without loss of generality $a < b$. The associated probability mass function would be $f(k) = p$, if $k = b$, and $f(k) = q = 1 - p$ if $k = a$. Then there will be a unique Bernoulli distribution for each fixed mean, variance pair. In particular, given p, $\mu = E[X]$ and $\sigma^2 = \text{Var}(X)$, find states a and b that produce the desired values of μ and σ^2.

5. (*Mathematica*) Referring to the roulette problem, Example 2, write a command that tracks the gambler's net winnings among n spins betting \$1 on red as n gets large. The command should compute and list plot the net winnings per spin. Run the command several times for large n. To what does the net winnings per spin seem to converge?

6. Suppose that the number of people among n who experience drowsiness when taking a certain blood pressure medication has the binomial distribution with success probability $p = .4$. Find the probability that no more than 3 among 10 such recipients of the medication become drowsy.

7. Find the second moment about 0 for the $b(n, p)$ distribution.

8. (*Mathematica*) Suppose that in front of a hotel in a large city, successive 20 second intervals constitute independent trials, in which either a single cab will arrive (with probability .2), or no cabs will arrive (with probability .8). Find the probability that at least 30 cabs arrive in an hour. What is the expected number of cabs to arrive in an hour?

9. (*Mathematica*) The 500 Standard & Poor's companies are often used to measure the performance of the stock market. One rough index simply counts the numbers of "advances," i.e., those companies whose stocks either stayed the same in value or rose. The term "declines" is used for those companies whose stocks fell in value. If during a certain bull market period 70% of the Standard & Poor's stocks are advances on average, what is the probability that on a particular day there are at least 300 advances? What are the expected value and variance of the number of advances?

10. For what value of the parameter p is the variance of the binomial distribution largest? (Assume that n is fixed.)

11. Prove that when $p = .5$, the p.m.f. of the binomial distribution is symmetric about p.

12. (*Mathematica*) Write a command in *Mathematica* to simulate 500 at bats of a baseball hitter whose theoretical average is .300 (that is, his probability of a hit on a given at bat is .3). Use the command to simulate several such 500 at bat seasons, making note of the empirical proportion of hits. What is the expected value of the number of hits, and what is the standard deviation of the batting average random variable (i.e., hits/at bats)?

13. Use linearity of expectation to devise an alternative method of proving the formula $E[X] = n\,p$ where $X \sim b(n, p)$. (Hint: See Exercise 2.)

14. (*Mathematica*) Unscrupulous promoters have booked 55 people on a bus trip to a shopping mall, knowing that the bus will only hold 52. They are counting on the fact that history indicates that on average 10% of the people who book reservations do not show up for the trip. The fine print on the reservations promises a double-your-money back refund of $100 if for any reason a reserved seat is not available for the ticket holder on the day of the trip. What is the probability that the promoters will have to issue at least one refund? What is the expected total refund that will be paid?

15. (*Mathematica*) A population of 100 people contains 20 who are regular consumers of a certain soft drink. For a product survey, samples of size 10 are taken under two procedures: (1) sampling in a batch without replacement; (2) sampling in sequence with replacement. Compute and compare the probability mass functions of the number of people in the sample who are regular consumers of this soft drink for the two procedures. Compute relative differences between corresponding probability masses for the two procedures for each state 0, 1, 2, ... , 10. (The *relative difference* between positive numbers a and b is the absolute difference $|a - b|$ divided by the smaller of a and b.)

16. Each day, a woman will visit one politically oriented on line chat room among four that she likes. She is equally likely to visit chat room 1 as chat room 2, and she is twice as likely to visit each of these as she is to visit each of chat rooms 3 and 4. In a 4 day period, find the probability that she visits chat room 1 at least 3 times.

17. (*Mathematica*) Write a *Mathematica* function which computes values of the multinomial probability mass function (see formula (6)). The input parameters should be a list of number of desired occurrences of each of the states $\{x_1, x_2, ..., x_k\}$, the number of trials n, and a list of multinomial probabilities $\{p_1, p_2, ..., p_k\}$. Use it to find the probability that in 20 rolls of a fair die, you roll 4 sixes, 7 fives, 6 fours, and 1 three. (*Mathematica* note: The function Multinomial will not apply directly to a sequence enclosed in list braces such as the x's above, but if you compose Multinomial with the *Mathematica* function Sequence using the syntax: Multinomial[Sequence@@xlist], then the list braces will be stripped off before feeding the list into Multinomial. Or, just use the Factorial function to make your own version of the multinomial coefficient.)

Sample Problems from Actuarial Exam P

18. A hospital receives 1/5 of its flu vaccine shipments from Company X and the remainder of its shipments from other companies. Each shipment contains a very large number of vaccine vials. For Company X's shipments, 10% of the vials are ineffective. For every other company, 2% of the vials are ineffective. The hospital tests 30 randomly selected vials from a shipment and finds that one vial is ineffective. What is the probability that this shipment came from Company X?

19. A large pool of adults earning their first driver's license includes 50% low-risk drivers, 30% moderate-risk drivers, and 20% high-risk drivers. Because these drivers have no prior driving record, an insurance company considers each driver to be randomly selected from the pool. This month, the insurance company writes 4 new policies for adults earning their first driver's license. What is the probability that these 4 will contain at least two more high-risk drivers than low-risk drivers?

20. A study is being conducted in which the health of two independent groups of ten policyholders is being monitored over a one-year period of time. Individual participants in the study drop out before the end of the study with probability .2 (independently of the other participants). What is the probability that at least 9 participants complete the study in one of the two groups, but not in both groups?

| 2.3 Geometric and Negative Binomial Random Variables

In the last section we learned about binomial experiments, which are closed-ended and sequential. There was a fixed number of trials, each of which could result only in two possible outcomes called success and failure. We now consider the case of an open-ended experiment which continues until some fixed number r of successes is reached. Consider for example an electrical switch which endures most of its strain when it changes from the off to the on state. There is no appreciable wear on the switch, but it has some low probability of switch failure each time it is switched on, which does not change with use. Suppose that we call a "success" a malfunction of the switch and a "failure" a switch-on without malfunction. Then the lifetime of the switch, measured in terms of number of times it is switched on without malfunction, is the number of "failures" required to reach the first "success" event. A random variable X which counts the number of Bernoulli trials strictly before the first success occurs (i.e., the number of failures) has what is called the *geometric distribution*. A natural generalization is a random variable which counts the number of failures prior to the rth success where r is a positive integer. The distribution of this random variable is known as the *negative binomial distribution*.

Activity 1   Devise other examples of experimental situations where the geometric or negative binomial distribution would be a natural model.

It is fairly easy to derive the p.m.f.'s of these two distributions. Consider first the geometric experiment which repeats dichotomous trials with success probability p until the first success occurs. The event that the number of failures is exactly k is the event that there are k failures in a row, each of probability $q = 1 - p$, followed by a success, which has probability p. By the independence of the trials, the number X of failures prior to the first success has the following *geometric p.m.f. with parameter p* (abbr. geometric(p)):

$$f(k) = P[X = k] = (1 - p)^k \cdot p, \quad k = 0, 1, 2, 3, \ldots \tag{1}$$

You can think about the negative binomial experiment in a similar way, although now the possible arrangements of successes and failures in the trials prior to the last one must be considered. In order for there to be exactly k failures prior to the rth success, the experiment must terminate at trial $k + r$, and there must be exactly $r - 1$ successes during the first $k + r - 1$ trials, and a success on trial $k + r$. The first $r - 1$ successes may occur in any positions among the first $k + r - 1$ trials. Therefore, the probability of this part of the event is given by a binomial probability with $k + r - 1$ trials, $r - 1$ successes and success probability p. This binomial probability is multiplied by the probability p of success on trial $k + r$. Therefore, the *negative binomial p.m.f. with parameters r and p* is

$$f(k) = P[X = k] = \binom{k + r - 1}{r - 1} p^{r-1} q^k \cdot p = \binom{k + r - 1}{r - 1} p^r (1 - p)^k,$$

$$k = 0, 1, 2, \ldots \tag{2}$$

where the random variable X counts the number of failures prior to the rth success.

The *Mathematica* distribution objects for these two distributions are

GeometricDistribution[p]   and   NegativeBinomialDistribution[r, p]

For example, the shape of the geometric(.5) p.m.f. is shown in Figure 9(a). A simulation of 500 observed values of a geometric(.5) random variable yields the histogram of the empirical distribution in Figure 9(b). Both graphs illustrate the exponential decrease of probability mass suggested by formula (1) as the state grows.

```
Needs["KnoxProb7`Utilities`"];
```

```
f[x_] := PDF[GeometricDistribution[0.5], x];
geomprobs = Table[f[x], {x, 0, 8}];
states = {0, 1, 2, 3, 4, 5, 6, 7, 8};
g1 = ProbabilityHistogram[
    states, geomprobs, BarColor → Gray];
geomdatalist = RandomInteger[
    GeometricDistribution[0.5], 500];
g2 = Histogram[geomdatalist, 9, "Probability",
    ChartStyle → Gray, PlotRange → All];
GraphicsRow[{g1, g2}, PlotRange → {0, .5}]
```

Figure 2.9 - (a) Histogram of geometric(.5) p.m.f.; (b) Histogram of 500 simulated observations from geometric(.5)

Activity 2   Show that the geometric p.m.f. in formula (1) defines a valid probability mass function.

Example 1   A telemarketer must call a list of cardholders of a certain lender to attempt to sell them credit card insurance, which pays off the debt in the event that the cardholder dies or becomes disabled and loses the ability to pay. Past experience indicates that only 20% of those called will agree to buy the insurance. The telemarketer would like to make 20 sales in a particular week. Find the probability that it will take at least 100 phone calls to achieve this. Comment on the variability of the distribution of the number of unsuccessful calls.

Each attempted phone call constitutes a Bernoulli trial, and if the success event is the event that the cardholder buys the insurance, then the success probability is given to be $p = .20$. Because there are to be at least 100 calls total, and 20 of them are to be successes, we are asking for the probability of the event that there will be at least 80 unsuccessful calls prior to the 20th success. The random variable X which is the number of unsuccessful calls has the negative binomial distribution with parameters $r = 20$ and $p = .2$. Therefore, we must compute:

$$P[X \geq 80] = 1 - P[X \leq 79]$$

---

1 - CDF[NegativeBinomialDistribution[20, .2], 79]

0.480021

Thus, our telemarketer can be about 48% sure that he will make at least 100 calls. Let us look at a connected dot plot of the probability masses to get an idea of how variable the number of unsuccessful calls X is. (We will talk about the variance of the distribution later.)

```
f[x_] :=
  PDF[NegativeBinomialDistribution[20, 0.2], x];
pmf = Table[{x, f[x]}, {x, 20, 140}];
ListPlot[pmf, Joined → True, PlotStyle → Black]
```

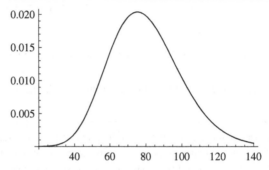

Figure 2.10 - Negative binomial(20,.2) probabilities

The distribution of X is nearly symmetric about a point that seems to be just less than 80, with a slight right skew. It appears that at least 2/3 of the probability weight is between 60 and 100, and there does not seem to be much weight at all to the right of 120. This tells the telemarketer that on repeated sales campaigns of this kind, most of the time between 60 and 100 unsuccessful calls will be made, and very rarely will he have to make more than 120 unsuccessful calls. ∎

Activity 3 Another kind of question that can be asked about the situation in Example 1 is: for what number c is it 95% likely that the number of unsuccessful calls X is less than or equal to c? Use *Mathematica* to do trial and error evaluation of the c.d.f. in search of this c.

Example 2 A snack machine dispenses bags of chips and candy. One of its ten dispenser slots is jammed and non-functioning. Assume that successive machine customers make a selection randomly. What is the probability that the first malfunction occurs strictly after the 20th customer? What is the probability that the third malfunction occurs on the 50th customer's selection?

Since one among ten slots is defective, the number of normal transactions (failures) prior to the first malfunction (success) has the geometric(.1) distribution. Consequently, the probability that the first malfunction happens for customer number 21 or higher is the probability that at least 20 normal transactions occur before the first malfunction, which is the complement of the probability that 19 or fewer normal transactions occur. This is computed as follows.

```
1 - CDF[GeometricDistribution[.1], 19]
```

0.121577

So it is highly probable, around 88%, that the first malfunction happens by the time the 20th customer uses the machine.

Similarly the number of good transactions until the third malfunction has the negative binomial distribution with parameters $r = 3$ and $p = .1$. For the second question we want the probability that there are exactly 47 good transactions, which is:

```
PDF[NegativeBinomialDistribution[3, .1], 47]
```

0.00831391

We have an opportunity to take this example in a different direction, one which indicates the power of probabilistic reasoning in drawing conclusions from data. Suppose we do not know how many among the ten dispenser slots are jammed, but we do know that malfunctions occurred on the 4th, 10th, 18th, and 24th customers among the first 24. Can the pattern of problems indicate a most likely guess for the number of jammed slots?

Because there exists at least one defect and not all slots are bad, our choices for the number of defective slots are 1, 2, 3, 4, 5, 6, 7, 8, and 9, under which the probabilities of a malfunction are, respectively, $p = .1, .2, .3, .4, .5, .6, .7, .8$, and $.9$. The number of normal transactions prior to the first malfunction clearly has the geometric(p) distribution, but we can also imagine restarting a geometric experiment each time a malfunction occurs, keeping track of the number of normal transactions prior to the next malfunction. We have four such experiments, yielding independent geometric(p) random variables X_1, X_2, X_3, X_4. The observed values of the X's, that is, the number of failures between successes, are 3, 5, 7, and 5. Among the candidate p values, which gives the highest probability to this list of observations? As a function of p, the probability of this list of observed values is

$$P[X_1 = 3, \ X_2 = 5, \ X_3 = 7, \ X_4 = 5] =$$
$$p(1 - p)^3 \, p(1 - p)^5 \, p(1 - p)^7 \, p(1 - p)^5$$

Summing exponents to simplify, we try to maximize this function of p over the set of possible p values above.

```
prob[p_] := p^4 (1 - p)^20
```

```
Table[{p, prob[p]}, {p, .1, .9, .1}] // TableForm
```

$$
\begin{array}{ll}
0.1 & 0.0000121577 \\
0.2 & 0.0000184467 \\
0.3 & 6.46317 \times 10^{-6} \\
0.4 & 9.35977 \times 10^{-7} \\
0.5 & 5.96046 \times 10^{-8} \\
0.6 & 1.42497 \times 10^{-9} \\
0.7 & 8.37177 \times 10^{-12} \\
0.8 & 4.29497 \times 10^{-15} \\
0.9 & 6.561 \times 10^{-21}
\end{array}
$$

From the table, we see that the p value which makes the observed data likeliest is the second one, $p = .2$; in other words, the data are most consistent with the assumption that exactly two slots are jammed. ■

Activity 4 Redo the last part of Example 2, assuming that the stream of malfunctions came in on customers 1, 3, 6, and 8. Find the value of p that makes this stream most likely. Do it again, assuming that the malfunctions were on customers 2, 7, 10, and 17.

We close the section by deriving the first two moments of these distributions.

Theorem 1. (a) The mean and variance of the geometric(p) distribution are

$$
\mu = \frac{(1-p)}{p}, \quad \sigma^2 = \frac{(1-p)}{p^2} \tag{3}
$$

(b) The mean and variance of the negative binomial(r, p) distribution are

$$
\mu = \frac{r(1-p)}{p}, \quad \sigma^2 = \frac{r(1-p)}{p^2} \tag{4}
$$

Proof. (a) By the definition of expected value,

$$
\begin{aligned}
E[X] &= \Sigma_{k=0}^{\infty} k \cdot (1-p)^k \cdot p \\
&= p(1-p) \Sigma_{k=0}^{\infty} k \cdot (1-p)^{k-1}
\end{aligned}
$$

The infinite series is of the form $\sum\limits_{k=0}^{\infty} k \cdot x^{k-1}$ for $x = 1 - p$. This series is the derivative of the series $\sum\limits_{k=0}^{\infty} x^k$ with respect to x, and the latter series has the closed form $1/(1-x)$. Hence,

$$
\sum_{k=0}^{\infty} k \cdot x^{k-1} = \frac{d}{dx}\left(\frac{1}{1-x}\right) = \frac{1}{(1-x)^2}
$$

Evaluating at $x = 1 - p$,

$$\mu = \mathrm{E}[X] = p(1 - p) \cdot \frac{1}{(1 - (1 - p))^2} = \frac{1 - p}{p}$$

The variance of the geometric distribution is the subject of Exercise 7.

(b) The number of failures X until the rth success can be thought of as the sum of the number of failures X_1 until the first success, plus the number of failures X_2 between the first and second successes, etc., out to X_r, the number of failures between successes $r - 1$ and r. Each of the X_i's has the geometric(p) distribution; hence by linearity of expectation $\mathrm{E}[X] = r \cdot \mathrm{E}[X_1]$, which yields the result when combined with part (a).

  The proof of the variance formula is best done using an important result that we have not yet covered: when random variables are independent of one another the variance of their sum is the sum of their variances. Then, reasoning as we did for the mean, $\mathrm{Var}(X) = r \cdot \mathrm{Var}(X_1)$, which yields the formula for the variance of the negative binomial distribution when combined with the formula in part (a) for the variance of the geometric distribution. We will prove this result on the variance of a sum in Section 2.6, but Exercise 13 gives you a head start. ■

Exercises 2.3

1. For the electrical switch discussed at the start of the section, how small must the switch failure probability be such that the probability that the switch will last at least 1000 trials (including the final trial where the switch breaks) is at least 90%?

2. Derive the c.d.f. of the geometric(p) distribution.

3. If X has the geometric(p) distribution, find $P[X > m + n \mid X > n]$.

4. (*Mathematica*) On a long roll of instant lottery tickets, on average one in 20 is a winner. Find the probability that the third winning ticket occurs somewhere between the 50th and 70th tickets on the roll.

5. (*Mathematica*) Simulate several random samples of 500 observations from the negative binomial distribution with parameters $r = 20$ and $p = .2$, produce associated histograms, and comment on the shape of the histograms as compared to the graph in Figure 10.

6. A certain home run hitter in baseball averages a home run about every 15 at bats. Find the probability that it will take him (a) at least 10 at bats to hit his first home run; (b) at least 25 at bats to hit his second home run; (c) at least 25 at bats to hit his second home run given that it took exactly 10 at bats to hit his first. (d) Find also the expected value and standard deviation of the number of at bats required to hit the second home run.

7. Derive the formula for the variance of the geometric(p) distribution. (Hint: it will be simpler to first find $\mathrm{E}[X(X - 1)]$.)

8. (*Mathematica*) A shoe store owner is trying to estimate the probability that customers who enter his store will purchase at least one pair of shoes. He observes one day that the 3rd, 6th, 10th, 12th, 17th, and 21st customers bought shoes. Assuming that the customers make their decisions independently, what purchase probability p maximizes the likelihood of this particular customer stream?

9. (*Mathematica*) Write a command that simulates a stream of customers as described in Exercise 8. It should take the purchase probability p as one argument, and the number of customers in total that should be simulated as another. It should return a list of coded observations like {B,N,N,B,...} indicating "buy" and "no buy."

10. Show in the case $r = 3$ that the negative binomial p.m.f. in formula (2) defines a valid probability mass function.

11. (*Mathematica*) A *random walk to the right* is a random experiment in which an object moves from one time to the next on the set of integers. At each time, as shown in the figure, either the object takes one step right to the next higher integer with probability p, or stays where it is with probability $1 - p$. If as below the object starts at position 0, and $p = .3$, what is the expected time required for it to reach position 5? What is the variance of that time? What is the probability that it takes between 15 and 30 time units to reach position 5?

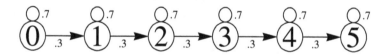

Exercise 11

12. (*Mathematica*) Study the dependence of the negative binomial distribution with parameter $r = 3$ on the success probability parameter p by comparing connected line graphs of the p.m.f. for several values of p. Be sure to comment on the key features of the graph of a p.m.f.: location, spread, and symmetry.

13. This exercise foreshadows the result to come later on the variance of a sum of independent random variables which was needed in the proof of the theorem about the variance of the negative binomial distribution. Suppose that two random variables X and Y are such that joint probabilities involving both of them factor into the product of individual probabilities, for example:

$$P[X = i, \ Y = j] = P[X = i] \, P[Y = j]$$

Let the expected value of any function $g(X, Y)$ of the two random variables be defined by the weighted average

$$E[g(X, Y)] = \sum_i \sum_j g(i, j) P[X = i, Y = j]$$

Write down and simplify an expression for $\text{Var}(X + Y) = E[\{(X + Y) - (\mu_x + \mu_y)\}^2]$ and show that it reduces to $\text{Var}(X) + \text{Var}(Y)$.

14. (*Mathematica*) Consider a basketball team which is estimated to have a 60% chance of winning any game, and which requires 42 wins to clinch a playoff spot. Find the expected value and standard deviation of the number of games that are required to do this. Think about the assumptions you are making when you do this problem, and comment on any possible difficulties.

Sample Problems from Actuarial Exam P

15. In modeling the number of claims filed by an individual under an automobile policy during a three-year period, an actuary makes the simplifying assumption that for all integers $n \geq 0$, $p_{n+1} = \frac{1}{5} p_n$, where p_n represents the probability that the policyholder files n claims during the period. Under this assumption, what is the probability that a policyholder files more than one claim during the period?

16. An insurance policy on an electrical device pays a benefit of $4000 if the device fails during the first year. The amount of the benefit decreases by $1000 each successive year until it reaches 0. If the device has not failed by the beginning of any given year, the probability of failure during that year is .4. What is the expected benefit under this policy?

2.4 Poisson Distribution

The last member of the collection of single-variable discrete distributions that we will look at is the *Poisson distribution*. It arises as a model for random variables that count the number of occurrences of an event in a given region of space or time, such as the number of users of a computer lab during a fixed time period, the number of dandelions in a yard, the number of dust particles in a room, the number of cars arriving to an intersection in a fixed time period, etc. Its state space is the non-negative integers $\{0, 1, 2, ...\}$, and the probability mass function that characterizes it is in formula (1) below. So, this distribution depends on a single parameter called μ, hence, we refer to it as the Poisson(μ) distribution.

In the cases of the other discrete distributions that we have studied, the mass function of the associated random variable was derivable from primitive assumptions about the randomness of an experiment. The Poisson distribution is a bit different, since it comes up as an idealization of a binomial experiment whose number of trials approaches infinity while the success probability becomes vanishingly small. To see the idea, consider some event that happens sporadically on a time axis, in such a way that the chance that it happens during a short time interval is proportional to the length of the interval, and the events that it happens in each of two disjoint time intervals are independent. Without loss of generality, let the time axis be the bounded interval $[0, 1]$, and also let μ be the proportionality constant, that is, P[event happens in $[a, b]] = \mu(b - a)$. Then, if the time interval $[0, 1]$ is broken up into n disjoint, equally sized subintervals $[0, 1/n], (1/n, 2/n], ..., ((n - 1)/n, 1]$,

$$p = \text{P[event happens in interval} ((j - 1)/n, \ j/n]] = \frac{\mu}{n}$$

for all subintervals. The number of occurrences during the whole interval $[0,1]$ is the sum of the number of occurrences in the subintervals; hence it has the binomial distribution with parameters n and $p = \mu/n$. But what do the probability masses approach as $n \longrightarrow \infty$? We will proceed to argue that they approach the following $Poisson(\mu)$ p.m.f.:

$$f(x) = \text{P}[X = x] = \frac{e^{-\mu} \mu^x}{x!}, \quad x = 0, \ 1, \ 2, \ ... \tag{1}$$

We can write out the binomial probabilities as

$$\begin{aligned}
\text{P[exactly } x \text{ successes]} &= \binom{n}{x} p^x (1 - p)^{n-x} \\
&= \frac{n!}{x! \, (n-x)!} \left(\frac{\mu}{n}\right)^x \left(1 - \frac{\mu}{n}\right)^{n-x} \tag{2} \\
&= \frac{\mu^x}{x!} \left(1 - \frac{\mu}{n}\right)^n \frac{n(n-1) \cdots (n-x+1)}{n^x} \left(1 - \frac{\mu}{n}\right)^{-x}
\end{aligned}$$

By a standard result from calculus, the factor $\left(1 - \frac{\mu}{n}\right)^n$ approaches $e^{-\mu}$ as $n \longrightarrow \infty$ In the activity below you will analyze the other factors to finish the proof that the binomial probabilities approach the Poisson probabilities as n becomes large and the success probability p becomes correspondingly small.

Activity 1 Show that the third and fourth factors in the bottom expression on the right side of (2) approach 1 as $n \longrightarrow \infty$. You will have to deal with the case $x = 0$ separately. Conclude that the binomial probability masses b(x; n, μ/n) approach the Poisson(μ) masses in formula (1) as n approaches infinity.

In summary, the Poisson distribution is an approximation for the binomial distribution with large n and small p, and the Poisson parameter relates to these by $\mu = n \, p$.

Mathematica has an object for the Poisson distribution, called

PoissonDistribution[μ]

which can be used in the usual way with the RandomInteger, PDF, and CDF functions. Let us use it to see how close the binomial distribution is to the Poisson distribution for large n

and small *p*. We take $n = 100$, $\mu = 4$, and $p = \mu/n = 4/100$. You will be asked to extend the investigation to other values of the parameters in the exercises. First is a table of the first few values of the Poisson and binomial p.m.f.'s, followed by superimposed connected list plots.

```
Poissonpmf[x_] := N[PDF[PoissonDistribution[4], x]];
Binomialpmf[x_] :=
  PDF[BinomialDistribution[100, .04], x];
TableForm[Join[{{"x", "Poisson", "binomial"}},
  Table[
    {x, Poissonpmf[x], Binomialpmf[x]}, {x, 0, 12}]]]
```

x	Poisson	binomial
0	0.0183156	0.0168703
1	0.0732626	0.070293
2	0.146525	0.144979
3	0.195367	0.197333
4	0.195367	0.199388
5	0.156293	0.159511
6	0.104196	0.105233
7	0.0595404	0.0588803
8	0.0297702	0.0285201
9	0.0132312	0.0121475
10	0.00529248	0.00460591
11	0.00192454	0.0015702
12	0.000641512	0.000485235

```
Poissonlist = Table[{x, N[Poissonpmf[x]]}, {x, 0, 12}];
Binomiallist =
  Table[{x, N[Binomialpmf[x]]}, {x, 0, 12}];
g1 = ListPlot[Poissonlist, Joined → True,
    PlotStyle → Black];
g2 = ListPlot[Binomiallist, Joined → True,
    PlotStyle → Gray];
Show[{g1, g2}]
```

Figure 2.11 - List plots of b(100,.04) and Poisson(4) p.m.f.'s

The connected line graphs hardly differ at all and, in fact, the largest absolute difference between probability masses is only around .004 which takes place at $x = 4$.

Activity 2 Show that the Poisson(μ) probabilities in formula (1) do define a valid probability mass function. Try to obtain a closed form for the c.d.f. of the distribution. Can you do it? (In Exercise 2 you are asked to use *Mathematica* to form a table of the c.d.f. for several values of μ.)

Example 1 We will suppose for this example that the number of ant colonies in a small field has the Poisson distribution with parameter $\mu = 16$. This is reasonable if the field can be broken up into a large number of equally sized pieces, each of which has the same small probability of containing a colony, independent of the other pieces. Then the probability that there are at least 16 colonies is

$$P[X \geq 16] = 1 - P[X \leq 15] = 1 - \sum_{k=0}^{15} \frac{e^{-16} 16^k}{k!}$$

Using *Mathematica* to evaluate this probability we compute

```
N[1 - CDF[PoissonDistribution[16], 15]]
```

```
0.533255
```

Next, suppose that we no longer know μ, but an ecology expert tells us that fields have 8 or fewer ant colonies about 90% of the time. What would be a good estimate of the parameter μ? For this value of μ, $P[X \leq 8] = .90$. We may write $P[X \leq 8]$ as a function of μ as below, and then check several values of μ in search of the one that makes this probability closest to .90. I issued a preliminary Table command to narrow the search from μ values between 2 and 10 to values between 5 and 6. The Table command below shows that to the nearest tenth, $\mu = 5.4$ is the parameter value we seek. ■

```
eightprob[mu_] := CDF[PoissonDistribution[mu], 8]
```

```
Table[N[{mu, eightprob[mu]}], {mu, 5, 6, .1}]
```

{{5., 0.931906}, {5.1, 0.925182},
 {5.2, 0.918065}, {5.3, 0.910554}, {5.4, 0.90265},
 {5.5, 0.894357}, {5.6, 0.885678}, {5.7, 0.876618},
 {5.8, 0.867186}, {5.9, 0.857389}, {6., 0.847237}}

The parameter μ of the Poisson distribution gives complete information about its first two moments, as the next theorem shows.

Theorem 1. If $X \sim \text{Poisson}(\mu)$, then $E[X] = \mu$ and $\text{Var}(X) = \mu$.

Proof. First, for the mean,

$$
\begin{aligned}
E[X] &= \Sigma_{k=0}^{\infty} k \cdot \frac{e^{-\mu} \mu^k}{k!} \\
&= e^{-\mu} \cdot \mu \cdot \Sigma_{k=1}^{\infty} \frac{\mu^{k-1}}{(k-1)!} \\
&= e^{-\mu} \cdot \mu \cdot e^{\mu} \\
&= \mu
\end{aligned}
$$

The first line is just the definition of expected value. In the second line we note that the $k = 0$ term is just equal to 0 and can be dropped, after which we cancel k with the k in $k!$ in the bottom, and then remove two common factors $e^{-\mu} \cdot \mu$. Changing variables to say $j = k - 1$ we see that the series in line 2 is just the Taylor series expansion of e^{μ}.

To set up the computation of the variance, we first compute $E[X(X - 1)]$ similarly to the above computation.

$$
\begin{aligned}
E[X(X-1)] &= \Sigma_{k=0}^{\infty} k \cdot (k-1) \frac{e^{-\mu} \mu^k}{k!} \\
&= e^{-\mu} \cdot \mu^2 \cdot \Sigma_{k=2}^{\infty} \frac{\mu^{k-2}}{(k-2)!} \\
&= e^{-\mu} \cdot \mu^2 \cdot e^{\mu} \\
&= \mu^2
\end{aligned}
$$

You are asked to complete the computation in the next activity. ∎

Activity 3 By expanding the expression $E[X(X - 1)]$ in the proof of Theorem 1 and using the computational formula for the variance, finish the proof.

Since μ is both the mean and the variance of the Poisson(μ) distribution, as μ grows we would expect to see the probability weight shifting to the right and spreading out. In Figure 12 we see a connected list plot of the probability mass function for three Poisson distributions with parameters 2 (leftmost), 5 (middle), and 8 (rightmost), and this dependence on μ is evident. We also see increasing symmetry as μ increases.

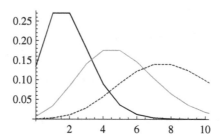

Figure 2.12 - Poisson p.m.f.'s, parameters 2 (dark), 5 (gray), and 8 (dashed)

In the electronic notebook, you can run the input cell below to produce an animation of the Poisson mass functions with parameters ranging from 2 through 8 to observe the shift to the right.

```
f[x_, mu_] := PDF[PoissonDistribution[mu], x]
Animate[ListPlot[Table[{x, f[x, mu]}, {x, 0, 15}],
    Joined → True], {mu, 2, 8}]
```

Poisson Processes

Probably the most important instance of the Poisson distribution is in the study of what is called the *Poisson process*. A Poisson process records the cumulative number of occurrences of some phenomenon as time increases. Therefore, such a process is not just a single random variable, but a family of random variables (X_t) indexed by a variable t usually thought of as time. We interpret X_t as the number of occurrences of the phenomenon in $[0, t]$. A family of random variables such as this is called generically a *stochastic process*. We will cover stochastic processes in more detail, and also say more about the Poisson process, in Chapter 6.

Figure 2.13 - Sample path of a Poisson process

Since occurrences happen singly and randomly, and X_t counts the total number of them, if for a fixed experimental outcome ω we plot $X_t(\omega)$ as a function of t we get a nondecreasing step function starting at 0, which jumps by exactly 1 at the random times $T_1(\omega)$, $T_2(\omega)$, $T_3(\omega)$, ... of occurrence of the phenomenon. Such a function, called a *sample path* of the process, is shown in Figure 13, for the case where the first three jumps occur at times $T_1 = 1.2$, $T_2 = 2.0$, $T_3 = 3.4$.

By our earlier discussion of the domain of application of the Poisson distribution, it is reasonable to assume that as long as the probability of an occurrence in a short interval is proportional to the length of the interval, and the numbers of occurrences on disjoint intervals are independent, X_t should have a Poisson distribution.

What should be the parameter of this Poisson distribution? Let λ be the expected number of occurrences per unit time, i.e., $\lambda = E[X_1]$. In a time interval of length t such as $[0, t]$ we should expect $\lambda \cdot t$ occurrences. Thus, $E[X_t] = \lambda t$ and since the expected value of a Poisson random variable is the same as its parameter, it makes sense that $X_t \sim$ Poisson(λt). The constant λ is called the *rate of the process*.

All of this can be put on a rigorous footing, but we will not do that until later. After we study a continuous distribution called the *exponential distribution*, we will be able to define a Poisson process constructively, by supposing that the times between successive occurrences $T_n - T_{n-1}$ are independent and have a common exponential distribution. For now though, we will simply assume that certain situations generate a Poisson process, which means a family (X_t) of random variables for which X_t has the Poisson(λt) distribution and whose sample paths are as in Figure 13.

Activity 4   Many examples of Poisson processes revolve around service situations, where the occurrence times are arrival times of customers. Give some examples of such situations, and discuss what the underlying assumptions must be in order that a Poisson process is a good model for the cumulative number of customers as a function of time.

Example 2 Suppose that cars travelling on an expressway arrive to a toll station according to a Poisson process with rate $\lambda = 5$ per minute. Thus, we are assuming that the numbers of cars arriving in disjoint time intervals are independent, and the probability of an arrival in a short time interval is proportional to the length of the interval. Let X_t be the total number of cars in $[0, t]$. Find: (a) $P[X_2 > 8]$; (b) $P[X_1 < 5 \mid X_{.5} > 2]$; (c) assuming that λ is no longer known, find the largest possible arrival rate λ such that the probability of 8 or more arrivals in the first minute is no more than .1.

To answer question (a), we note that since $\lambda = 5$ and the time interval is 2 minutes, $X_2 \sim$ Poisson(10); hence by complementation of the c.d.f., the probability is

```
N[1 - CDF[PoissonDistribution[10], 8]]
```

0.66718

It is therefore about 67% likely that more than 8 cars will come in two minutes, which is intuitively reasonable given the average arrival rate of 5 per minute.

For question (b), we have that $X_1 \sim$ Poisson(5) and $X_{.5} \sim$ Poisson(2.5). In order for both $\{X_{.5} > 2\}$ and $\{X_1 < 5\}$ to occur, either $X_{.5} = 3$ or $X_{.5} = 4$, in which case the number of arrivals in [.5, 1] is determined. Thus,

$$P[X_1 < 5 \mid X_{.5} > 2] = \frac{P[X_1 < 5 \cap X_{.5} > 2]}{P[X_{.5} > 2]}$$

$$= \frac{P[X_{.5} = 3, X_1 - X_{.5} \le 1] + P[X_{.5} = 4, X_1 - X_{.5} = 0]}{P[X_{.5} > 2]}$$

Now the random variable $X_1 - X_{.5}$ in the numerator of the last expression is the number of arrivals in the time interval (.5, 1], which is independent of $X_{.5}$, the number of arrivals in [0, .5]. The intersection probabilities in the numerator therefore factor into the product of the individual event probabilities. But also, we may think of restarting the Poisson process at time .5, after which it still satisfies the Poisson process hypotheses so that $X_1 - X_{.5}$ has the same distribution as $X_{.5}$, which is Poisson(2.5). By these observations, we are left to compute

$$P[X_1 < 5 \mid X_{.5} > 2] = \frac{P[X_{.5} = 3]\,P[X_{.5} \le 1] + P[X_{.5} = 4]\,P[X_{.5} = 0]}{P[X_{.5} > 2]}$$

In *Mathematica*, this is the following:

```
f[x_] := PDF[PoissonDistribution[2.5], x]
F[x_] := CDF[PoissonDistribution[2.5], x]
f[3] * F[1] + f[4] * f[0]
─────────────────────────
        1 - F[2]
```

0.158664

For part (c), we now know only that $X_1 \sim$ Poisson($\lambda \cdot 1$). The probability of 8 or more arrivals is the same as $1 - P[X_1 \le 7]$, and we would like this probability to be less than or equal to .1. We can define the target probability as a function of λ, and then compute it for several values of λ. A preliminary Table command narrows the search to values of λ in [4, 5]. The Table command below indicates that the desired rate λ is around 4.6. ∎

```
prob[lambda_] :=
  1 - CDF[PoissonDistribution[lambda], 7];
Table[{lambda, N[prob[lambda]]}, {lambda, 4, 5, .1}]
```

```
{{4., 0.0511336}, {4.1, 0.0573121},
 {4.2, 0.0639433}, {4.3, 0.0710317}, {4.4, 0.0785794},
 {4.5, 0.0865865}, {4.6, 0.095051}, {4.7, 0.103969},
 {4.8, 0.113334}, {4.9, 0.123138}, {5., 0.133372}}
```

Exercises 2.4

1. Assume that the number of starfish X in an ocean region has the Poisson(20) distribution. Find (a) $P[X = 20]$; (b) $P[X > 20]$; (c) $P[\mu - \sigma < X < \mu + \sigma]$, where μ and σ are the mean and standard deviation of the distribution of X.

2. (*Mathematica*) Use *Mathematica* to form a table of cumulative probabilities $F(x) = P[X \leq x]$ for each of the four Poisson distributions with parameters $\mu = 1, 2, 3$, and 4. For all distributions, begin at $x = 0$ and end at $x = 10$.

3. (*Mathematica*) Suppose that the sample 3, 2, 4, 3, 4, 5, 6, 2 comes from a Poisson(μ) distribution. What value of μ makes this sample most likely to have occurred? (Justify your answer well.) Compare this value of μ to the sample mean $\overline{X}$.

4. Find the third moment about 0 for the Poisson(μ) distribution.

5. If $X \sim$ Poisson(μ), find an expression for the expectation: $E[X(X - 1)(X - 2) \cdots (X - k)]$.

6. (*Mathematica*) Write a function of n, p, and k that finds the largest absolute difference between the values of the b(n,p) probability mass function and the Poisson(np) mass function, among states $x = 0, 1, 2, \ldots, k$. Use it to study the change in the maximum absolute difference for fixed $p = 1/10$ as values of n increase from 20 to 100 in multiples of 10. Similarly, fix $n = 200$ and study the maximum absolute difference as p takes on the values 1/2, 1/3, 1/4, 1/5, 1/6, 1/7, 1/8, 1/9, 1/10.

7. Suppose that the number of traffic tickets issued during a week in a particular region has the Poisson distribution with mean 10. Find the probability that the number of tickets next week will be at least 2 standard deviations below the average.

8. If the number of defects in a length of drywall has the Poisson distribution, and it is estimated that the proportion with no defects is .3, estimate the proportion with either 1 or 2 defects.

9. If (X_t) is a Poisson process with rate 2/min., find the probability that X_3 exceeds its mean by at least 2 standard deviations.

10. Suppose that customer arrivals to a small jewelry store during a particular period in the day form a Poisson process with rate 12/hr. (a) Find the probability that there will be more than 12 customers in a particular hour; (b) Given that there are more than 12 customers in the first hour, find the probability that there will be more than 15 customers in that hour.

11. If outside line accesses from within a local phone network form a Poisson process with rate 1/min., find the joint probability that the cumulative number of accesses is equal to 2 at

time 1 minute, 3 at time 2 minutes, and 5 at time 3 minutes.

12. Find a general expression for $P[X_s = j \cap X_{t+s} = k]$ for a Poisson process (X_t) of rate λ.

13. (*Mathematica*) Suppose that emergency 911 calls come in to a station according to a Poisson process with unknown rate λ per hour. Find the value of λ that is most likely to have resulted in a stream of calls at times: .1, .3, .6, 1.0, 1.1, 1.4, 1.8, 2.0, 2.2, 2.3, 2.8, 3.6. (Hint: consider the numbers of calls in successive intervals of length .5: [0, .5], (.5, 1.0], ...)

Sample Problems from Actuarial Exam P

14. An actuary has discovered that policyholders are three times as likely to file two claims as to file four claims. If the number of claims filed has a Poisson distribution, what is the variance of the number of claims filed?

15. A baseball team has scheduled its opening game for April 1. If it rains on April 1, the game is postponed and will be played on the next day that it does not rain. The team purchases insureance against rain. The policy will pay $1000 for each day, up to 2 days, that the opening game is postponed. The insurance company determines that the number of consecutive days of rain beginning on April 1 is a Poisson random variable with mean .6. What is the standard deviation of the amount the insurance company will have to pay?

2.5 Joint, Marginal, and Conditional Distributions

In this section we would like to take the ideas of conditional probability and independence that were introduced in Sections 1.5 and 1.6 into the domain of random variables and their distributions. This transition should be a very straightforward one. We already encountered the idea of joint distributions of random variables in Section 2.2 on the multinomial distribution, and we would like to explore this idea further, introduce the ideas of marginal and conditional distributions, and use those to define independence of random variables.

Earlier we studied a contingency table for the Peotone airport problem, and a similar example here should motivate the ideas well. Suppose that a researcher in cognitive psychology is conducting an experiment on memory. She exposes to several subjects four numbers between 1 and 100 on flash cards, and after fifteen minutes she asks the subjects to recall the numbers. The experimenter records how many of the numbers the subjects correctly recalled. Then the same subjects are presented four more numbers between 1 and 100 orally. After another fifteen minutes, the experimenter again records how many of them each subject got right. At the end of the experiment, she classifies each subject as to how many of the numbers that were shown visually and how many that were spoken were correctly remembered. Assume that the frequencies are in the table below. I have computed the total number of subjects in each row and column for your convenience.

Y = visual correct	0	1	2	3	4	row sum
0	4	6	10	8	3	31
1	2	5	7	12	6	32
2	1	4	10	15	6	36
3	2	6	9	14	10	41
4	1	3	5	11	11	31
col sum	10	24	41	60	36	171

X = oral correct

Consider the experiment of picking a subject randomly from this group of 171 subjects. Let X and Y, respectively, denote the oral and visual scores of the subject. Then, for instance,

$$P[X = 0, \ Y = 0] = \frac{4}{171}, \ P[X = 0, \ Y = 1] = \frac{6}{171}, \ P[X = 0, \ Y = 2] = \frac{10}{171},$$

etc. The complete list of the probabilities

$$f(x, y) = P[X = x, \ Y = y] \tag{1}$$

over all possible values (x, y) is called the *joint probability mass function* of the random variables X and Y. Joint mass functions for more than two discrete random variables are defined similarly (see Exercise 14).

Activity 1.   Referring to the memory experiment, what is $P[X = 0]$? $P[X = 1]$? $P[X = 2]$? $P[X = 3]$? $P[X = 4]$? What do these probabilities add to?

The activity above suggests another kind of distribution that is embedded in the joint distribution described by f. Working now with Y using the column sums we find:

$$P[Y = 0] = \frac{10}{171}; \ P[Y = 1] = \frac{24}{171}; \ P[Y = 2] = \frac{41}{171};$$
$$P[Y = 3] = \frac{60}{171}; \ P[Y = 4] = \frac{36}{171}$$

It is easy to check that these probabilities sum to 1; hence, the function $q(y) = P[Y = y]$ formed in this way is a valid probability mass function. Notice that to compute each of these probabilities, we add up joint probabilities over all x values for the fixed y of interest, for example,

$$\begin{aligned}
P[Y = 0] &= P[X = 0, \ Y = 0] + P[X = 1, \ Y = 0] + P[X = 2, \ Y = 0] \\
&\quad + P[X = 3, \ Y = 0] + P[X = 4, \ Y = 0] \\
&= \frac{4}{171} + \frac{2}{171} + \frac{1}{171} + \frac{2}{171} + \frac{1}{171} \\
&= \frac{10}{171}
\end{aligned}$$

The Law of Total Probability justifies this procedure.

In general, the *marginal probability mass function* of Y is

$$q(y) = P[Y = y] = \sum_x P[X = x, Y = y] = \sum_x f(x, y) \tag{2}$$

where f is the joint p.m.f. of X and Y. Similarly, the *marginal probability mass function* of X is obtained by adding the joint p.m.f. over all values of y:

$$p(x) = P[X = x] = \sum_y P[X = x, Y = y] = \sum_y f(x, y) \tag{3}$$

The idea of a *conditional distribution of one discrete random variable given another* is analogous to the discrete conditional probability of one event given another. Using the memory experiment again as our model, what is the conditional probability that a subject gets 2 visual numbers correct given that the subject got 3 oral numbers correct? The condition on oral numbers restricts the sample space to subjects in the line numbered 3 in the table, which has 41 subjects. Among them, 9 got 2 visual numbers correct. Hence,

$$P[Y = 2 \mid X = 3] = \frac{9}{41}$$

It should be easy to see from this example that for jointly distributed discrete random variables, $P[Y = y \mid X = x]$ is a well defined conditional probability of one event given another. This leads us to the definition of the *conditional probability mass function of Y given $X = x$*:

$$q(y \mid x) = P[Y = y \mid X = x] = \frac{P[X = x, \ Y = y]}{P[X = x]} = \frac{f(x, y)}{p(x)} \tag{4}$$

Reasoning as above, the full conditional p.m.f. of Y given $X = 3$ in the memory experiment is

$$P[Y = 0 \mid X = 3] = \frac{2}{41}, \ P[Y = 1 \mid X = 3] = \frac{6}{41}, \ P[Y = 2 \mid X = 3] = \frac{9}{41},$$
$$P[Y = \ 3 \mid X = 3] = \frac{14}{41}, \ \ P[Y = 4 \mid X = 3] = \frac{10}{41}$$

Similarly, the *conditional probability mass function of X given $Y = y$* is

$$p(x \mid y) = P[X = x \mid Y = y] = \frac{P[X = x, \ Y = y]}{P[Y = y]} = \frac{f(x, y)}{q(y)} \tag{5}$$

Formulas (4) and (5) tie together the three kinds of distributions: joint, marginal, and conditional. The conditional mass function is the quotient of the joint mass function and the marginal mass function of the random variable being conditioned on.

The concept of independence also carries over readily to two or more discrete random variables. In the case of two discrete random variables X and Y, we say that they are *independent* of each other if and only if for all subsets A and B of their respective state spaces,

$$P[X \in A, Y \in B] = P[X \in A] \cdot P[Y \in B] \tag{6}$$

(Exercise 15 gives the analogous definition of independence of more than two random variables.) Alternatively, we could define independence by the condition

$$P[Y \in B \mid X \in A] = P[Y \in B] \tag{7}$$

for all such sets A and B, with the added proviso that $P[X \in A] \neq 0$. (How does (7) imply the factorization criterion (6), and how is it implied by (6)?)

For example, from the memory experiment table, $P[Y = 4 \mid X = 4] = 11/31$, since there are 31 subjects such that $X = 4$, and 11 of them remembered 4 numbers that were visually presented. However, $P[Y = 4]$ is only 36/171, so that the chance that 4 visual numbers are remembered is increased by the occurrence of the event that 4 auditory numbers were remembered. Even this one violation of condition (7) is enough to show that the random variables X and Y are dependent (that is, not independent). But in general, do you really have to check all possible subsets A of the X state space ($2^5 = 32$ of them here) with all possible subsets B of the Y state space (again 32 of them) in order to verify independence? Fortunately the answer is no, as the following important activity shows.

Activity 2 Show that X and Y are independent if and only if their joint p.m.f. $f(x, y)$ factors into the product of the marginal p.m.f.'s $p(x) \cdot q(y)$.

Thus, in our example instead of checking $32 \cdot 32 = 1024$ possible combinations of sets A and B, we need only check the factorization of f for $5 \cdot 5 = 25$ possible combinations of states x and y.

Independence for more than two random variables is explored in the exercises. The basic idea is that several random variables are independent if and only if any subcollection of them satisfies factorization of intersection probabilities as in formula (6). This also follows if and only if the joint p.m.f. factors into the product of the marginals.

We will now look at a series of examples to elaborate on the ideas of joint, marginal, and conditional distributions and independence.

Example 1 In a roll of two fair 12-sided dice, let X and Y be the observed faces. What is the joint distribution of X and Y? Find a general formula for $P[X + Y = z]$, $z = 2, 3, \ldots, 24$.

We assume that X and Y are independent random variables, each with marginal p.m.f. $f(x) = 1/12$, $x = 1, 2, \ldots, 12$. By the result of Activity 2, the joint p.m.f. of these two random variables is the product of their marginals:

$$f(x, y) = p(x) \cdot q(y) = \frac{1}{12} \cdot \frac{1}{12} = \frac{1}{144}, \quad x, y = 1, 2, \ldots, 12$$

Now let $Z = X + Y$ be the total of the two observed faces. In order for Z to equal z, X must assume one of its possible values x, and Y must equal $z - x$. Then the distribution of Z is, by the law of total probability,

$$g(z) = P[Z = z] = \sum_{x=1}^{z-1} P[X = x, Y = z - x]$$

We need to consider separately the cases where $z \le 12$ and $z > 12$, because when $z \le 12$ each joint probability in the sum above is non-zero, but when $z > 12$ only terms such that both x and $z - x$ are in $\{1, 2, \ldots, 12\}$ are non-zero. In the case $z \le 12$,

$$
\begin{aligned}
g(z) = P[Z = z] &= \sum_{x=1}^{z-1} P[X = x, Y = z - x] \\
&= \sum_{x=1}^{z-1} \frac{1}{144} \\
&= \frac{1}{144}(z - 1), \quad z = 2, 3, \ldots, 12
\end{aligned}
$$

When $z > 12$, we need both $1 \le x \le 12$ and $1 \le z - x \le 12 \Longrightarrow x \ge z - 12$. Therefore,

$$
\begin{aligned}
g(z) = P[Z = z] &= \sum_{x=z-12}^{12} P[X = x, Y = z - x] \\
&= \sum_{x=z-12}^{12} \frac{1}{144} \\
&= \frac{1}{144}(12 - (z - 12) + 1) \\
&= \frac{1}{144}(25 - z), \quad z = 13, 14, \ldots, 24
\end{aligned}
$$

You should convince yourself that this is an appropriate generalization of the distribution of the sum of the faces of two ordinary 6-sided dice. ∎

Activity 3 In the previous example, find the conditional mass function of the sum Z given that $X = x$ for each of the x values 2, 6, and 12.

Example 2 This example concerns a generalization of the hypergeometric distribution that we studied earlier. Suppose that the faculty at Old VineCovered College is sharply divided along political lines, with 45 of the 100 faculty members who are political liberals, 30 who are centrists, and 25 who are political conservatives. The Dean of the College is appointing a committee to revise the general education curriculum by selecting 8 faculty members at random, in a batch and without replacement, from the 100. Let us find the joint distribution of the numbers of liberals, centrists, and conservatives on the committee, the marginal distributions of each political group, and the probability that the liberals will have a majority.

In order to have, say i liberals, j centrists, and k conservatives on the committee, where $i + j + k$ is required to be 8, we must select i liberals in a batch without replacement from the 45 liberals, and similarly j centrists from the 30 and k conservatives from the 25. By combinatorics, since there are $\binom{100}{8}$ possible equally likely committees,

$$f(i, j, k) = P[i \text{ liberals, } j \text{ centrists, } k \text{ conservatives}]$$

$$= \frac{\binom{45}{i}\binom{30}{j}\binom{25}{k}}{\binom{100}{8}}, \quad i, j, k \in \{0, 1, \ldots, 8\}, \quad i + j + k = 8$$

Now the marginals may be found by using the definition in (2) of marginal p.m.f.'s and summing out over the other random variable states. Try this in the activity that follows this example. But we can also work by combinatorial reasoning. In order to have exactly i liberals, we must select them from the subgroup of 45, and then select any other $8 - i$ people from the 55 non-liberals to fill out the rest of the committee. Therefore, the marginal p.m.f. of the number of liberals on the committee is

$$p(i) = P[i \text{ liberals}] = \frac{\binom{45}{i}\binom{55}{8-i}}{\binom{100}{8}} \tag{8}$$

Similarly, the other marginals are

$$q(j) = P[j \text{ centrists}] = \frac{\binom{30}{j}\binom{70}{8-j}}{\binom{100}{8}},$$

$$r(k) = P[k \text{ conservatives}] = \frac{\binom{25}{k}\binom{75}{8-k}}{\binom{100}{8}}$$

All of these marginal distributions are of the hypergeometric family.

By the way, it is easy to check that

$$f(i, j, k) \neq p(i) \cdot q(j) \cdot r(k)$$

so that the numbers of the three political types in the sample are not independent random variables. Intuitively, knowing the number of liberals, for instance, changes the distribution of the number of centrists (see also Exercise 5).

Now the probability that the liberals will have a majority is the probability that there are at least five liberals on the committee. It is easiest to use the marginal distribution in formula (8) for the number of liberals, and then to let *Mathematica* do the computation:

$$P[\text{liberals have majority}] \; = \; \sum_{i=5}^{8} \frac{\dbinom{45}{i}\dbinom{55}{8-i}}{\dbinom{100}{8}}$$

```
N[1 - CDF[HypergeometricDistribution[8, 45, 100], 4]]
```

0.251815

We see that the probability of a liberal majority occurring by chance is only about 1/4. ■

Activity 4 In the above example, find $p(i)$, the marginal p.m.f. of the number of liberals by setting $k = 8 - i - j$ in the joint p.m.f. and then adding the joint probabilities over the possible j values.

Example 3 If two random variables X and Y are independent and we observe instances of them $(X_1, Y_1), (X_2, Y_2), \dots , (X_n, Y_n)$, what patterns would we expect to find?

We will answer the question by simulating the list of n pairs. For concreteness, let X have the discrete p.m.f. $p(1) = 1/4$, $p(2) = 1/4$, $p(3) = 1/2$ and let Y have the discrete p.m.f. $q(1) = 5/16$, $q(2) = 5/16$, $q(3) = 3/8$. The program below simulates two numbers at a time from *Mathematica*'s random number generator and converts them to observations of X and Y, respectively. If the random number generator works as advertised, the values of X and Y that are simulated should be (or appear to be) independent. We will then tabulate the number of times the pairs fell into each of the 9 categories $(1, 1)$, $(1, 2)$, $(1, 3)$, $(2, 1)$, $(2, 2)$, $(2, 3)$, $(3, 1)$, $(3, 2)$, $(3, 3)$, and study the pattern.

The first two commands simulate individual observations from the distributions described in the last paragraph, the third uses them to simulate a list of pairs, and the last tabulates frequencies of the nine states and presents them in a table.

```
SimX[] := Module[{rand}, rand = RandomReal[];
    If[rand < 1/4, 1, If[1/4 ≤ rand < 1/2, 2, 3]]]
```

```
SimY[] := Module[{rand}, rand = RandomReal[];
    If[rand < 5/16, 1, If[5/16 ≤ rand < 10/16, 2, 3]]]
```

```
SimXYPairs[numpairs_] :=
 Table[{SimX[], SimY[]}, {numpairs}]
```

```
XYPairFrequencies[numpairs_] :=
    Module[{simlist, freqtable, nextpair, x, y},
        simlist = SimXYPairs[numpairs];
  (* generate data *)
    freqtable = Table[0, {i, 1, 3}, {j, 1, 3}];
  (* initialize table *)
    Do[nextpair = simlist[[i]];
  (* set up next x,y pair *)
        x = nextpair[[1]]; y = nextpair[[2]];
        freqtable[[x, y]] = freqtable[[x, y]] + 1,
  (* increment the frequency table *)
            {i, 1, Length[simlist]}];
        (* output the table *)
    TableForm[{{" ", "Y=1", "Y=2", "Y=3"},
                        {"X=1", freqtable[[1, 1]],
    freqtable[[1, 2]], freqtable[[1, 3]]},
        {"X=2", freqtable[[2, 1]],
    freqtable[[2, 2]], freqtable[[2, 3]]},
                {"X=3", freqtable[[3, 1]],
    freqtable[[3, 2]], freqtable[[3, 3]]}}]]
```

```
SeedRandom[165 481];
XYPairFrequencies[1000]
```

	Y=1	Y=2	Y=3
X=1	83	78	85
X=2	85	73	88
X=3	162	152	194

Above is a sample run. There is some variability in category frequencies, even when the sample size is 1000, but a few patterns do emerge. In each X row, the $Y = 1$ and $Y = 2$ frequencies are about equal, and the $Y = 3$ frequency is a bit higher. Similarly, in each Y column, the $X = 1$ frequency is close to that of $X = 2$, and their frequencies are only around half that of $X = 3$. If you look back at the probabilities assigned to each state, you can see why you should have expected this. In general, if independence holds, the distribution of data into columns should not be affected by which row you are in, and the distribution of data into rows should not be affected by which column you are in. ■

Activity 5 What characteristics of simulated contingency tables like the one in the last example would you expect in the case where the random variables are not independent?

Example 4 Suppose that a project is composed of two tasks that must be done one after the other, and the task completion times are random variables X and Y. The p.m.f. of X is discrete uniform on the set $\{3, 4, 5\}$, and the conditional probability mass functions for Y given each of the three possible values for X are as in the table below. Find the probability distribution of the total completion time for the project, and find the probability that the project will be finished within 6 time units.

	$Y = 1$	$Y = 2$	$Y = 3$
$X = 3$	1/2	1/4	1/4
$X = 4$	1/3	1/3	1/3
$X = 5$	1/4	1/4	1/2

The possible values of the project completion time $X + Y$ are $\{4, 5, 6, 7, 8\}$. Itemizing cases and using the multiplication rule for conditional probability, we get:

$$P[X + Y = 4] = P[X = 3, Y = 1] = P[Y = 1 \mid X = 3] \, P[X = 3] = \frac{1}{2} \cdot \frac{1}{3} = \frac{1}{6} = \frac{6}{36},$$

$$P[X + Y = 5] = P[X = 3, Y = 2] + P[X = 4, Y = 1] = \frac{1}{4} \cdot \frac{1}{3} + \frac{1}{3} \cdot \frac{1}{3} = \frac{7}{36},$$

$$P[X + Y = 6] = P[X = 3, Y = 3] + P[X = 4, Y = 2] + P[X = 5, Y = 1]$$
$$= \frac{1}{4} \cdot \frac{1}{3} + \frac{1}{3} \cdot \frac{1}{3} + \frac{1}{4} \cdot \frac{1}{3} = \frac{10}{36},$$

$$P[X + Y = 7] = P[X = 4, Y = 3] + P[X = 5, Y = 2] = \frac{1}{3} \cdot \frac{1}{3} + \frac{1}{4} \cdot \frac{1}{3} = \frac{7}{36},$$

$$P[X + Y = 8] = P[X = 5, Y = 3] = \frac{1}{2} \cdot \frac{1}{3} = \frac{1}{6} = \frac{6}{36}$$

The probability that the completion time is less than or equal to 6 is the total of the first three of these, that is, $6/36 + 7/36 + 10/36 = 23/36$. ■

Exercises 2.5

1. In the memory experiment example at the beginning of the section, find the conditional p.m.f. of Y given $X = 0$, and the conditional p.m.f. of X given $Y = 2$.

2. Suppose that X and Y have the joint distribution in the table below. Find the marginal distributions of X and Y. Are X and Y independent random variables?

		Y		
	1	2	3	4
1	1/16	1/16	1/16	1/16
X 2	1/32	1/32	1/16	1/16
3	1/16	1/16	1/32	1/32
4	1/8	1/8	1/16	1/16

3. Argue that two discrete random variables X and Y cannot be independent unless their joint state space is a Cartesian product of their marginal state spaces, i.e., $E = E_x \times E_y = \{(x, y) : x \in E_x, y \in E_y\}$.

Exercise 4

4. Suppose that a joint p.m.f. $f(x, y)$ puts equal weight on all the integer grid points marked in the diagram. Find: (a) the marginal distribution of X; (b) the marginal distribution of Y; (c) the conditional distribution of X given $Y = 1$; (d) the conditional distribution of Y given $X = 1$.

5. In Example 2, find the conditional distribution of the number of centrists on the committee given that the number of liberals is 3.

6. Show that if two discrete random variables X and Y are independent, then their *joint cumulative distribution function*, defined by $F(x, y) = P[X \le x, Y \le y]$, factors into the product of the marginal c.d.f.'s $F_x(x) = P[X \le x]$ and $F_y(y) = P[Y \le y]$.

7. In the contingency table below, subjects were classified according to their age group and their opinion about what should be done with a government budget surplus. One subject is drawn at random; let X denote the age category and let Y denote the opinion category for this subject.

(a) Compute the marginal distribution of X;

(b) Compute the marginal distribution of Y;

(c) Compute the conditional distribution of Y given that X is the $21 - 35$ age category;

(d) Compute the conditional distribution of X given that Y is the Lower Taxes opinion category;

(e) Considering the data in the table to be a random sample of American voters, do the age group and opinion variables seem independent?

	Save Social Security	Reduce National Debt	Lower Taxes	Increase Defense Spending
21 – 35	22	10	63	15
36 – 50	46	20	85	60
50 – 65	89	54	70	41
over 65	106	32	15	20

8. (*Mathematica*) Write a *Mathematica* program to simulate observations (X, Y) having the joint density in Exercise 4.

9. Show that if two discrete random variables X and Y are independent under the intersection form of the definition, then for all x such that $p(x) \neq 0$, $q(y \mid x) = q(y)$, and for all y such that $q(y) \neq 0$, $p(x \mid y) = p(x)$, where $p(x)$ and $q(y)$ are the marginal p.m.f.'s of X and Y, and $p(x \mid y)$ and $q(y \mid x)$ are the conditional p.m.f.'s.

10. Suppose that in baseball a pitcher throws a fastball (coded by 0), a curve (1), or a slider (2) with probabilities 1/2, 1/4, and 1/4, respectively. Meanwhile the hitter guesses fastball (coded by 0 again), curve (1), or slider (2) with probabilities 5/8, 1/8, and 1/4, respectively. The hitter's guess is independent of the pitcher's choice. The probabilities that the hitter will hit safely under each possible pair of pitcher pitches and hitter guesses are in the table below. Find the probability that the hitter hits safely.

		Y (hitter)		
		0	1	2
X (pitcher)	0	.4	.05	.2
	1	.1	.3	.25
	2	.25	.2	.4

11. A machine can either be in a working state ($X = 1$) or malfunctioning ($X = 0$). A diagnostic test either reports that the machine is working ($Y = 1$), or that it is malfunctioning ($Y = 0$). Assume that the marginal distribution of X is $p(1) = .9$, $p(0) = .1$, assume that the conditional distribution of Y given $X = 0$ is $q(0 \mid 0) = .95$, $q(1 \mid 0) = .05$, and assume that the conditional distribution of Y given $X = 1$ is $q(0 \mid 1) = .10$, $q(1 \mid 1) = .90$. Find the conditional distribution of X given $Y = 1$, and the conditional distribution of X given $Y = 0$.

12. Show that the conditional p.m.f. $q(y \mid x)$ is always a valid p.m.f. for each fixed x such that $p(x) \neq 0$.

13. The number of failures until the first win in a particular slot machine has the geometric distribution with parameter 1/5. If successive plays on the machine are independent of one another, use reasoning similar to Example 1 to compute the probability mass function of the number of failures until the second win.

14. The *joint probability mass function* of many discrete random variables $X_1, X_2, \ldots, X_k$ is the function

$$f(x_1, x_2, \ldots, x_k) = P[X_1 = x_1, \ X_2 = x_2, \ \ldots, \ X_k = x_k]$$

Joint marginal distributions of subgroups of the X's can be found by summing out over all values of the states x for indices not in the subgroup. If a joint mass function $f(x_1, x_2, x_3)$ puts equal probability on all corners of the unit cube $[0,1] \times [0,1] \times [0,1]$, find the joint marginals of X_1 and X_2, X_2 and X_3, and X_1 and X_3. Find the one variable marginals of X_1, X_2, and X_3.

15. Discrete random variables $X_1, X_2, \ldots, X_k$ are *mutually independent* if for any subsets B_1, $B_2, \ldots, B_k$ of their respective state spaces

$$P[X_1 \in B_1, \ X_2 \in B_2, \ \ldots, \ X_k \in B_k] = P[X_1 \in B_1] \cdot P[X_2 \in B_2] \cdots P[X_k \in B_k]$$

(a) Argue that if the entire group of random variables is mutually independent, then so is any subgroup.

(b) Show that if $X_1, X_2, \ldots, X_k$ are mutually independent, then their joint p.m.f. (see Exercise 14) factors into the product of their marginal p.m.f.'s.

(c) Show that if $X_1, X_2, \ldots, X_k$ are mutually independent, then

$$P[X_1 \in B_1, \ X_2 \in B_2 \mid X_3 \in B_3 \ldots, \ X_k \in B_k] = P[X_1 \in B_1] \cdot P[X_2 \in B_2]$$

Sample Problems from Actuarial Exam P

16. An insurance company determines that N, the number of claims received in a week, is a random variable with $P[N = n] = \frac{1}{2^{n+1}}$, where $n \geq 0$. The company also determines that the number of claims received in a given week is independent of the number of claims received in any other week. Determine the probability that exactly seven claims will be received during a given two-week period.

17. A car dealership sells 0, 1, or 2 luxury cars on any day. When selling a car, the dealer also tries to persuade the customer to buy an extended warranty for the car. Let X denote the number of luxury cars sold in a given day, and let Y denote the number of extended warranties sold. Suppose that the joint probability mass function of X and Y is

$$P[X = 0, Y = 0] = 1/6, \ P[X = 1, Y = 0] = 1/12, \ P[X = 1, Y = 1] = 1/6,$$
$$P[X = 2, Y = 0] = 1/12, \ P[X = 2, Y = 1] = 1/3, \ P[X = 2, \ Y = 2] = 1/6.$$

What is the variance of X?

2.6 More on Expectation

Now that we know about joint, marginal, and conditional distributions, there are some essential matters that we need to finish involving expected value of discrete random variables. Of particular importance in this section are the results on expected value and variance of a linear combination of random variables, and the definitions of the summary measures of association between random variables called the *covariance* and *correlation*. These ideas have a great deal of impact on sampling, and its use in making inferences about the population from which the sample is taken, because they reveal important properties of the mean of a random sample.

Recall from the last section the joint probability mass function of two random variables X and Y, defined by

$$f(x, y) = P[X = x, Y = y]$$

For several random variables $X_1, X_2, \ldots, X_n$ the joint p.m.f. is defined analogously:

$$f(x_1, x_2, \ldots, x_n) = P[X_1 = x_1, X_2 = x_2, \ldots, X_n = x_n] \tag{1}$$

In the single variable case we defined the expectation of a function g of a random variable X as the weighted average of the possible states $g(x)$, weighted by their probabilities $f(x)$. The analogous definition for many random variables follows.

Definition 1. The *expected value* of a function $g(X_1, X_2, \ldots, X_n)$ of random variables whose joint p.m.f. is as in (1) is:

$$E[g(X_1, X_2, \ldots, X_n)] = \sum_{x_1} \cdots \sum_{x_n} g(x_1, x_2, \ldots, x_n) \cdot f(x_1, x_2, \ldots, x_n) \tag{2}$$

where the multiple sum is taken over all possible states $(x_1, x_2, \ldots, x_n)$.

Example 1 Let X be the total number of heads in two flips of a fair coin and let Y be the total number of heads in two further flips, which we assume are independent of the first two. Then $X + Y$ is the total number of heads among all flips. Each of X and Y have the binomial distribution with parameters $n = 2$ and $p = 1/2$, and by the independence assumption,

$$
\begin{aligned}
E[X + Y] &= \sum_{x=0}^{2} \sum_{y=0}^{2} (x + y) f(x, y) \\
&= \sum_{x=0}^{2} \sum_{y=0}^{2} (x + y) \binom{2}{x} .5^2 \binom{2}{y} .5^2 \\
&= .5^4 (0 \cdot 1 \cdot 1 + 1 \cdot 1 \cdot 2 + 2 \cdot 1 \cdot 1 + 1 \cdot 2 \cdot 1 + 2 \cdot 2 \cdot 2 \\
&\qquad + 3 \cdot 2 \cdot 1 + 2 \cdot 1 \cdot 1 + 3 \cdot 1 \cdot 2 + 4 \cdot 1 \cdot 1) \\
&= .5^4 \cdot 32 = 2
\end{aligned}
$$

Notice that the result is just $1 + 1$, that is, the expected value of X plus the expected value of Y. This is not a coincidence, as you will see in the next theorem. ∎

Theorem 1. If X and Y are discrete random variables with finite means, and c and d are constants, then

$$E[c \cdot X + d \cdot Y] = c \cdot E[X] + d \cdot E[Y] \tag{3}$$

Proof. Let $f(x, y)$ be the joint p.m.f. of the two random variables and let $p(x)$ and $q(y)$ be the marginals of X and Y. By formula (2) the expectation on the left is

$$
\begin{aligned}
E[c \cdot X + d \cdot Y] &= \textstyle\sum_x \sum_y (c \cdot x + d \cdot y) f(x, y) \\
&= \textstyle\sum_x c \cdot x \sum_y f(x, y) + \sum_y d \cdot y \sum_x f(x, y) \\
&= c \textstyle\sum_x x \cdot p(x) + d \cdot \sum_y y \cdot q(y) \\
&= c \cdot E[X] + d \cdot E[Y] \quad \blacksquare
\end{aligned}
$$

It is easy to extend the property in formula (3), called *linearity of expectation*, to many random variables. (See Exercise 5.) Think about the following question, to which we will return shortly after a theorem about the variance of a linear combination of random variables.

Activity 1 What does Theorem 1 imply about the simple arithmetical average of two random variables X and Y?

Theorem 2. If X and Y are independent discrete random variables with finite variances, and c and d are constants, then

$$\text{Var}(c \cdot X + d \cdot Y) = c^2 \cdot \text{Var}(X) + d^2 \cdot \text{Var}(Y) \tag{4}$$

Proof. By Theorem 1, the mean of $c \cdot X + d \cdot Y$ is $c \cdot \mu_x + d \cdot \mu_y$, where the μ's are the individual means of X and Y. By the definition of variance,

$$\text{Var}(c \cdot X + d \cdot Y) = E\left[\left((c \cdot X + d \cdot Y) - \left(c \cdot \mu_x + d \cdot \mu_y\right)\right)^2 \right]$$

Group the terms involving X together, and group those involving Y to get

$$\text{Var}(c \cdot X + d \cdot Y) = E\left[\left(c(X - \mu_x) + d(Y - \mu_y)\right)^2 \right]$$

Expansion of the square yields

$$
\begin{aligned}
\text{Var}(c \cdot X + d \cdot Y) = c^2 \cdot E\left[(X - \mu_x)^2\right] + d^2 \cdot E\left[(Y - \mu_y)^2\right] \\
+ 2 \cdot c \cdot d \cdot E\left[(X - \mu_x)(Y - \mu_y)\right]
\end{aligned} \tag{5}
$$

But the last term in the sum on the right equals zero (see Activity 2), which proves (4). $\blacksquare$

Activity 2   Use the independence assumption in the last theorem to show that $E\big[(X - \mu_x)(Y - \mu_y)\big] = 0$. Generalize: show that if X and Y are independent, then $E[g(X)\,h(Y)] = E[g(X)]\,E[h(Y)]$.

Theorem 2 also extends easily to the case of linear combinations of many random variables.

Example 2 Let X_1, X_2, ..., X_n be a random sample drawn in sequence and with replacement from some population with mean μ and variance σ^2. The *mean of the sample* is the simple arithmetical average of the sample values

$$\overline{X} = \frac{X_1 + X_2 + \ldots + X_n}{n} \tag{6}$$

If the sample is in a *Mathematica* list then *Mathematica* can calculate the sample mean with the command Mean[datalist] as follows.

```
sample = {3, 2, 4, 6, 5};
Mean[sample]
```

4

In the study of statistics, one of the most fundamental ideas is that because a random sample is subject to chance influences, so is a statistic based on the sample such as $\overline{X}$. So, $\overline{X}$ is a random variable, and you can ask what its mean and variance are. The previous theorems allow us to answer the question. By linearity of expectation,

$$\begin{aligned}
E[\overline{X}] &= E\left[\frac{X_1 + X_2 + \ldots + X_n}{n}\right] \\
&= \frac{1}{n}\left(E[X_1] + E[X_2] + \cdots + E[X_n]\right) \\
&= \frac{1}{n}n\mu = \mu
\end{aligned} \tag{7}$$

So $\overline{X}$ has the same mean value as each sample value X_i. It is in this sense an accurate estimator of μ. Also, by Theorem 2,

$$\begin{aligned}
\mathrm{Var}\,(\overline{X}) &= \mathrm{Var}\left(\frac{X_1 + X_2 + \ldots + X_n}{n}\right) \\
&= \frac{1}{n^2}\left(\mathrm{Var}\,(X_1) + \cdots + \mathrm{Var}\,(X_n)\right) \\
&= \frac{1}{n^2}n\sigma^2 = \frac{\sigma^2}{n}
\end{aligned} \tag{8}$$

As the sample size n grows, the variance of $\overline{X}$ becomes small and, in this sense, it becomes a more precise estimator of μ as n increases.

Let us use *Mathematica* to see these properties in action. First, we will build a simulator of a list of sample means from a given discrete distribution. The arguments of the next command are the number of sample means we want to simulate, the distribution from which we are simulating, and the size of each random sample. The RandomInteger function is used to create a sample of the given size, then Mean is applied to the sample to produce a sample mean.

```
SimulateSampleMeans[nummeans_,
    distribution_, sampsize_] :=
    Table[Mean[RandomInteger[
    distribution, sampsize]], {nummeans}]
```

Now we will simulate and plot a histogram of 100 sample means of random samples of size 20 from the geometric distribution with parameter 1/2.

```
Needs["KnoxProb7`Utilities`"];
SeedRandom[4532];
```

```
sample = SimulateSampleMeans[
    100, GeometricDistribution[0.5], 20];
Histogram[sample, 10, ChartStyle → Gray]
```

Figure 2.14 - Sample histogram of 100 means of samples of size 20 from geometric(.5)

This histogram is the observed distribution of the sample mean $\overline{X}$ in our simulation. Recall that the mean of the underlying geometric distribution is $\frac{1/2}{1-1/2} = 1$, and we do see that the center of the frequency histogram is roughly at $x = 1$.

To see the effect of increasing the sample size of each random sample, let's try quadrupling the sample size from 20 to 80.

```
SeedRandom[7531];
sample = SimulateSampleMeans[
    100, GeometricDistribution[.5], 80];
Histogram[sample, 10, ChartStyle → Gray]
```

Figure 2.15 - Sample histogram of 100 means of samples of size 80 from geometric(.5)

Again the center point is around 1, but whereas the range of observed values in the case that the sample size was 20 extended from around .4 to 1.8, now it only extends from around .7 to 1.4, about half as wide. The theory explains this phenomenon. According to Theorem 2, since the variance of the geometric(.5) distribution is $\frac{1-1/2}{1/2^2} = 2$, the standard deviation of $\overline{X}$ in the first experiment is $\sigma/\sqrt{n} = \sqrt{2}/\sqrt{20}$, and in the second experiment it is $\sqrt{2}/\sqrt{80} = \sqrt{2}/(2\cdot\sqrt{20})$, which is exactly one half of the standard deviation in the first experiment. ∎

Activity 3 Try simulating 100 sample means of samples of size 10 from the Poisson(4) distribution. Predict the histogram you will see before actually plotting it. Then try samples of size 90. What happens to the standard deviation of the sample mean?

Covariance and Correlation

The assumption of independence in Theorem 2 on the variance of a linear combination is a rather severe one. (Note that in Theorem 1 on the mean, we did not need to assume that the random variables in the combination were independent.) As we know, when sampling occurs without replacement, successive sampled values are not independent. And yet, it would be interesting to know about the variability of the sample mean. Can anything be said in the dependent case about the variance of the combination?

First we need a detour to introduce the concepts of covariance and correlation.

Definition 2. If X and Y are discrete random variables with means μ_x and μ_y, then the *covariance* between X and Y is

$$\sigma_{xy} = \text{Cov}(X, Y) = \text{E}\left[(X - \mu_x)(Y - \mu_y)\right] \tag{9}$$

Furthermore, if X and Y have finite standard deviations σ_x and σ_y, then the *correlation* between X and Y is

$$\rho_{xy} = \text{Corr}(X, Y) = \frac{\text{Cov}(X, Y)}{\sigma_x \sigma_y} \tag{10}$$

Example 3 To get an idea of how these two expectations measure dependence between X and Y, consider the two simple discrete distributions in Figure 16.

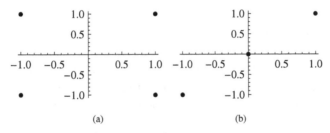

(a)   (b)

Figure 2.16 - Two discrete distributions

In part (a) of the figure, we put equal probability weight of 1/4 on each of the corners $(-1,-1)$, $(-1,1)$, $(1,1)$, and $(1,-1)$. It is easy to check that both μ_x and μ_y are zero. The covariance for the distribution in (a) is therefore

$$\sigma_{xy} = \text{E}[X\,Y] = \tfrac{1}{4}(-1)(-1) + \tfrac{1}{4}(-1)(1) + \tfrac{1}{4}(1)(1) + \tfrac{1}{4}(1)(-1) = 0$$

and so the correlation is 0 as well. If you look at the defining formula for covariance, you see that it will be large and positive when Y tends to exceed its mean at the same time X exceeds its mean. But for this joint distribution, when X exceeds its mean (i.e., when it has the value 1), Y still will either equal 1 or -1 with equal probability, which is responsible for the lack of correlation. However in Figure 16(b), suppose we put equal probability of 1/3 on each of the states $(-1,-1)$, $(0,0)$, and $(1,1)$. Again, it is easily checked that both μ_x and μ_y are zero. This time the covariance is

$$\sigma_{xy} = \frac{1}{3}(-1)(-1) + \frac{1}{3}(0)(0) + \frac{1}{3}(1)(1) = \frac{2}{3}$$

The random variables X and Y actually have the same marginal distribution (check this) which puts a probability weight of 1/3 on each of -1, 0, and 1, and so their common variance is also easy to compute as $\sigma^2 = 2/3$. Thus, the correlation between X and Y is

$$\rho = \frac{\sigma_{xy}}{\sigma_x \sigma_y} = \frac{2/3}{\sqrt{2/3}\,\sqrt{2/3}} = 1$$

For the distribution in Figure 16(b), X and Y are perfectly (positively) correlated. (Notice that when X is bigger than its mean of 0, namely when $X = 1$, Y is certain to be greater than its mean; in fact it can only have the value 1.) It turns out that this is the most extreme case. In Exercise 12 you are led through a proof of the important fact that:

> If ρ is the correlation between random variables X and Y,
> then $|\rho| \le 1$ (11)

and in addition it is true that ρ equals positive 1 or negative 1 if and only if, with certainty, Y is a linear function of X. ∎

Remember also the result of Activity 2. If X and Y happen to be independent, then $E\big[(X - \mu_x)(Y - \mu_y)\big] = E[X - \mu_x]\,E\big[Y - \mu_y\big] = 0$. This case represents the other end of the spectrum. Statistical independence implies that the covariance (and correlation) are zero. Perfect linear dependence implies that the correlation is 1 (or -1 if Y is a decreasing linear function of X).

Another set of results in the exercises (Exercise 11) about covariance and correlation is worth noting. If two random variables are both measured in a different system of units, that is if X is transformed to $aX + b$ and Y is transformed to $cY + d$, where a and c are positive, then

$$Cov(aX + b, cY + d) = a \cdot c \cdot Cov(X, Y),$$
$$Corr(aX + b, cY + d) = Corr(X, Y)$$

(12)

In particular, the covariance changes under the change of units, but the correlation does not. This makes the correlation a more standardized measure of dependence than the covariance. One way of understanding this is to note that

$$\rho_{xy} = \frac{Cov(X, Y)}{\sigma_x \sigma_y} = \frac{E\big[(X - \mu_x)(Y - \mu_y)\big]}{\sigma_x \sigma_y} = E\Big[\frac{X - \mu_x}{\sigma_x} \cdot \frac{Y - \mu_y}{\sigma_y}\Big]$$

so that the correlation is the expected product of standardized differences of the two random variables. You can check that each of $\frac{X - \mu_x}{\sigma_x}$ and $\frac{Y - \mu_y}{\sigma_y}$ have mean zero and variance 1, and each measures the number of standard deviation units by which the random variables differ from their respective means.

Finally, expansion of the expectation that defines the covariance yields the useful simplifying formula:

$$
\begin{aligned}
\text{Cov}(X, Y) &= \text{E}\big[(X - \mu_x)(Y - \mu_y)\big] \\
&= \text{E}[X \cdot Y] - \mu_x \text{E}[Y] - \mu_y \text{E}[X] + \mu_x \mu_y \\
&= \text{E}[X \cdot Y] - \mu_x \mu_y
\end{aligned}
$$

Example 4 Recall the psychology experiment on memory from the beginning of Section 2.5. If once again we let X be the number of orally presented numbers that were remembered, and Y be the number of visually presented numbers that were remembered, then we can compute the covariance and correlation of X and Y. The arithmetic is long, so we will make good use of *Mathematica*. First we define the marginal distributions of the two random variables, which are the marginal totals divided by the overall sample size. We also introduce the list of states for the two random variables

```
xmarginal = {31, 32, 36, 41, 31} / 171;
ymarginal = {10, 24, 41, 60, 36} / 171;
xstates = {0, 1, 2, 3, 4};
ystates = {0, 1, 2, 3, 4};
```

The marginal means are the dot products of the marginal distributions and the state lists:

```
{mux, muy} = {xmarginal.xstates, ymarginal.ystates}
```

$$
\left\{ \frac{39}{19}, \frac{430}{171} \right\}
$$

The variances can be found by taking the dot product of the marginal mass lists with the square of the state lists minus the means:

```
{sigsqx, sigsqy} =
  {xmarginal.((xstates - mux) * (xstates - mux)),
     ymarginal.((ystates - muy) * (ystates - muy))}
```

$$
\left\{ \frac{2030}{1083}, \frac{38\,084}{29\,241} \right\}
$$

By computational formula (13), it now suffices to subtract the product of marginal means from the expected product. For the latter, we need the full dot product of the joint probabilities with the product of X and Y states, which is below.

```
jointmassfn = {4, 6, 10, 8, 3, 2, 5, 7, 12, 6, 1, 4, 10,
        15, 6, 2, 6, 9, 14, 10, 1, 3, 5, 11, 11} / 171;
stateprods = {0, 0, 0, 0, 0, 0, 1, 2, 3, 4, 0, 2,
        4, 6, 8, 0, 3, 6, 9, 12, 0, 4, 8, 12, 16};
expectedprod = jointmassfn.stateprods
```

$$\frac{943}{171}$$

Hence, the covariance and correlation are

```
covar = N[expectedprod - mux * muy]
```

0.353032

```
corr = N[ ──────covar──────── ]
          √sigsqx * √sigsqy
```

0.225946

These two random variables are therefore rather weakly correlated. ∎

Example 5 Suppose that an investor puts half of his wealth into an asset with a random rate of return R_1 whose mean is .05 and whose standard deviation is .03, and the other half into a second asset with random rate of return R_2 whose mean is .06 and whose standard deviation is .05. If the correlation between the two rates of return is $-.5$, find the mean and variance of the total investment rate of return.

The total rate of return on the investment is the linear combination $R = .5\,R_1 + .5\,R_2$ of the rates of return on the two assets. By linearity, the mean rate of return is

$$\mu_R = E[.5\,R_1 + .5\,R_2] = .5\,\mu_1 + .5\,\mu_2 = 5\,(.05) + .5\,(.06) = .055$$

The variance of the rate of return is

$$
\begin{aligned}
\text{Var}\,(R) &= E\big[\,(R - \mu_R)^2\,\big] \\
&= E\big[((.5\,R_1 + .5\,R_2) - (.5\,\mu_1 + .5\,\mu_2))^2\big] \\
&= E\big[((.5\,R_1 - .5\,\mu_1) + (.5\,R_2 - .5\,\mu_2))^2\big] \\
&= (.5)^2\,E\big[(R_1 - \mu_1)^2\big] + (.5)^2\,E\big[(R_2 - \mu_2)^2\big] + 2\,(.5)\,(.5)\,E[(R_1 - \mu_1)\,(R_2 - \mu_2)] \\
&= (.5)^2\,\text{Var}(R_1) + (.5)^2\,\text{Var}(R_2) + 2\,(.5)\,(.5)\,\text{Cov}(R_1, R_2)
\end{aligned}
$$

Notice that we have computed in a similar way to the derivation leading to formula (5). Since $\text{Cov}(R_1, R_2) = \rho\, \sigma_1\, \sigma_2$, we compute that the variance and standard deviation of the total rate of return are:

```
VarR = .25 * (.03)² + .25 * (.05)² +
    2 * (.25) * (-.5) * (.03) * (.05)
σR = √VarR
```

0.000475

0.0217945

It is a very interesting fact that the negative correlation between the asset rates of return has allowed the overall rate of return to have a smaller standard deviation, about .022, than either of the individual assets. This illustrates the value of diversification. ∎

It is time to turn back to the distributional properties of the sample mean in the case of sampling without replacement. To set up this study, look at the variance of a linear combination of two random variables which are not assumed to be independent. By formula (5),

$$\text{Var}(c\cdot X + d\cdot Y) = c^2 \cdot \text{Var}(X) + d^2 \cdot \text{Var}(Y) + 2\, c\cdot d\cdot \text{Cov}(X, Y) \tag{14}$$

Extend this result to three random variables by doing the next activity.

Activity 4 Find an expression analogous to (14) for $\text{Var}(c\cdot X + d\cdot Y + e\cdot Z)$.

The proof of Theorem 2 can be generalized (see Exercise 6) to give the formula

$$\text{Var}\left(\sum_{i=1}^{n} c_i\, X_i\right) = \sum_{i=1}^{n} c_i^2\, \text{Var}(X_i) + 2 \sum_{i=1}^{n}\sum_{j=1}^{i-1} c_i\, c_j\, \text{Cov}(X_i, X_j) \tag{15}$$

In particular, for the sample mean $\overline{X} = \frac{1}{n}\sum_{i=1}^{n} X_i$ each coefficient c_i equals $1/n$ and, if the X_i's are identically distributed with variance σ^2, then we obtain

$$\begin{aligned}
\text{Var}(\overline{X}) &= \sum_{i=1}^{n} \frac{1}{n^2}\sigma^2 + 2 \sum_{i=1}^{n}\sum_{j=1}^{i-1} \frac{1}{n^2}\text{Cov}(X_i, X_j) \\
&= \frac{\sigma^2}{n} + \frac{2}{n^2} \sum_{i=1}^{n}\sum_{j=1}^{i-1} \text{Cov}(X_i, X_j)
\end{aligned} \tag{16}$$

Since σ^2/n is the variance of X in the independent case, we can see that if the pairs (X_i, X_j) are all positively correlated (i.e., their covariance is positive), then the variance of $\overline{X}$ in the dependent case is greater than the variance in the independent case and, if the pairs are all negatively correlated, the dependent variance is less than the independent variance.

Example 6 Suppose that a population can be categorized into type 0 individuals and type 1 individuals. This categorization may be done on virtually any basis: male vs. female, at risk for diabetes vs. not, preferring Coke vs. Pepsi, etc. We draw a random sample of size n in sequence of these 0's and 1's: $X_1, X_2, \ldots, X_n$. Notice that in this case, since the X's only take on the values 0 and 1, the sample mean $\overline{X}$ is also the sample proportion of 1's, which should be a good estimator of the population proportion of 1's. This is because each X_i has the Bernoulli(p) distribution, whose mean is p and whose variance is $p(1-p)$ (formulas (2) and (3) of Section 2.2). Thus, by formula (7),

$$E[\overline{X}] = p$$

and so $\overline{X}$ is an accurate estimate of p. If the sample is drawn with replacement, then the X's will be independent, hence formula (8) gives in the independent case

$$\mathrm{Var}(\overline{X}) = \frac{p(1-p)}{n}$$

Therefore, $\overline{X}$ becomes an increasingly precise estimator of p as the sample size n increases.

Now let's look at the case of sampling in sequence without replacement. Suppose our population has finite size M, and L individuals in it are of type 1, hence $p = L/M$. Each X_i is marginally Bernoulli(p), and so the equation $E[\overline{X}] = p$ still holds. But the sample values are now correlated. To compute the covariance between X_i and X_j, note that the probability that both of them equal 1 is

$$\frac{L}{M} \cdot \frac{L-1}{M-1}$$

since we require $X_i = 1$, and once that is given, there are $L - 1$ type 1's remaining to select from the $M - 1$ population members who are left which can make $X_j = 1$. Hence,

$$E[X_i \cdot X_j] = 1 \cdot P[X_i = 1, \ X_j = 1] = \frac{L}{M} \cdot \frac{L-1}{M-1}$$

and by computational formula (13),

$$\mathrm{Cov}(X_i, X_j) = \frac{L}{M} \cdot \frac{L-1}{M-1} - \frac{L}{M} \cdot \frac{L}{M} = \frac{L}{M} \cdot \frac{L-M}{M(M-1)}$$

after simplification. This can be rewritten as

$$\mathrm{Cov}(X_i, X_j) = -\frac{p(1-p)}{M-1}$$

if we use the facts that $p = L/M$ and $1 - p = (M - L)/M$. Notice that we are in one of the cases cited above, where all of the X pairs are negatively correlated. Finally, substitution into (16) gives

$$\begin{aligned}
\operatorname{Var}(\overline{X}) &= \frac{p(1-p)}{n} - \frac{2}{n^2} \sum_{i=1}^{n} \sum_{j=1}^{i-1} \frac{p(1-p)}{M-1} \\
&= \frac{p(1-p)}{n} - \frac{2}{n^2} \frac{(n-1)n}{2} \frac{p(1-p)}{M-1} \\
&= \frac{p(1-p)}{n} \left(1 - \frac{n-1}{M-1}\right)
\end{aligned}$$

One interesting thing about formula (17) is that as the sample size n approaches the population size M, the variance of $\overline{X}$ decreases to zero, which is intuitively reasonable since sampling is done without replacement, and the more of the population we sample, the surer we are of $\overline{X}$ as an estimator of p.

Figure 2.17 - Variance of the sample mean as a function of sample size, with (thin) and without (thick) replacement

In Figure 17 is a *Mathematica* plot of the variance of $\overline{X}$ as a function of n for fixed $M = 1000$ and $L = 500$ (hence $p = 1/2$). The thin curve is the variance of the sample mean in the independent case, shown for comparison. You can open up the closed cell in the electronic version of the text, and try a few different plot domains to get a complete look at the graphs. Here we see that for sample sizes in the range of 100 to 500, both variances are getting small, but the variance for the non-replacement case is less than that for the replacement case, making it a more precise estimator of p. ∎

Conditional Expectation

We would like to introduce one more notion involving expectation: conditional expectation of a random variable given (a) an event; and (b) the value taken on by another random variable. Conditional expectation has come to occupy an important place in applications of probability involving time-dependent random processes, such as queues and financial objects, which are discussed in Chapter 6.

Just as the conditional probability of an event given another is a revised probability that assumes the second event occured, the conditional expectation of a random variable is a revised average value of that random variable given some information. Beginning with the idea of conditional expectation given an event, recall that Exercise 8 of Section 1.5 established that if A is an event of positive probability, then the function $Q[B] = P[B \mid A]$ is a probability measure as a function of events B. If Y is a random variable on the same sample space in which A and B reside, we can define the conditional expectation of Y given A as follows:

Definition 3. Let Y be a discrete random variable mapping a sample space Ω to a state space E, and let A be a fixed event of positive probability. Denote by P_A the conditional probability measure on Ω given A described above. Then the *conditional expectation of Y given A* is

$$E[Y \mid A] = \sum_{y \in E} y \cdot P_A[Y = y] \tag{18}$$

By considering outcomes individually instead of combining those outcomes ω such that $Y(\omega) = y$ into one set, the expression for $E[Y \mid A]$ can also be written

$$E[Y \mid A] = \sum_{\omega \in \Omega} Y(\omega) \cdot P_A[\omega] \tag{19}$$

Suppose for example that Y is the random variable that we looked at near the beginning of Section 2.1, which takes two possible values $\{0, 1\}$ on six possible equally likely outcomes $\{a, b, c, d, e, f\}$, as shown in Figure 18. Consider the conditional expectation of Y given the event $A = \{d, e, f\}$. Since the original outcomes were equally likely, the conditional probability measure $P_A[B] = \frac{P[A \cap B]}{P[A]} = \frac{P[A \cap B]}{1/2}$ will place equal likelihood of 1/3 on each of the outcomes in $A = \{d, e, f\}$. (Why?). Therefore,

$$E[Y \mid A] = \frac{1}{3} \cdot 0 + \frac{2}{3} \cdot 1 = \frac{2}{3}.$$

Notice that the unconditional expectation of Y is 1/3, so knowledge that A occurred changes the average value of Y. This computation should help to clarify the intuition that the conditional expectation of a random variable given an event is the average value that the random variable takes on over the outcomes that are in the event.

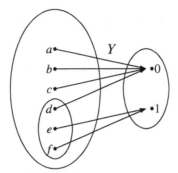

Figure 2.18 - Conditional expectation of a random variable given a set of outcomes

The next result is an interesting alternative characterization of the conditional expectation reminiscent of the definition of conditional probability.

Theorem 3. Let Y be a discrete random variable and let A be an event of positive probability. Denote by I_A the indicator random variable for A, that is,

$$I_A(\omega) = \begin{cases} 1 & \text{if } \omega \in A \\ 0 & \text{otherwise} \end{cases}$$

Then,

$$E[Y \mid A] = \frac{E[Y \cdot I_A]}{P[A]} \tag{20}$$

Proof. The conditional probability measure P_A gives no probability weight to outcomes ω that are not in A, hence $Y(\omega) P_A[\omega] = Y(\omega) \cdot I_A(\omega) \cdot P_A[\omega] = \frac{Y(\omega) \cdot I_A(\omega) \cdot P[\omega \cap A]}{P[A]}$. For every outcome ω, $I_A(\omega) \cdot P[\omega \cap A] = I_A(\omega) \cdot P[\omega]$, since both sides are $P[\omega]$ if $\omega \in A$ and zero if not. Thus,

$$\begin{aligned}
E[Y \mid A] &= \sum_{\omega \in \Omega} Y(\omega) \cdot P_A[\omega] \\
&= \sum_{\omega \in \Omega} \frac{Y(\omega) \cdot I_A(\omega) \cdot P[\omega]}{P[A]} \\
&= \frac{1}{P[A]} \sum_{\omega \in \Omega} Y(\omega) \cdot I_A(\omega) \cdot P[\omega] \\
&= \frac{E[Y \cdot I_A]}{P[A]} \quad \blacksquare
\end{aligned}$$

Activity 5 Use formula (20) to show that conditional expectation is linear, i.e.,
$E[c_1 Y_1 + c_2 Y_2 \mid A] = c_1 E[Y_1 \mid A] + c_2 E[Y_2 \mid A]$.

In Exercise 16 you are asked to prove a new version of the law of total probability for conditional expectation, as follows: If sets B_1, B_2, ..., B_n form a partition of the sample space, then

$$E[Y] = \sum_{i=1}^{n} E[Y \mid B_i] P[B_i] \tag{21}$$

So the expected value of a random variable can be found by conditioning its expectation on the occurrence of B_i, then un-conditioning by weighting by $P[B_i]$ and summing.

Most often we want to compute conditional expectations of random variables given observed values of other random variables, that is $E[Y \mid X = x]$. As long as both X and Y are defined on the same sample space Ω the set $A = \{\omega \in \Omega : X(\omega) = x\}$ is a perfectly good event, and the definition of $E[Y \mid A]$ above applies, so that we do not need to give a new definition. But it is more common to give an equivalent definition using conditional mass functions. The following theorem does that a bit more generally, giving a formula for the conditional expectation of functions of Y.

Theorem 4. Let X and Y be discrete random variables with joint p.m.f. $f(x, y)$, and let $q(y \mid x)$ be the conditional p.m.f. of Y given $X = x$. Then the conditional expectation of a function g of Y given $X = x$ is

$$E[g(Y) \mid X = x] = \sum_{y} g(y) q(y \mid x) \tag{22}$$

Proof. With $A = \{X = x\}$, we have that

$$
\begin{aligned}
E[g(Y) \mid X = x] &= E[g(Y) \mid A] \\
&= \sum_{\omega \in \Omega} g(Y(\omega)) \cdot P_A[\omega] \\
&= \sum_{\omega \in \Omega} g(Y(\omega)) I_A(\omega) \cdot P_A[\omega] \\
&= \sum_{\omega \in \Omega} \frac{g(Y(\omega)) \cdot I_{\{X=x\}}(\omega) \cdot P[\omega \cap \{X=x\}]}{P[X=x]}
\end{aligned}
$$

Reorder the terms in the last sum so that all outcomes for which $Y(\omega) = y$ are grouped together, and add over the possible states y of Y. For such outcomes, $g(Y(\omega)) = g(y)$, and $\sum_{\omega:Y(\omega)=y} I_A(\omega) \cdot P[\omega \cap \{X = x\}] = P[X = x, Y = y]$. Therefore,

$$
\begin{aligned}
E[g(Y) \mid X = x] &= \sum_{y} \frac{g(y)}{P[X=x]} \sum_{\omega:Y(\omega)=y} I_A(\omega) \cdot P[\omega \cap \{X = x\}] \\
&= \sum_{y} \frac{g(y)}{P[X=x]} P[X = x, Y = y]
\end{aligned}
$$

In view of the definition of $q(y \mid x)$, this completes the proof. ∎

By Theorem 4, the conditional expectation is not really a new quantity; it is just expectation as defined earlier (the weighted average of the values $g(y)$) except that the conditional p.m.f. of Y is used as the weighting function instead of the marginal p.m.f. of Y. It follows that the conditional expectation may well depend on the observed X value x.

The most important functions g of Y whose expectations we calculate are the moments. The *conditional mean of Y given X = x* is

$$\mu_{Y|x} = E[Y \mid X = x] = \sum_y y \cdot q(y \mid x) \tag{23}$$

and the *conditional variance of Y given X = x* is

$$\sigma_{Y|x}^2 = \text{Var}(Y \mid X = x) = E\left[(Y - \mu_{Y|x})^2 \mid X = x\right]$$
$$= \sum_y (y - \mu_{Y|x})^2 \cdot q(y \mid x) \tag{24}$$

Notice that it is the conditional mean $\mu_{Y|x}$, not the marginal mean μ_y that appears in the square in the conditional variance. The following is an important activity to check your understanding of the definition of conditional variance.

Activity 6 Show that $\sigma_{Y|x}^2$ is not in general the same as $E\left[(Y - \mu_y)^2 \mid X = x\right]$. Show also the computational rule $\sigma_{Y|x}^2 = E\left[Y^2 \mid X = x\right] - \mu_{Y|x}^2$.

Example 7 Let random variables X and Y have the joint mass function $f(x, y)$ shown in the table below. Find the conditional mean and variance $\mu_{Y|0} = E[Y \mid X = 0]$ and $\text{Var}(Y \mid X = 0)$. Compute also the conditional mean and variance of X given $Y = 0$.

		Y	
		0	1
X	−1	1/6	0
	0	1/3	1/3
	1	1/6	0

Notice that the marginal probability $p(0) = P[X = 0] = 1/3 + 1/3 = 2/3$. Thus, the conditional probability mass function of Y given $X = 0$ is

$$q(0 \mid 0) = \frac{f(0, 0)}{p(0)} = \frac{1/3}{2/3} = \frac{1}{2}, \quad q(1 \mid 0) = \frac{f(0, 1)}{p(0)} = \frac{1/3}{2/3} = \frac{1}{2}$$

From this it is easy to calculate:

$$E[Y \mid X = 0] = \frac{1}{2} \cdot 0 + \frac{1}{2} \cdot 1 = \frac{1}{2}$$

$$\mathrm{Var}(Y \mid X = 0) = \frac{1}{2}\cdot\left(0 - \frac{1}{2}\right)^2 + \frac{1}{2}\cdot\left(1 - \frac{1}{2}\right)^2 = \frac{1}{4}$$

For the second part of the question, observe that the marginal probability $q(0) = P[Y = 0] = 1/6 + 1/3 + 1/6 = 2/3$. Therefore, the conditional mass function of X given $Y = 0$ is

$$p(-1 \mid 0) = \frac{f(-1, 0)}{q(0)} = \frac{1/6}{2/3} = \frac{1}{4},$$

$$p(0 \mid 0) = \frac{f(0, 0)}{q(0)} = \frac{1/3}{2/3} = \frac{1}{2},$$

$$p(1 \mid 0) = \frac{f(1, 0)}{q(0)} = \frac{1/6}{2/3} = \frac{1}{4}$$

Thus,

$$E[X \mid Y = 0] = \frac{1}{4}\cdot(-1) + \frac{1}{2}\cdot 0 + \frac{1}{4}\cdot 1 = 0$$

$$\mathrm{Var}(X \mid Y = 0) = \frac{1}{4}\cdot(-1 - 0)^2 + \frac{1}{2}\cdot(0 - 0)^2 + \frac{1}{4}\cdot(1 - 0)^2 = \frac{1}{2}. \quad \blacksquare$$

The activity below lists another important property of conditional expectation, analogous to the relationship between independence and conditional probability.

Activity 7 Show that if X and Y are independent, then $E[Y \mid X = x] = E[Y]$.

Example 8 Recall Example 1.5-3, in which a Xerox machine could be in one of four states of deterioration labeled 1, 2, 3, and 4 on each day. The matrix of conditional probabilities of machine states tomorrow given today is reproduced here for your convenience.

		\<br\>1	tomorrow\<br\>2	\<br\>3	\<br\>4
	1	3/4	1/8	1/8	0
today	2	0	3/4	1/8	1/8
	3	0	0	3/4	1/4
	4	0	0	0	1

Now that we know about random variables and conditional distributions, we can introduce X_1, X_2, and X_3 to represent the random states on Monday through Wednesday, respectively. Now let us suppose that the state numbers are proportional to the daily running costs when the machine is in that state. Thus, we can find for example the expected total cost through Wednesday, given that $X_1 = 1$ was the initial state on Monday. The conditional distribution of X_2 given $X_1 = 1$ is, from the table,

$$P[X_2 = 1 \mid X_1 = 1] = 3/4, \; P[X_2 = 2 \mid X_1 = 1] = 1/8, \; P[X_2 = 3 \mid X_1 = 1] = 1/8$$

Also, you can check that by the law of total probability, the conditional distribution of X_3 given $X_1 = 1$ is (see Figure 11 of Chapter 1)

$$P[X_3 = 1 \mid X_1 = 1] = \frac{9}{16}, \quad P[X_3 = 2 \mid X_1 = 1] = \frac{6}{32},$$
$$P[X_3 = 3 \mid X_1 = 1] = \frac{13}{64}, \quad P[X_3 = 4 \mid X_1 = 1] = \frac{3}{64}$$

Therefore, the expected total cost from Monday through Wednesday given that the machine was in state 1 on Monday is

$$\begin{aligned}
E[X_1 + X_2 + X_3 \mid X_1 = 1] &= 1 + E[X_2 \mid X_1 = 1] + E[X_3 \mid X_1 = 1] \\
&= 1 + \left(1 \cdot \tfrac{3}{4} + 2 \cdot \tfrac{1}{8} + 3 \cdot \tfrac{1}{8}\right) \\
&\quad + \left(1 \cdot \tfrac{9}{16} + 2 \cdot \tfrac{6}{32} + 3 \cdot \tfrac{13}{64} + 4 \cdot \tfrac{3}{64}\right) \\
&= \tfrac{273}{64} = 4.265625 \qquad \blacksquare
\end{aligned}$$

One other important property of conditional expectation is worth our attention. To set it up, observe that formula (22) implies that $E[Y \mid X = x]$ is a function of x; call it $h(x)$ say. We write the composition $h(X)$ as $E[Y \mid X]$, thinking of it as the expectation of Y given the random variable X. The function $h(X)$ is random, and its expectation simplifies this way:

$$E[h(X)] = E[E[Y \mid X]] = E[Y] \tag{25}$$

To prove this, note that if $p(x)$ denotes the marginal mass function of X, then by Theorem 4,

$$\begin{aligned}
E[E[Y \mid X]] = E[h(X)] &= \textstyle\sum_x h(x)\, p(x) \\
&= \textstyle\sum_x \sum_y y\, q(y \mid x)\, p(x) \\
&= \textstyle\sum_x \sum_y y\, f(x, y) \\
&= E[Y]
\end{aligned}$$

This property appears similar to, and is in fact referred to as, the *law of total probability for expectation*. We can find the expected value of a random variable Y by conditioning on X, then "un-conditioning," which here means taking the expected value of the conditional expectation with respect to X.

Example 9 Let us check formula (25) in the setting of the two random variables X and Y of Example 7. Given $X = -1$, Y can have only one value, namely 0, hence $E[Y \mid X = -1] = 0$. Given $X = 0$, the values $Y = 0$ and $Y = 1$ occur with equal probability of 1/2, so $E[Y \mid X = 0] = 1/2$, as we already knew from Example 7. Given $X = 1$, the only possible value for Y is $Y = 0$, hence $E[Y \mid X = 1] = 0$. The expression $E[E[Y \mid X]]$ is

$$E[E[Y\mid X]] = E[Y\mid X = -1]\,P[X = -1] + E[Y\mid X = 0]\,P[X = 0]$$
$$+ E[Y\mid X = 1]\,P[X = 1]$$
$$= 0 + \frac{1}{2}\cdot\frac{2}{3} + 0$$
$$= \frac{1}{3}$$

Since the marginal mass function of Y is $q(0) = 1/6 + 1/3 + 1/6 = 2/3$, $q(1) = 0 + 1/3 + 0 = 1/3$, the mean of Y is $E[Y] = 0\cdot\frac{2}{3} + 1\cdot\frac{1}{3} = \frac{1}{3}$, which agrees with $E[E[Y\mid X]]$. ■

Exercises 2.6

1. Rederive the results for the mean and variance of the $b(n, p)$ distribution using another method, namely by observing that the number of successes Y is equal to the sum of random variables $X_1 + X_2 + \cdots + X_n$ where $X_i = 1$ or 0, respectively, according to whether the ith trial is a success or failure.

2. Suppose that the pair of random variables (X, Y) has a distribution that puts equal probability weight on each of the points in the integer triangular grid shown in the figure. Find (a) $E[X + Y]$; (b) $E[2X - Y]$.

Exercise 2

3. If X_1, X_2, X_3, X_4, and X_5 have the multinomial distribution with parameters $n = 20$, $p_1 = 1/5$, $p_2 = 1/10$, $p_3 = 1/10$, $p_4 = 2/5$, and $p_5 = 1/5$, find $E[X_1 + X_2 + X_3]$. Without computation, what is $E[X_1 + X_2 + X_3 + X_4 + X_5]$?

4. If (X_t) is a Poisson process of van arrivals with rate 2 per hour and 6 individuals in each van, find the expected value and variance of the number of individuals arriving in either the first or third hours.

5. Extend the linearity of expectation (Theorem 1) to linear combinations of n random variables X_1, X_2, ... , X_n.

6. Prove formula (15), the general expression for the variance of a linear combination $\text{Var}(\sum_i c_i X_i)$. To what does the formula reduce when the X_i's are independent?

7. (*Mathematica*) Write *Mathematica* commands which take as input a finite list $\{x_1, x_2, \ldots, x_m\}$ of states of a random variable X, a list $\{y_1, y_2, \ldots, y_n\}$ of states of a random variable Y, and a matrix $\{\{p11, p12, \ldots, p1n\}, \ldots, \{pm1, pm2, \ldots, pmn\}\}$ of joint probabilities of states (i.e., $pij = P[X = i, Y = j]$), and return: (a) the mean of X; (b) the mean of Y; (c) the variance of X; (d) the variance of Y; (e) the covariance of X and Y; (f) the correlation of X and Y.

8. Find the covariance and correlation between the random variables X and Y in Exercise 2.

9. If X and Y, respectively, are the numbers of successes and failures in a binomial experiment with n trials and success probability p, find the covariance and correlation between X and Y.

10. (*Mathematica*) We saw in the section that the probability distribution of the sample mean $\overline{X}$ of a random sample of size n clusters more and more tightly around the distributional mean μ as $n \longrightarrow \infty$. Here let us look at a different kind of convergence, studying a time series of the sample mean as a function of n, $\overline{X}(n) = (X_1 + X_2 + \cdots + X_n)/n$ as we add more and more observations.

Write a *Mathematica* command that successively simulates a desired number of observations from a discrete distribution, updating $\overline{X}$ each time m new observations have been generated (where m is also to be an input parameter). The command should display a connected list plot of that list of means. Run it several times for several distributions and larger and larger values of n. Report on what the sequence of means seems to do.

11. Show that if a, b, c, and d are constants with $a, c > 0$, and if X and Y are random variables, then

$$\text{Cov}(aX + b, cY + d) = a \cdot c \cdot \text{Cov}(X, Y)$$
$$\text{Corr}(aX + b, cY + d) = \text{Corr}(X, Y)$$

12. Show that the correlation ρ between two random variables X and Y must always be less than or equal to 1 in magnitude, and furthermore, show that if X and Y are perfectly linearly related, then $|\rho| = 1$. (The latter implication is almost an equivalence but we don't have the tools to see that as yet. To show that $|\rho| \leq 1$, carry out the following plan. First reduce to the case where X and Y have mean 0 and variance 1 by considering the random variables $(X - \mu_x)/\sigma_x$ and $(Y - \mu_y)/\sigma_y$. Then look at the discriminant of the non-negative valued quadratic function of t: $E[(W + tZ)^2]$, where W and Z are random variables with mean 0 and variance 1.)

13. (*Mathematica*) If the population in Example 6 is of size 1000, and sampling is done without replacement, use a graph to find out how large must the sample be to guarantee that the standard deviation of $\overline{X}$ is no more than .03. (Hint: you do not know a priori what p is, but at most how large can $p(1 - p)$ be?)

14. (*Mathematica*) Consider the problem of simulating random samples taken from the finite population {1, 2, ..., 500}, and computing their sample means. Plot histograms of the sample means of 50 random samples of size 100 in both the cases of sampling with replacement and sampling without replacement. Comment on what you see, and how it relates to the theoretical results of this section.

15. Suppose that random variable X maps eight outcomes in a sample space to the integers as follows: On outcomes ω_1, ω_2 X has the value 0; on outcomes ω_3, ω_4, ω_5 X has the value 1; and on outcomes ω_6, ω_7, ω_8 X has the value 2. The probability measure on the sample space gives equal likelihood to outcomes ω_1, ω_2, ω_3, ω_4, which are each twice as likely as each of ω_5, ω_6, ω_7, ω_8. Find the conditional expectation of X given the event $B = \{\omega_1, \omega_2, \omega_8\}$.

16. As in Example 8, compute the expected total cost through Wednesday given that the initial state on Monday was 2.

17. Find $E[Y \mid X = 1]$ and $E[Y \mid X = 2]$ for the random variables X and Y whose joint distribution is given in Exercise 2.5-2.

18. Compute $h(x) = E[Y^2 \mid X = x]$ for each possible x, if X and Y have the joint distribution in Exercise 2. Use this, together with formula (20), to find $E[Y^2]$.

19. Suppose that random variables X and Y have the joint density shown in the table below. Compute the conditional mean and variance $\mu_{Y|1} = E[Y \mid X = 1]$ and $\text{Var}(Y \mid X = 1)$. Demonstrate the equation $E[E[Y \mid X]] = E[Y]$ for these random variables.

		Y	
	0	1	2
0	1/6	0	0
1	1/6	1/8	0
X 2	1/6	1/8	1/24
3	0	1/8	1/24
4	0	0	1/24

20. Theorem 4 can be extended slightly to include the case where g is a function of both X and Y; simply replace the random variable X by its observed value x as follows: $E[g(X, Y) \mid X = x] = E[g(x, Y) \mid X = x] = \sum_y g(x, y) q(y \mid x)$. Use this to show that $E[h(X) g(Y) \mid X = x] = h(x) E[g(Y) \mid X = x]$, that is, given the value of X, functions of X may be factored out of the conditional expectation as if they were constants.

21. Show that if Y is a discrete random variable and $B_1, B_2, \ldots, B_n$ is a partition of the sample space, then $E[Y] = \sum_{i=1}^{n} E[Y \mid B_i] P[B_i]$.

Sample Problems from Actuarial Exam P

22. An insurance policy pays a total medical benefit consisting of two parts for each claim. Let X represent the part of the benefit that is paid to the surgeon, and let Y represent the part that is paid to the hospital. The variance of X is 5000, the variance of Y is 10,000, and the variance of the total benefit $X + Y$ is 17,000. Due to increasing medical costs, the company that issues the policy decides to increase X by a flat amount of 100 per claim and to increase Y by 10% per claim. Calculuate the variance of the total benefit after these revisions have been made.

23. Let X denote the size of a surgical claim and let Y denote the size of the associated hospital claim. An actuary is using a model in which $E[X] = 5$, $E[X^2] = 27.4$, $E[Y] = 7$, $E[Y^2] = 51.4$, and $\text{Var}(X + Y) = 8$. Let $C_1 = X + Y$ denote the size of the combined claims before the application of a 20% surcharge on the hospital portion of the claim, and let C_2 denote the size of the combined claims after the application of that surcharge. Calculate $\text{Cov}(C_1, C_2)$.

24. An actuary determines that the annual numbers of tornadoes in counties P and Q are jointly distributed as follows:

County Q

County P	0	1	2	3
0	.12	.06	.05	.02
1	.13	.15	.12	.03
2	.05	.15	.10	.02

Calculate the conditional variance of the annual number of tornadoes in county Q, given that there are no tornadoes in county P.

CHAPTER 3

CONTINUOUS PROBABILITY

3.1 From The Finite to the (Very) Infinite

In this chapter we begin the transition from discrete probability, in which sample spaces have either finitely many or countably many outcomes, to continuous probability, in which there is a continuum of possible outcomes. This area of study continues in the next chapter when we review some of the most useful probability distributions and the properties of continuous random variables.

Calculus and set theory are the tools for adapting what we have done so far to the more complicated continuous setting. Remember that when the sample space of a random experiment is finite or countable, to construct a probability it is enough to make an assignment $P[\omega]$ of positive probability to each outcome $\omega \in \Omega$. To be valid, these outcome probabilities must be non-negative and must sum to 1. In continuous probability, we are usually interested in sample spaces $\Omega \subset \mathbb{R}$ which are intervals. But the problem is that we cannot assign strictly positive probability to all points in an interval $[a, b]$ without violating the condition that the sample space has probability 1. To see this, observe that in order for $P[\,[a, b]\,]$ to be less than or equal to 1, we can have at most 1 point in $[a, b]$ of probability in $(1/2, 1]$, at most 2 points of probability in $(1/3, 1/2]$, at most 3 points of probability in $(1/4, 1/3]$, etc. The set of points in $[a, b]$ of strictly positive probability is the disjoint union as n ranges from 2 to ∞ of the sets of points whose probability is in $(1/n, 1/(n-1)]$. Since the set of points of positive probability is therefore a countable union of finite sets, it is at most countable. In particular it follows that the set of points of positive probability on Ω can contain no intervals.

Another problem concerns what we mean by events. A typical sampling experiment in discrete probability is to sample two items, in sequence and with replacement, from the universe $\{a, b, c\}$. The sample space for this experiment is

$$\Omega = \{(a, a), (a, b), (a, c), (b, a), (b, b), (b, c), (c, a), (c, b), (c, c)\}$$

which has 9 elements. Each of the $2^9 = 512$ subsets of the sample space is an event. The analogous problem in the continuous domain of sampling two real numbers has all of $\mathbb{R}^2$ as its sample space. The set of all subsets of $\mathbb{R}^2$ is a large breed of infinity indeed, and it also includes some pathological sets whose properties are incompatible with the way we would like to formulate continuous probability. So in the rest of this book we will restrict our attention to a smaller, more manageable class of events: the so-called *Borel sets*, which are described shortly.

In working with events previously, we were accustomed to taking unions, intersections, and complements of events and expecting the resulting set to be an event. This is the idea behind the following definition.

Definition 1. A σ-*algebra* $\mathcal{H}$ of a set Ω is a family of subsets of Ω which contains Ω itself and is closed under countable union and complementation; i.e., if $A_1, A_2, A_3, \ldots \in \mathcal{H}$ then $\bigcup_i A_i \in \mathcal{H}$ and if $A \in \mathcal{H}$ then $A^c \in \mathcal{H}$.

The definition does not specifically refer to closure under countable intersections, but the next activity covers that case.

Activity 1 Use DeMorgan's laws to argue that if $\mathcal{H}$ is a σ-algebra then $\mathcal{H}$ is also closed under countable intersection.

Example 1 Although the σ-algebra concept is most appropriate for continuous probability models, it sheds light on the idea to construct one in the discrete case. Consider a sample space $\Omega = \{1, 2, 3, 4, 5, 6\}$. Find (a) the smallest σ-algebra containing the event $\{1, 2\}$; and (b) the smallest σ-algebra containing both of the events $\{1, 2\}$ and $\{3, 4, 5\}$.

For part (a) observe that the power set, that is the family of all subsets of Ω is certainly closed under countable union and intersection, and it contains the whole space Ω Therefore it is a σ-algebra containing $\{1, 2\}$, but it may not be the smallest one. Since Ω and $\{1, 2\}$ are in the σ-algebra that we would like to construct, so must be their complements, which are $\emptyset$ and $\{3, 4, 5, 6\}$. Complementation of any of the four subsets of Ω that we have listed so far does not give us any new candidates, so we turn to unions. The empty set unioned with any set is that set, and Ω unioned with any set is just Ω so we find no new candidates for the σ-algebra that way. But also $\{1, 2\} \bigcup \{3, 4, 5, 6\} = \Omega$, which is already in our σ-algebra, so no new subsets can be found. The smallest σ-algebra containing $\{1, 2\}$ is therefore

$$\mathcal{H}_a = \{\Omega, \emptyset, \{1, 2\}, \{3, 4, 5, 6\}\}.$$

In part (b) there is a second subset of Ω to consider. Both $\{3, 4, 5\}$ and its complement $\{1, 2, 6\}$ must be adjoined to the previous family $\mathcal{H}$ Are any more subsets necessary to include? To maintain closure under union, the set $\{1, 2\} \bigcup \{3, 4, 5\} = \{1, 2, 3, 4, 5\}$ must be included, as must be its complement $\{6\}$. The smallest σ-algebra containing both events $\{1, 2\}$ and $\{3, 4, 5\}$ must at least contain all of the following subsets:

$$\mathcal{H}_b = \{\Omega, \emptyset, \{1, 2\}, \{3, 4, 5, 6\}, \{3, 4, 5\}, \{1, 2, 6\}, \{1, 2, 3, 4, 5\}, \{6\}\}.$$

Here we have listed out the subsets by pairs, with complementary sets grouped together, so clearly the complement of each set in $\mathcal{H}_b$ is also in $\mathcal{H}_b$. The following activity verifies that $\mathcal{H}_b$ is indeed a σ-algebra. ∎

> **Activity 2** Check that the union of any pair of events in $\mathcal{H}_b$ above is again in $\mathcal{H}_b$. Why is this enough to show that $\mathcal{H}_b$ is closed under countable union for this example?

The underlying theoretical structure of probability is wrapped up in the following definition.

Definition 2. A *probability space* is a triple $(\Omega, \mathcal{H}, P)$ where Ω is a set called the *sample space*, $\mathcal{H}$ is a σ-algebra of subsets of Ω that we call *events*, and P is a probability measure on Ω that maps events to $[0, 1]$.

It is perhaps needlessly pedantic to be so formal in the discrete case in which Ω is finite or countable, $\mathcal{H}$ is usually just the power set of Ω, and P is able to be defined simply by its action on outcomes. But as we move to continuous probability there is more to be said about these three objects.

To get started, let the sample space be $\Omega = \mathbb{R}$. (The construction for higher dimensional sample spaces or interval subsets of $\mathbb{R}$ is analogous.) Certainly intervals should qualify as events. For definiteness, let us use intervals of the form $(a, b]$ as our most primitive building blocks, where a is allowed to be $-\infty$ Notice that the intersection of any two of these is either empty or is another interval of this form.

If we start with a generating family of sets such as the collection of all intervals $\mathcal{L} = \{(a, b]\}$, we can talk about the family of all σ-algebras $\mathcal{H}$ that contain $\mathcal{L}$ This family does have a member, since the power set of Ω is always a σ-algebra, and it contains Ω and $\mathcal{L}$ Exercise 4 asks you to show that an arbitrary intersection of σ-algebras is also a σ-algebra. Therefore the intersection of all σ-algebras containing $\mathcal{L}$ is again a σ-algebra, and it must be the smallest σ-algebra containing $\mathcal{L}$ in terms of set containment. The sets in this σ-algebra are the Borel sets.

Definition 3. The *Borel subsets* of the real line are the sets comprising the smallest σ-algebra containing all intervals $(a, b]$.

Example 2 Let us look more concretely at examples of members of the σ-algebra of Borel sets on the real line. All primitive half-open, half-closed intervals such as $(9, 11]$ and $(-\infty, 3]$ are Borel sets. Since the family of Borel sets is closed under complementation, open rays such as (b, ∞), which are complementary to closed rays $(-\infty, b]$ are Borel sets. Singleton sets $\{b\}$ are also Borel sets, because they can be expressed as countable intersections of intervals $(a_n, b]$ where the sequence of left endpoints a_n increases to b. Similarly, a closed ray of the form $[a, \infty)$ is a Borel set, because it is expressible as a countable intersection of intervals (a_n, ∞) where the left endpoints a_n increase to a. Exercise 6 asks you to show that intervals of all other possible forms (a, b), $[a, b]$, $[a, b)$, for a and b finite, and $(-\infty, b)$ are Borel sets as well. Also, any finite or countable union of intervals of any form is a Borel set, hence sets like $(0, 1) \cup [8, \infty)$ are Borel sets. We will rarely see any set more complicated than a finite, disjoint union of intervals. In higher dimensions, the σ-algebra of Borel sets is the smallest σ-algebra containing all generalized rectangles $[a_1, b_1) \times [a_2, b_2) \times \cdots$, which is sufficiently rich to include most events of interest. ∎

So, in cases where the sample space Ω is the real line, the family of events $\mathcal{H}$ will be the collection of Borel subsets of Ω If the sample space is a subset of $\mathbb{R}$ then the events are the Borel sets contained in that subset. The main advantage to limiting attention to the Borel sets is that we can compute Riemann integrals over these sets in working with continuous probability models.

Next we must look at how to characterize the probability of events in the continuous case. Whatever we do should be consistent with the definition of a probability measure in Section 1.2, which requires: (i) $P[\Omega] = 1$; (ii) For all $A \in \mathcal{H}$ $P[A] \geq 0$; and (iii) If A_1, A_2, $\cdots$ is a sequence of pairwise disjoint events then $P[A_1 \cup A_2 \cup \cdots] = \sum_i P[A_i]$. If

these axioms are satisfied, Propositions 1-5 in Section 1.2, which give the main properties of probability, still hold without further proof, because they were proved assuming only the axioms, not the finiteness of Ω We will discover that many of the other properties of probability, random variables, and distributions will carry over as well.

Continuous probabilities can often be thought of as limits of discrete probabilities where outcomes are measured to higher and higher degrees of accuracy. To motivate the idea, consider a sample space of numerical outcomes in $\Omega = [0, 1]$ rounded down to the nearest tenth. The histogram might look as in Figure 1(a) if the distribution of probability among outcomes is taken to be uniform for simplicity. Each of the 10 outcomes receives equal probability of .1. Now suppose that outcomes are rounded down to the nearest .05. The possible outcomes are now $\{0, .05, .1, .15, ... , .9, .95\}$ of which there are 20, hence each would receive probability $1/20 = .05$ under a uniform model. This is sketched in Figure 1(b). One can imagine continuing the process; if outcomes were rounded down to the nearest .01, then each of the 100 resulting outcomes $\{0, .01, .02, ... , .99\}$ would receive probability .01.

(a)   (b)

Figure 3.1 - Two discrete uniform probability measures

As more precision of measurement is used, the number of outcomes increases and their individual probabilities decrease, but one thing is consistent: the *probability density*. In general a scaled probability histogram called a *density histogram* can be drawn, by setting the bar height for an interval from x to $x + \Delta x$ equal to the density, that is the probability weight at x divided by the length of the interval Δx. If we do this, then the bar area, which is the bar height times Δx, is equal to the probability of outcome x. For instance, for the probability measure in Figure 1(a) with weights of .1 on each outcome spaced at an interval $\Delta x = .1$, the probability density is $.1/.1 = 1$. In Figure 1(b), the probability density is $.05/.05 = 1$ again. The two density histograms are in Figure 2. Notice the resemblance to the continuous uniform distribution of probability on [0, 1] that we studied in Chapter 1. Probabilities of sets of outcomes are total rectangle areas in the density histograms; for instance, in part (a), the probability of the set of outcomes {.4, .5} is the total of the areas of the rectangles above those outcomes, which is $1 \cdot .1 + 1 \cdot .1 = .2$. In part (b), the probability that the outcome takes a value between .2 and .4 inclusive is the total of the five rectangle areas between these two bounds, each of which is $.05 \cdot 1$, hence the total probability is $5 \cdot .05 = .25$.

(a)   (b)

Figure 3.2 - Two discrete probability density histograms

Example 3 The idea of using a continuous probability model as an approximation to a discrete model with many states each of which has small probability is more than just a theoretical device. The growing field of financial mathematics makes heavy use of a model for random asset price motions called the *binomial branch model*. In this, the price of an asset begins at some value x at time 0, and moves to a later value at a later time that we will call time 1 through a sequence of steps. In each step, the process has just two possible next values: "up" with probability p, or "down" with probability $1 - p$. Figure 3 shows two such steps, in which from the starting value x the asset can either go up to position 2 or down to position 3; then in the next step if it is at position 2 it can either go up to position 4 or down to 5, and if it is at position 3 it can go up to 5 or down to 6.

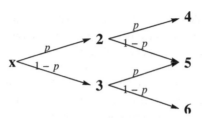

Figure 3.3 - Binomial branch process

Suppose that a binomial branch process is normalized so that $x = 0$, $p = 1/2$, and there are n time steps between time 0 and time 1. In each step the value either goes up by a step size of $\sqrt{1/n}$ or down by the same amount. After n steps, if there have been k ups and $n - k$ downs, then the value of the asset is $0 + k\sqrt{1/n} - (n - k)\sqrt{1/n} = (2k - n)\sqrt{1/n}$ for $k = 0, 1, 2, \ldots, n$. These $n + 1$ values are the outcomes in the sample space Ω of possible asset values at time 1. They are separated from each other by a distance of $\Delta x = 2\sqrt{1/n}$. The probability that the outcome corresponding to k ups happens is the binomial probability:

$$\binom{n}{k}(.5)^k (.5)^{n-k}.$$

These probabilities form a valid probability measure on the sample space $\Omega = \left\{(2k - n)\sqrt{1/n} : k = 0, 1, 2, \ldots, n\right\}$.

```
AssetPriceDistribution[numsteps_] :=
 Module[{timestep, stepsize,
    stateprobs, statelist, densitylist},
  timestep = 1 / numsteps;
  stepsize = Sqrt[timestep];
  stateprobs =
    Table[PDF[BinomialDistribution[numsteps, .5], k],
     {k, 0, numsteps}];
  densitylist = stateprobs / (2 * stepsize);
  statelist = Table[
     (2 k - numsteps) * stepsize, {k, 0, numsteps}];
  ProbabilityHistogram[statelist, densitylist,
    Ticks → {statelist, Automatic}, BarColor -> Gray]]
```

The *Mathematica* command above lets you plot the density histogram for this discrete probability model. First the time step $1/n$ and the asset price step size $\sqrt{1/n}$ are computed. The binomial probabilities are assembled into a list, and divided by $\Delta x = 2\sqrt{1/n}$ to get the densities. The states $(2k - n)\sqrt{1/n}$ are assembled into a list, and the probability density histogram is computed.Let us see how these density histograms behave as the number of time steps n becomes large. The closed cell below contains the graph of a continuous function called the standard normal density function (about which we will have much to say later) that superimposes well onto the density histogram. The graphics output for the plot is named g5. In the electronic file you can manipulate the value of the number of steps ($n = 12$ is shown here) in Figure 4 to see that the densities converge to this function. The fact that probabilities of outcomes are areas of rectangles in the density histogram suggests that as the number of steps approaches infinity, one should be able to find probabilities involving the asset price at time 1 by finding areas under this standard normal density function. ∎

```
Manipulate[Show[
    {AssetPriceDistribution[n], g5}, BaseStyle -> 8],
    {n, 2, 20, 1}, ControlPlacement → Top]
```

Figure 3.4 - Binomial branch probability densities

The following examples illustrate ways of defining probability measures in continuous sample spaces.

Example 4 Let $f(x) = \frac{3}{4}\left(1 - x^2\right)$ and let the sample space Ω be the interval $[-1, 1]$. Let us argue that the following defines a probability measure on the Borel subsets of Ω and find several example event probabilities.

$$P[E] = \int_E f(x)\,d\,x \tag{1}$$

The probability measure is constructed to find probabilities of events (Borel sets) by integration, that is, by finding the area under the graph of f corresponding to the event on the x-axis. It can be shown in more advanced work that the integral in (1) is well-defined on the family of Borel sets which we are taking as our events. The idea here is to think of the function f as determining how densely the probability weight is concentrated around each x. Adding density $f(x)$ times infinitesimal length $d\,x$ over all points x in event E gives the total probability weight of E. For our building block events $(a, b]$ we have

$$P[\,(a, b]\,] = \int_a^b f(x)\,d\,x \tag{2}$$

First, here is a *Mathematica* definition of our function f, and a computation and display of the probability associated with the interval $(0,.5]$, which is the shaded area under the density function in Figure 5.

```
f[x_] := 0.75 * (1 - x²);  ∫₀·⁵ f[x] dx
```

0.34375

```
PlotContsProb[f[x], {x, -1, 1}, {0, 0.5},
  Ticks -> {{-1, 0, 0.5, 1}, Automatic},
  AxesOrigin -> {0, 0}, PlotRange -> All,
  PlotStyle -> Black, ShadingStyle -> Gray]
```

Figure 3.5 - Shaded area = P[(0, .5]]

To display the area, we have used the command

PlotContsProb[densityfn, domain, between]

in KnoxProb7`Utilities`, which takes the density function, a plot domain, and another pair called *between* which designates the endpoints of the shaded interval. Notice that the probability weight of <u>every</u> singleton point $\{a\}$ is 0 in this model, since $\int_a^a f(x)\,dx = 0$. Only sets that have length will receive positive probability. As a further example, the probability of the event $[-.3, .2]$ is

$$\int_{-.3}^{.2} \mathbf{f[x]}\ \mathbf{dx}$$

0.36625

Activity 3 Why is $\int_a^b f(x)\,dx$ also equal to $P[\,(a,\ b)\,]$, $P[\,[a,\ b]\,]$, and $P[\,[a,\ b)\,]$?

Continuing with Example 4, we are obliged to check whether P defined by (1) is a valid probability measure. First observe that the integral of f over all of $\Omega = [-1, 1]$ is equal to 1, as required by the first axiom of probability:

$$\int_{-1}^{1} \mathbf{f[x]}\ \mathbf{dx}$$

1.

It is clear from the fact that $f(x) = \frac{3}{4}\left(1 - x^2\right)$ that the integrand in the defining expression for $P[E]$ is non-negative on Ω, hence $P[E] \geq 0$ for all events E, establishing axiom 2 of probability. The third axiom of probability, that is, the additivity over countable disjoint unions, is actually tricky to show in general, because of the fact that the Borel events can be quite complicated. We will omit the proof. But at least for unions of finitely many disjoint intervals the additivity property of the integral, shown in calculus, yields the conclusion of the third axiom without difficulty, as follows.

$$P[(a_1, b_1] \cup (a_2, b_2] \cup [(a_3, b_3]] = \int\limits_{(a_1,b_1]\cup(a_2,b_2]\cup(a_3,b_3]} f(x)\,d\,x$$

$$= \int_{a_1}^{b_1} f(x)\,d\,x + \int_{a_2}^{b_2} f(x)\,d\,x + \int_{a_3}^{b_3} f(x)\,d\,x$$

$$= P[(a_1, b_1]] + P[(a_2, b_2]] + P[(a_3, b_3]] \quad \blacksquare$$

Example 5 Suppose that an electronic device on which there is a maintenance plan survives for a uniformly distributed time (See Section 1.3) in the real interval [0, 4]. It will be replaced at the earlier of time 3 and its death time. Find a probability measure appropriate for the experiment of observing the time of replacement.

The possible outcomes ω of the experiment belong to the sample space $\Omega = [0, 3]$. But the actual death time of the device is assumed to be such that the probability that it lies in a Borel subset E of [0, 4] is the length of the subset divided by 4, which is the same as $\int_E \frac{1}{4}\,d\,x$. There is a non-zero probability that the replacement occurs exactly at time 3, namely:

$$P[\text{replace at time 3}] = P[\text{device survives at least 3 time units}]$$
$$= \int_3^4 \frac{1}{4}\,d\,x = \frac{1}{4} \tag{3}$$

Hence the outcome $\omega = 3$ should receive positive probability weight 1/4. For Borel subsets $E \subset [0, 3)$ we have

$$P[E] = P[\text{replace at time } t \in E] = P[\text{device dies at time } t \in E]$$
$$= = \int_E \frac{1}{4}\,d\,x = \frac{1}{4} \cdot \text{length}\,(E) \tag{4}$$

Together, formulas (3) and (4) define our probability measure. Axiom 1 of probability is satisfied, since

$$P[\Omega] = P[3] + P[[0, 3)] = \frac{1}{4} + \frac{1}{4} \cdot \text{length}([0, 3)) = \frac{1}{4} + \frac{3}{4} = 1.$$

Axiom 2 is clearly satisfied, since the point 3 has positive probability, and for Borel subsets of [0, 3) formula (4) must always give a non-negative result. Do Activity 4 below to help convince yourself that the third axiom holds. $\blacksquare$

Activity 4 Argue that P is Example 5 does satisfy the additivity axiom of probability for two disjoint events A_1 and A_2. (Hint: since the two events are disjoint, at most one of them can contain the point 3.)

Example 6 Consider the sample space $\Omega = \{(x, y) \in \mathbb{R}^2 : x^2 + y^2 \le 4\}$, that is, the disk of radius 2 in the plane. Let a set function be defined on the Borel subsets of Ω by

$$P[E] = \int\int_E c \cdot \left(4 - x^2 - y^2\right) dx\, dy$$

Find the value of c that makes this a valid probability measure and compute the probability of the unit square $P[[0,\ 1] \times [0,\ 1]]$.

In this continuous model, we are defining probabilities as volumes bounded by the surface graph of a two-variable function and the $x - y$ coordinate plane. Notice that the condition $x^2 + y^2 \le 4$ forces the integrand to be non-negative over Ω as long as $c > 0$. Hence axiom 2 of probability will hold, and axiom 3 will also follow from the additivity of the double integral over disjoint sets. What we simply need to do is to choose c so that the double integral over all of Ω is 1. By symmetry this double integral is 4 times the integral over the piece of the disk in the first quadrant:

```
(* P[Ω] = *)

4 * ∫₀² ∫₀^√(4-x²) c * (4 - x² - y²) dy dx
```

8 c π

Setting this equal to 1 yields $c = 1/(8\pi)$. The probability of the square is the volume underneath the surface shown in Figure 6, which is computed below as $5/(12\pi)$. ∎

```
(* P[[0,1] × [0,1]] = *)
∫₀¹ ∫₀¹ 1/(8 π) * (4 - x² - y²) dy dx
```

$$\frac{5}{12\ \pi}$$

```
Plot3D[1/(8 π) (4 - x² - y²), {x, 0, 1}, {y, 0, 1},
    PlotStyle → Gray, Lighting → "Neutral",
    BaseStyle → {8, FontFamily → "Times"}]
```

Figure 3.6 - Probability modeled as volume

Exercises 3.1

1. Devise an example of a random experiment and associated probability space in which a countable number of points receive positive probability and as well an interval of real numbers receives probability through integration. (Hint: a randomized experiment can be constructed which randomly chooses between two other experiments.) Carefully describe the action of the probability measure.

2. Let $\Omega = \{1, 2, 3, 4, 5\}$ be a sample space. Find (a) the smallest σ-algebra containing the events $\{2, 3\}$ and $\{1, 5\}$; and (b) the smallest σ-algebra containing both of the events $\{1, 2\}$ and $\{1, 2, 3\}$.

3. Explain why the set of all integers is a Borel set. Explain why any at most countable set of points on the real axis is a Borel set.

4. Prove that an arbitrary intersection (not necessarily a countable one) of σ-algebras is a σ-algebra. Conclude that the intersection of all σ-algebras containing a collection of sets $\mathcal{L}$ is the smallest σ-algebra containing $\mathcal{L}$

5. What is the smallest σ-algebra containing a singleton set $\{x\}$, for $x \in \Omega$?

6. Show that all of the intervals (a, b), $[a, b]$, $(-\infty, b)$, $[a, b)$ are Borel sets for any real numbers $a < b$.

7. Find a constant c such that $P[E] = \int_E c/x^2 \, dx$ is a valid probability measure on $\Omega = [1, \infty)$. Then find $P[(2, 4) \cup (8, \infty)]$.

8. Find a constant c such that $P[E] = \int_E c \log(x) \, dx$ is a valid probability measure on $\Omega = [1, e]$, and then compute $P[[2, e]]$ and $P[[1, 2]]$.

9. Find a constant c such that $P[E] = \int_E \int \frac{c}{x \cdot y} \, dx \, dy$ is a valid probability measure on $\Omega = [1, 2] \times [1, 2]$, and then compute $P[[1.5, 2] \times [0, 1.5]]$.

10. *(Mathematica)* Consider a binomial branch process as in Example 3 with 6 time steps, normalized initial price 0 and $p = 1/2$. Each time step, the price changes by $\pm \sqrt{1/6}$. Find and plot the probability mass function of the asset value at time 1, and use it to find the exact

probability that the asset value at time 1 is at least 1. Then superimpose the graph of the following function onto a probability density histogram and use numerical integration to find the approximate area underneath that function corresponding to the interval $[1, \infty)$ in order to continuously approximate the same probability. Does your picture suggest a better interval over which to integrate in order to improve the approximation?

```
f[x_] := PDF[NormalDistribution[0, 1], x]
```

11. A random variable X mapping a sample space Ω with a σ-algebra $\mathcal{H}$ to $\mathbb{R}$ is called *measurable with respect to* $\mathcal{H}$ if for every Borel subset B of $\mathbb{R}$, the inverse image $X^{-1}(B) = \{\omega : X(\omega) \in B\}$ is in $\mathcal{H}$ Consider the collection of all sets $\mathcal{H}_x = \{X^{-1}(B)\}$ as B ranges through all Borel sets in $\mathbb{R}$ Show that $\mathcal{H}_x$ is a σ-algebra with respect to which X is measurable, and in fact it is the smallest such σ-algebra.

3.2 Continuous Random Variables and Distributions

Next we would like to study the notion of random variables in the continuous setting. Random variables, as we know, attach a numerical value to the outcome of a random phenomenon. In the discrete case we used the probability mass function to characterize the likelihood $p(x) = P[X = x]$ that a random variable X takes on each of its possible states x. Then for subsets B of the state space E,

$$P[X \in B] = \sum_{x \in B} p(x) = \sum_{x \in B} P[X = x]$$

The continuous case will come out similarly, with an integral replacing the discrete sum. But why are we concerned with continuous random variables? There are many interesting experiments which give rise to measurement variables that are inherently continuous, that is, they take values in intervals on the line rather than at finitely many discrete points. For instance, the pressure on an airplane wing, the amount of yield in a chemical reaction, the measurement error in an experimental determination of the speed of light, and the time until the first customer arrives to a store are just a few among many examples of continuous variables. To make predictions about them, and to understand their central tendency and spread, we must have a way of modeling their probability laws.

We have already set up the machinery to characterize the probability distribution of a continuous random variable in Section 3.1: integration of a continuous function (thought of as a density of probability) over a set is the means for calculating the probability of that set. But now we consider probability density functions on the state spaces E of random variables rather than on the sample spaces Ω. From here on, Ω will be pushed to the background.

Definition 1. A random variable X with state space $E \subset \mathbb{R}$ is said to have the *probability density function* (abbr. p.d.f.) f if, for all Borel subsets B of E,

$$P[X \in B] = \int_B f(x)\,dx \tag{1}$$

In order for the axioms of probability to be satisfied, to be a valid density function f must be non-negative and $\int_E f(x)\,dx = 1$. Figure 7 gives the geometric meaning of formula (1). The probability that X will fall into set B is the area under the density curve on the subset B of the real axis.

```
Needs["KnoxProb7`Utilities`"]
```

Figure 3.7 - Shaded area is $P[X \in [.5,\ 1.5]]$ **Figure 3.8** - Shaded area is $F(-.7) = P[X \le -.7]$

The *cumulative distribution function* of a continuous random variable will be prominently featured from here on. The definition of the c.d.f. F is the same as it was before, though now it is computed as an integral:

$$F(x) = P[X \le x] = \int_{-\infty}^{x} f(t)\,dt \tag{2}$$

It totals all probability weight to the left of and including point x (see Figure 8). By the Fundamental Theorem of Calculus, differentiating both sides of (2) with respect to x gives the relationship

$$F'(x) = f(x) \tag{3}$$

Thus, the c.d.f. also characterizes the probability distribution of a continuous random variable, since if it is given then the density function is determined.

Activity 1 From calculus, the differential equation $F'(x) = f(x)$ only determines F up to an additive constant, given f. In our setting, what further conditions are present which guarantee that each density f has exactly one c.d.f.?

From past work we know one named continuous density function so far: *the continuous uniform density on [a,b]* has value $1/(b-a)$ on the interval $[a, b]$, and vanishes outside the interval. For example, if f is the continuous uniform density on $[0, 4]$, then

$$f(x) = \begin{cases} 1/4 & \text{if } x \in [0, 4] \\ 0 & \text{otherwise} \end{cases}$$

and hence the associated c.d.f. is

$$F(x) = \int_0^x \frac{1}{4}\,dt = \frac{1}{4}x \text{ if } x \in [0, 4]$$

For $x < 0$, $F(x) = P[X \le x] = 0$, and for $x > 4$, $F(x) = P[X \le x] = 1$, since X must take a value in $[0, 4]$. The graphs of the density and c.d.f. are shown in Figure 9. Note according to (3) that the density at each point x is the slope of the c.d.f. graph at that x.

```
f[x_] := Which[x < 0, 0, 0 ≤ x ≤ 4, 1/4, x > 4, 0];
F[x_] := Which[x < 0, 0, 0 ≤ x ≤ 4, 0.25` x, x > 4, 1];
Plot[{f[x], F[x]}, {x, -1, 5},
  PlotStyle → {Black, Gray}]
```

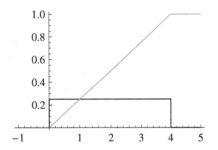

Figure 3.9 - Uniform(0,4) density (black), and c.d.f. (gray)

Mathematica's kernel contains definitions of many of the most common continuous distributions that we will use in this chapter and the next. The only one that we are acquainted with at this point is

Uniform Distribution[{a, b}]

which is the continuous uniform distribution on $[a, b]$. As in the discrete case, the functions

PDF[distribution, x]   and   CDF[distribution, x]

can be applied to distributions objects. PDF returns the density function, CDF returns the cumulative distribution function, and x is the desired argument of each function.

For example, for the uniform distribution on [0, 4] we can define the density and c.d.f. as follows:

```
f[x_] := PDF[UniformDistribution[{0, 4}], x];
F[x_] := CDF[UniformDistribution[{0, 4}], x];
```

Then we can calculate a probability like $P[X \in (1, 3]]$ in either of two ways in *Mathematica*; by integrating the density from 1 to 3, or by computing

$$P[1 < X \le 3] = P[X \le 3] - P[X \le 1] = F(3) - F(1).$$

```
∫₁³ f[x] dx
F[3] - F[1]
```

$$\frac{1}{2}$$

$$\frac{1}{2}$$

Example 1 If we simulate large random samples from a continuous uniform distribution, will a histogram of the data show the same flatness as the density function?

Consider for instance the uniform distribution on [0, 10]. Let us simulate 100 values from this distribution using the RandomReal command, and plot a histogram of the result with 10 rectangles. Two replications of this experiment are shown in Figure 10.

```
SeedRandom[63 471]
```

```
uniflist1 =
  RandomReal[UniformDistribution[{0, 10}], 100];
uniflist2 = RandomReal[
    UniformDistribution[{0, 10}], 100];
GraphicsRow[{Histogram[uniflist1,
    10, ChartStyle → Gray],
    Histogram[uniflist2, 10, ChartStyle → Gray]}]
```

Figure 3.10 - Two simulations of 100 uniform(0,10) observations

Because the distribution of probability is uniform, we expect to see 1/10th of the observations in each category, but in repeated runs it is typical to see a variation of relative frequencies ranging from a low of around .05 to a high of around .15. The moral of the story is that you should not expect histograms of random samples to look exactly like the density functions, although if there are drastic departures of data from the density, one should suspect that the density is not appropriate for the population being sampled from. Do the next activity to study the effect of increasing the sample size. ■

Activity 2 Do several more simulations like the one above, changing the sample size to 200, 500, and then 1000 sample values. What do you observe? Do you see any changes if you change the distribution to uniform(0, 1)?

Example 2 Consider the triangular density function sketched in Figure 11. (It could be appropriate for example to model the distribution of a random variable X that records the amount of roundoff error in the computation of interest correct to the nearest cent, where X is measured in cents and takes values between negative and positive half a cent.) Verify that this function is a valid density and find a piecewise defined formula for its cumulative distribution function.

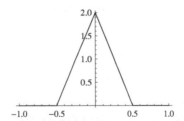

Figure 3.11 - A triangular density function

From the graph, f is clearly non-negative on its state space [−.5, .5] and its formula is

$$f(x) = \begin{cases} 2 + 4x & \text{if } -.5 \le x \le 0 \\ 2 - 4x & \text{if } 0 < x \le .5 \\ 0 & \text{otherwise} \end{cases}$$

The area under f corresponding to all of the state space is the area of the triangle, which is $(1/2) \cdot 1 \cdot 2 = 1$. Therefore f is a valid density function. For values of x between $-.5$ and 0 the c.d.f. is

$$F(x) = \int_{-.5}^{x} (2 + 4t)\,dt = 2t + 2t^2 \,|_{-.5}^{x} = 2x^2 + 2x + \frac{1}{2}$$

Notice that $F(0) = 1/2$, which is appropriate because the density is symmetric about 0, and therefore half of its overall probability of 1 lies to the left of $x = 0$. For values of x between 0 and $.5$,

$$F(x) = \frac{1}{2} + \int_{0}^{x} (2 - 4t)\,dt = \frac{1}{2} + \left(2t - 2t^2 \,|_{0}^{x}\right) = -2x^2 + 2x + \frac{1}{2}$$

The final form for F is

$$F(x) = \begin{cases} 0 & \text{if} \quad x < -.5 \\ 2x^2 + 2x + \frac{1}{2} & \text{if} \quad -.5 \le x < 0 \\ -2x^2 + 2x + \frac{1}{2} & \text{if} \quad 0 \le x < .5 \\ 1 & \text{if} \quad x \ge .5 \end{cases}$$

In the interest rounding example then, the probability that the rounding error is between $-.2$ and $.2$ is $F(.2) - F(-.2)$. This turns out to be $.82 - .18 = .64$ after calculation. ∎

Example 3 The *Weibull distribution* is a familiar one in reliability theory, because it is used as a model for the distribution of lifetimes. It depends on two parameters called α and β, which control the graph of its density function. The state space of a Weibull random variable X is the half line $[0, \infty)$, and the formula for the c.d.f. is

$$F(x) = P[X \le x] = 1 - e^{-(x/\beta)^\alpha}, \quad x \ge 0 \tag{4}$$

Mathematica's syntax for this distribution is

WeibullDistribution[α, β]

Below we construct the density and c.d.f. functions, and plot the two together in the case $\alpha = 2$, $\beta = 3$. Again as in formula (3), the density graph is clearly the derivative graph of the c.d.f.

```
g[x_, α_, β_] := PDF[WeibullDistribution[α, β], x];
G[x_, α_, β_] := CDF[WeibullDistribution[α, β], x];
Plot[{g[x, 2, 3], G[x, 2, 3]},
  {x, 0, 8}, PlotStyle → {Black, Gray}]
```

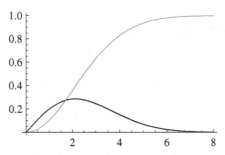

Figure 3.12 - Weibull(2, 3) p.d.f. (black), c.d.f. (gray)

In the electronic version of the text, you can experiment with various combinations of parameter values α and β using the Manipulate command below to see the effect on the Weibull density (see also Exercise 8).

```
Manipulate[
  Plot[g[x, α, β], {x, 0, 8}, PlotRange → {0, 1.6},
   AxesOrigin → {0, 0}], {α, 1, 4}, {β, 1, 4}]
```

If the lifetimes (in thousands of miles) of auto shock absorbers of a particular kind have a Weibull distribution with parameters $\alpha = 2$, $\beta = 30$, then for example the proportion of them that last at least 30,000 miles is $P[X \geq 30] = 1 - F(30)$, which is about 37%, as shown below. ∎

```
F[x_] := CDF[WeibullDistribution[2, 30], x];
N[1 - F[30]]
```

```
0.367879
```

Activity 3 Find the equation of the Weibull density function in general. Use *Mathematica* to plot it in the case where $\alpha = 2$, $\beta = 30$ as in the last example, and write a short paragraph about the shape and location of the density, and what the graph implies about the distribution of shock absorber lifetimes.

Joint, Marginal, and Conditional Distributions

In this subsection we carry over the ideas of joint, marginal, and conditional probability distributions from discrete to continuous probability. The adaptation is straightforward. In addition, we take another look at independence of random variables.

To characterize the joint distribution of two continuous random variables X and Y, we combine the ideas of the single variable density function and the two variable discrete probability mass function.

Definition 2. Random variables X and Y are said to have *joint probability density function* $f(x, y)$ if for any Borel subset B of $\mathbb{R}^2$,

$$P[(X, Y) \in B] = \int\int_B f(x, y)\, dx\, dy \tag{5}$$

The most familiar case of (5) occurs when B is a rectangular set, $B = [a, b] \times [c, d]$. Then (5) reduces to:

$$P[a \leq X \leq b,\ c \leq Y \leq d] = \int_a^b \int_c^d f(x, y)\, dy\, dx \tag{6}$$

For more complicated Borel subsets of $\mathbb{R}^2$, the techniques of multivariable calculus can be brought to bear to find the proper limits of integration. One final comment before we do an example computation: the extension of (5) and (6) to more than two jointly distributed random variables is simple. For n random variables X_1, X_2, ..., X_n, the density f becomes a function of the n-tuple $(x_1, x_2, ..., x_n)$ and the integral is an n-fold iterated integral. We will not deal with this situation very often, except in the case where the random variables are independent, where the integral becomes a product of single integrals.

Example 4 Consider the experiment of randomly sampling a pair of numbers X and Y from the continuous uniform $(0, 1)$ distribution. Suppose we record the pair (U, V), where $U = \min(X, Y)$ and $V = \max(X, Y)$. First let us look at a simulation of many replications of this experiment. Then we will pose a theoretical model for the joint distribution of U and V and ask some questions about them.

The next *Mathematica* command is straightforward. Given the distribution to be sampled from and the number of pairs to produce, it repeatedly simulates an X, then a Y, and then appends the min and max of the two to a list, which is plotted.

```
SimPairs[dist_, numpairs_] :=
 Module[{X, Y, UVlist}, UVlist = {};
  Do[X = RandomReal[dist]; Y = RandomReal[dist];
  AppendTo[UVlist, {Min[X, Y], Max[X, Y]}],
  {i, 1, numpairs}];
  ListPlot[UVlist, AspectRatio → 1, AxesLabel → {u, v},
  PlotStyle → {Black, PointSize[0.02`]}]]]
```

Here is a sample run of SimPairs; in the e-version of the text you should try running it again a few times, and increasing the number of pairs to see that this run is typical.

```
SimPairs[UniformDistribution[{0, 1}], 500]
```

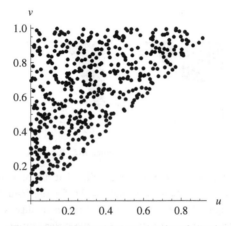

Figure 3.13 - 500 simulated sorted uniform(0, 1) pairs

Notice that points seem to be randomly spread out within the triangle whose vertices are (0, 0), (0, 1), and (1, 1), which is described by the inequalities $0 \le u \le v \le 1$. No particular region appears to be favored over any other. A moment's thought about how U and V are defined shows that indeed both can take values in the interval [0, 1], but the minimum U must be less than or equal to the maximum V; hence the state space of the pair (U, V) is the aforementioned triangle. But both of the data points from which this pair originated were uniformly distributed on [0, 1], which combined with the empirical evidence of Figure 13 suggests to us that a good model for the joint distribution of U and V would be a constant density over the triangle. In order for that density to integrate to 1, its value must be 1 divided by the area of the triangle, i.e., 1/(1/2) = 2. So we model the joint density of U and V as

$$f(u, v) = 2 \text{ if } 0 \le u \le v \le 1$$

and $f = 0$ otherwise. (We will deal with the problem of finding the distribution of transformed random variables analytically later in the book.)

We can ask, for example, for the probability that the larger of the two sampled values is at least 1/2. This is written as $P[V \geq .5]$ and is computed by integrating the joint density f over the shaded region in Figure 14. The easiest way to do this is to integrate first on u, with fixed v between .5 and 1, and u ranging from 0 to v. (You could also integrate on v first, but that would require splitting the integral on u into two parts, since the boundaries for v change according to whether u is less than .5 or not.)

$$\int_{.5}^{1}\int_{0}^{v} 2 \, du \, dv$$

0.75

Figure 3.14 - Region of integration for $P[V \geq .5]$ **Figure 3.15** - Region of integration for $P[V > 2U]$

As another example, let us find the probability that the maximum exceeds twice the minimum, i.e., $P[V > 2U]$. The boundary line $v = 2u$ for the inequality cuts through the state space as shown in Figure 15, and the solution of the inequality is the shaded area. This time the limits of integration are easy regardless of the order you use. Here we integrate v first, from $2u$ to 1, then u from 0 to .5. ∎

$$\int_{0}^{.5}\int_{2u}^{1} 2 \, dv \, du$$

0.5

Activity 4   Give geometric rather than calculus based arguments to calculate the two probabilities in Example 4.

Marginal probability density functions are defined in the continuous case much as marginal p.m.f.'s are in the discrete case. Instead of summing the joint mass function, we integrate the joint density over one of the variables.

Definition 3. Let X and Y be continuous random variables with joint density $f(x, y)$. The *marginal density function* of X is

$$p(x) = \int_{E_y} f(x, y)\, d y \tag{7}$$

where the integral is taken over all Y states that are possible for the given x. (We may integrate over the entire state space E_y of Y under the understanding that f may vanish for some y values, depending on x.) Similarly the *marginal density* of Y is

$$q(y) = \int_{E_x} f(x, y)\, d x \tag{8}$$

where the integral is taken over all X states that are possible for the given y.

Notice that the marginal densities completely characterize the individual probability distributions of the two random variables. For example, the probability that X lies in a Borel subset B of its state space is

$$P[X \in B] = P\big[X \in B, Y \in E_y\big] = \int_B \int_{E_y} f(x, y)\, d y\, d x = \int_B p(x)\, d x$$

where E_y is the state space of Y. You can do a similar derivation for Y to conclude that $P[Y \in B] = \int_B q(y)\, d y$. So, $p(x)$ and $q(y)$ are the probability density functions of X and Y in the one-variable sense.

In Example 4, for instance, where the joint density $f(u, v)$ was constantly equal to 2 on the triangle, the marginal density of the smaller sampled value U is

$$p(u) = \int_u^1 2\, d v = 2\,(1 - u), \quad 0 \le u \le 1$$

and the marginal density of the larger sampled value V is

$$q(v) = \int_0^v 2\, d u = 2\, v, \quad 0 \le v \le 1$$

(Why does it make intuitive sense that p should be a decreasing function and q an increasing function?)

Example 5 Let X and Y have a joint density of the form

$$f(x, y) = c(x^2 + y^2), \quad x, y \in [0, 1]$$

and $f = 0$ otherwise. Find the two marginal densities, find $P[X > 1/2, Y > 1/2]$, and find both $P[X > 1/2]$ and $P[Y > 1/2]$.

First, the coefficient c must be found such that the double integral of f over the rectangle $[0, 1] \times [0, 1]$ is 1, so that f is a valid joint density. We have

$$1 = \int_0^1 \int_0^1 c(x^2 + y^2)\, dy\, dx = c \int_0^1 \int_0^1 (x^2 + y^2)\, dy\, dx$$

hence c is the reciprocal of the last double integral. We compute this integral below in *Mathematica*.

$$\int_0^1 \int_0^1 \left(x^2 + y^2\right) dx\, dy$$

$$\frac{2}{3}$$

Therefore $c = 3/2$. In Figure 16 is a graph of the joint density function. If we repeatedly sampled from this joint density, the data points that we would see would be more numerous as we move farther away from the origin.

```
f[x_, y_] := 1.5 * (x² + y²);
Plot3D[f[x, y], {x, 0, 1},
  {y, 0, 1}, AxesLabel → {x, y, "f(x,y)"},
  PlotStyle → Gray, Lighting → "Neutral",
  BaseStyle → {8, FontFamily → "Times"}, Ticks →
    {{.2, .4, .6, .8}, {.2, .4, .6, .8}, {0, 3}}]
```

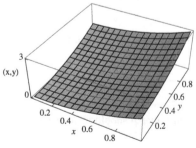

Figure 3.16 - Joint density $f(x, y) = \frac{3}{2}(x^2 + y^2)$

The marginal density of X is

$$p(x) = \frac{3}{2} \int_0^1 x^2 + y^2 \, d\,y = \frac{3}{2}\left(x^2 y + \frac{y^3}{3}\Big|_0^1\right) = \frac{3}{2}\left(x^2 + \frac{1}{3}\right), \quad x \in [0, 1]$$

and by symmetry the marginal of Y has the same form,

$$q(y) = \frac{3}{2}\left(y^2 + \frac{1}{3}\right), \quad y \in [0, 1]$$

We compute the requested probabilities in *Mathematica* below, by appropriately integrating the joint and marginal densities.

```
(* P[X > 1/2, Y > 1/2] = *)
3
- * ∫₁/₂¹ ∫₁/₂¹ (x² + y²) dx dy
2
```

$\dfrac{7}{16}$

```
(* P[X > 1/2] = *)
3
- * ∫₁/₂¹ (x² + 1 / 3) dx
2
```

$\dfrac{11}{16}$

The probability that $Y > 1/2$ is also 11/16, since X and Y have the same distribution. Notice that since $(11/16)(11/16) \neq 7/16$, the joint probability that both random variables exceed 1/2 does not factor into the product $P[X > 1/2] \cdot P[Y > 1/2]$. Thus, the event that X exceeds 1/2 is not independent of the event that Y exceeds 1/2. We will explore independence of two continuous random variables more thoroughly in a few more paragraphs. ∎

In the case where there are more than two jointly distributed random variables, the marginal distribution of each of them is obtained by integrating the joint density over all possible values of all random variables in the group other than the one you are computing the marginal p.d.f. for. For example, if X, Y, and Z have joint density $f(x, y, z)$ then the marginal density of Y would be

$$q(y) = \int_{E_z}\int_{E_x} f(x, y, z)\, d\,x\, d\,z$$

Activity 5 Carefully formulate the notion of a joint marginal distribution of a subgroup of a larger group of jointly distributed random variables. Write an integral expression for the joint marginal density of X and Y when random variables X, Y, Z, and W have a joint density function f.

There is an analogue of the idea of a conditional distribution of one random variable given another in the world of continuous random variables too. Remember that for discrete random variables, the expression

$$q(y \mid x) = P[Y = y \mid X = x] = \frac{P[Y = y, \, X = x]}{P[X = x]} = \frac{f(x, y)}{p(x)}$$

made perfect sense from basic definitions of probability. When X and Y are continuous random variables with joint density $f(x, y)$, the ratio of probabilities in the third expression above no longer makes sense, because both numerator and denominator are 0. But the ratio of joint density function to marginal density function in the fourth term does make sense, and we will take that as our definition of the conditional density, by analogy with the discrete case. (Actually, in Exercise 23 you will be led to make a better argument for the legitimacy of this definition by taking a conditional c.d.f. as the starting point.)

Definition 4. Let X and Y be continuous random variables with joint density $f(x, y)$. Denote by $p(x)$ and $q(y)$ the marginal density functions of X and Y, respectively. Then the *conditional density of Y given X = x* is given by

$$q(y \mid x) = \frac{f(x, y)}{p(x)} \tag{9}$$

and the *conditional density of X given Y = y* is

$$p(x \mid y) = \frac{f(x, y)}{q(y)} \tag{10}$$

Example 6 Consider again the random variables of Example 5. The joint density is

$$f(x, y) = \frac{3}{2}\left(x^2 + y^2\right), \quad x, \, y \in [0, 1]$$

and the two marginal densities are

$$p(x) = \frac{3}{2}\left(x^2 + \frac{1}{3}\right), \quad x \in [0, 1] \text{ and } q(y) = \frac{3}{2}\left(y^2 + \frac{1}{3}\right), \quad y \in [0, 1]$$

Therefore, by formula (9) the conditional density of y given x is

$$q(y \mid x) = \frac{f(x, y)}{p(x)} = \frac{(x^2 + y^2)}{\left(x^2 + \frac{1}{3}\right)}, \quad y \in [0, 1]$$

The density of Y does depend on what x value is observed. For example, here is a plot of the marginal density of Y, together with two of the conditionals, with $X = 1/4$, and $X = 3/4$.

```
q[y_] := 3/2 (y² + 1 / 3)

qygivenx[y_, x_] := (x² + y²) / (x² + 1 / 3)
```

```
Plot[{q[y], qygivenx[y, 1/4], qygivenx[y, 3/4]}, {y, 0, 1},
  PlotStyle → {Black, Gray, Dashing[{0.01, 0.01}]}]
```

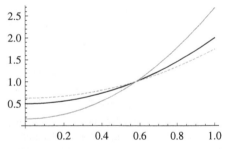

Figure 3.17 - Marginal of Y (black), conditional of Y given $X = 1/4$ (gray), $X = 3/4$ (dashed)

In the electronic text you can use the Manipulate command below to see how the conditional density changes as x changes.

```
Manipulate[Plot[{q[y], qygivenx[y, x]}, {y, 0, 1},
  PlotRange → {0, 3}, PlotStyle → {Thickness[0.01`],
  Dashing[{0.01`, 0.01`}]}], {x, 0, 1}]
```

If we wanted to compute the probability that Y exceeds 1/2 given an observed X value of 1/2, we would integrate the conditional density of Y as follows:

$$P[Y > 1/2 \mid X = 1/2] = \int_{1/2}^{1} q(y \mid 1/2)\, d\,y$$

which we can do in *Mathematica*. ∎

$$\int_{1/2}^{1} \frac{(1\,/\,2)^{2} + \mathbf{y}^{2}}{(1\,/\,2)^{2} + 1\,/\,3}\, \mathbf{dy}$$

$$\frac{5}{7}$$

Activity 6 Referring to Example 6, compute $P[X \in [1/3, 2/3] \mid Y = 3/4]$.

The last idea that we will take up in this section is the independence of continuous random variables. Here the definition expressed by formula (6) of Section 2.5 serves the purpose again without change. We repeat it here for your convenience:

Definition 5. Two continuous random variables X and Y are *independent* of each other if and only if for all Borel subsets A and B of their respective state spaces,

$$P[X \in A,\ Y \in B] = P[X \in A] \cdot P[Y \in B] \tag{11}$$

All of the comments made in Section 2.5 about independent random variables have their analogues here. An equivalent definition of independence would be $P[Y \in B \mid X \in A] = P[Y \in B]$ for all Borel subsets A and B such that $P[X \in A]$ is not zero. X and Y are also independent if and only if the joint probability density function $f(x, y)$ factors into the product of the marginal densities. In fact, X and Y are independent if and only if the joint c.d.f. $F(x, y) = P[X \leq x, Y \leq y]$ factors into the product $P[X \leq x]\,P[Y \leq y]$. The corresponding definition of independence for many random variables would amend formula (11) in the natural way: the simultaneous probability that all of the random variables fall into sets is the product of the marginal probabilities. Again in the many variable context, independence is equivalent to both the factorization of the joint density, and the factorization of the joint c.d.f. into the product of the marginals.

Activity 7 Show that if the joint p.d.f. of two continuous random variables factors into the product of the two marginal densities, then the random variables are independent.

Independence often arises in the following context: a *random sample* taken in sequence and with replacement from a population represented by a probability density f is a sequence $X_1, X_2, \ldots, X_n$ of independent random variables each of which has density f. Hence by what we said in the last paragraph, the joint density of the random variables X_i is

$$g(x_1, x_2, ..., x_n) = f(x_1) f(x_2) \cdots f(x_n) \tag{12}$$

Thus,

$$\begin{aligned}
P[X_1 \in B_1, ..., X_n \in B_n] &= \int_{B_1 \times \cdots \times B_n} g(x_1, ..., x_n)\, d\, x_1 \cdots d\, x_n \\
&= \int_{B_1} f(x_1)\, d\, x_1 \cdots \int_{B_n} f(x_n)\, d\, x_n
\end{aligned} \tag{13}$$

Example 7   What do independent continuous random variables look like? Example 3 introduced the Weibull distribution, which was governed by two parameters, α and β. Figure 12 showed the graph of the density and c.d.f. in the case $\alpha = 2$, $\beta = 3$. The density was asymmetrical, with a long right tail and a peak at around 2. First let us see what we should expect to observe if we simulate random samples (X, Y) of size $n = 2$ from the Weibull(2, 3) distribution. We can slightly alter the earlier command SimPairs to suit our needs. The change is that instead of tabulating the min and max of the raw data pair, we tabulate the raw data pair itself. The distinct calls to RandomReal should make the X and Y observations in a pair independent of one another, and different pairs should also be independent.

```
SimPairs[dist_, numpairs_] :=
 Module[{X, Y, pairlist},
  pairlist = {};
  Do[X = RandomReal[dist];
   Y = RandomReal[dist];
   AppendTo[pairlist, {X, Y}], {i, 1, numpairs}];
  ListPlot[pairlist, AspectRatio → 1,
   PlotStyle → {Black, PointSize[0.015]}]]]
```

In Figure 18 are the results of simulating 2000 such independent and identically distributed (X_i, Y_i) pairs using the SimPairs command.

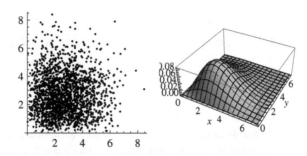

Figure 3.18 - 2000 simulated Weibull(2, 3) pairs   **Figure 3.19** - 2-dimensional Weibull(2, 3) joint density

The fact that X and Y have the same distribution shows up in the fact that the point cloud looks about the same in the orientation of Figure 18 as it would if the roles of X and Y were interchanged and the graph was stood on its side. The independence of X and Y is a little harder to see, but you can pick it out by observing that regardless of the vertical level, the dots spread out horizontally in about the same way, with a heavy concentration of points near 2, and a few points near 0 and out at large values between 6 and 8. Similarly, regardless of the horizontal position, the points spread out from bottom to top in this way. We expect this to happen because under independence, the observed value of one of the variables should not affect the distribution of the other.

The graph of the joint density in Figure 19 sheds even more light on this phenomenon. Our point cloud in Figure 18 is most dense exactly where the joint density in Figure 19 reaches its maximum, around (2, 2). Also the point cloud is sparse where the density is low, and the long right tail extending to 6 and beyond is visible in both graphs. But the joint density graph shows the independence property a little better. By examining the grid lines you can see that the relative density of points in the x direction is independent of the y value, and the relative density of points in the y direction is independent of the x value. The identical distribution of the X and Y variables is seen in Figure 19 by the fact that the cross-sections look the same regardless of whether you slice the surface perpendicularly to the x-axis or to the y-axis. ■

Exercises 3.2

1. Show that $f(x) = \dfrac{1}{\pi(1+x^2)}$ is a valid probability density function on $\Omega = \mathbb{R}$. (It is referred to as the Cauchy density.) Then find P[$[0, \infty)$] and P[$[-1, 1]$].

2. (*Mathematica*) Plot a graph of the density function $f(x) = c\, x^4\, e^{-3x}$ on $[0, \infty)$ for the appropriate choice of c. Compute P[$(3, 4]^c$].

3. Consider the function

$$f(x) = \begin{cases} Ce^{-kx} & \text{if } x \geq 0 \\ 0 & \text{otherwise} \end{cases}$$

where C and k are positive constants. Find C such that f is a valid probability density function. Find the cumulative distribution function associated to this distribution.

4. Find the cumulative distribution function of the uniform distribution on $[a, b]$. If X has this distribution, find $P\left[X > \frac{a+b}{2}\right]$.

5. For each of the following random variables, state whether it is best modeled as discrete or continuous, and explain your choice.
   (a) The error in measuring the height of a building by triangulation
   (b) The number of New Year's Day babies born in New York City next year
   (c) The relative humidity in Baltimore on a certain day at a certain time

6. The time T (in days) of a server malfunction in a computer system has the probability density function

$$f(t) = \begin{cases} \frac{1}{100} t e^{-t/10} & \text{if } t \geq 0 \\ 0 & \text{otherwise} \end{cases}$$

Verify that this is a valid density and compute the probability that the malfunction occurs after the tenth day.

7. The c.d.f. of a continuous random variable X is

$$F(x) = \begin{cases} 0 & \text{if } x < 1 \\ (1/4)x - 1/4 & \text{if } 1 \leq x < 2 \\ (1/2)x - 3/4 & \text{if } 2 \leq x < 3 \\ (1/4)x & \text{if } 3 \leq x < 4 \\ 1 & \text{if } x \geq 4 \end{cases}$$

Find a piecewise expression for the density function $f(x)$ and graph both f and F. Explain why it does not matter what value you set for f at the break points 1, 2, 3, and 4. If X has this probability distribution, find $P[1.5 \leq X \leq 3.5]$.

8. (*Mathematica*) By setting several different values for α and fixing β, graph several versions of the Weibull(α, β) density on the same set of axes. Describe the dependence of the density on the α parameter. Do the same for the β parameter.

9. Find the location of the maximum value of the Weibull(α, β) density in terms of α and β.

10. If a random variable X has the uniform(0, 1) distribution and $Y = 2X - 1$, what possible values can Y take on? Find $P[Y \leq y]$ (i.e., the c.d.f. of Y) for all such values y by reexpressing the inequality in terms of X. Use this to find the probability density function of Y.

11. (*Mathematica*) Suppose that the time D (in years) until death of an individual who holds a new 25 year $40,000 life insurance policy has the Weibull distribution with parameters $\alpha = 1.5$, $\beta = 20$. The policy will expire and not pay off anything if the policy holder survives beyond 25 years; otherwise, if the holder dies within 25 years it pays the $40,000 face value to his heirs. What is the probability that the heirs will receive the $40,000? What is the expected value of the amount that the insurance company will pay?

12. (*Mathematica*) The value V of an investment at a certain time in the future is a random variable with probability density function

$$f(v) = \frac{1}{v} \frac{1}{\sqrt{2\pi}} e^{-(\log v - 4)^2/2}, \ v > 0$$

Find the probability that the value will be at least 40. Find the *median* v^* of the value distribution, that is, the point such that $P[V \le v^*] = .5$.

13.   If X, Y, and Z are three independent uniform(0, 1) random variables, find $P\left[X + Y + Z \le \frac{1}{2}\right]$.

14. For the random variables U and V of Example 4, find $P\left[V - U \ge \frac{1}{4}\right]$.

15. (*Mathematica*) If X and Y are independent uniform(−1, 1) random variables, build a simulator of pairs (W, Z) where $W = X − Y$ and $Z = X + Y$. By looking at a large sample of such pairs and noting the construction of W and Z, what seems to be the state space of (W, Z)? Does a uniform model seem reasonable? Use your model to find $P[Z \ge 0]$.

16. (*Mathematica*) Two random variables X and Y have joint density

$$f(x, y) = \begin{cases} \frac{1}{2} e^{-y} & \text{if } x \in [0, 2], \ y \in [0, \infty) \\ 0 & \text{otherwise} \end{cases}$$

Sketch the graph of this joint density, and find $P[Y > X]$.

17. For the random variables X and Y of Exercise 16, find the marginal densities. Can you conclude anything about X and Y?

18. For the random variables U and V of Example 4, find the conditional densities $q(v \mid u)$ and $p(u \mid v)$ for all u and v. Compute also $P\left[V > \frac{3}{4} \mid U = \frac{1}{2}\right]$.

19. Explain why the random variables X and Y of Example 6 are not independent of each other. However, is there any case in which the marginal density $q(y)$ equals the conditional density $q(y \mid x)$?

20. (*Mathematica*) If X and Y have joint density of the form

$$f(x, y) = \frac{c}{x + y}, \quad x, y \in [1, 2]$$

find the marginal densities of X and Y, and compute $P\left[Y > \frac{5}{4} \mid X = \frac{3}{2}\right]$ and $P\left[X < \frac{3}{2} \mid Y = \frac{6}{5}\right]$.

21. If X_1, X_2, and X_3 are a random sample (in sequence and with replacement) from the uniform(0, 3) distribution, find (a) $P[X_1 > 1, X_2 < 1, X_3 > 2]$; (b) $P[\min(X_1, X_2, X_3) > 1]$

22. (*Mathematica*) Does the following random sample appear to have been sampled from the distribution whose density function is

$$f(x) = \frac{27}{2} x^2 e^{-3x}, \quad x > 0 ?$$

Use graphics to explain why or why not.

6.21, 3.97, 6.34, 5.31, 6.52, 4.57, 4.03, 5.99, 4.99, 5.54, 4.69, 5.09, 4.13, 5.32, 3.16, 4.16, 4.49, 5.88, 6.23, 7.42, 3.98, 5.29, 5.22, 6.4, 5.25, 5.75, 4.4, 1.25, 5.45, 3.55, 5.83, 3.48, 3.57, 4.95, 5.06, 4.55, 3.06, 5.36, 6.6, 4.57, 6.39, 5.61, 5.72, 4.55, 3.67, 3.79, 4.84, 6.05, 5.99, 6.37, 6.58, 4.48, 5.52, 5.89, 7.19, 2.9, 4.1, 5.96, 5.56, 5.6, 4.72, 6.19, 4.14, 3.76, 4.17, 4.13, 4.05, 6.44, 5.23, 5.88, 5.4, 7.23, 2.82, 5.05, 5.57, 4.74, 4.73, 3.78, 6.0, 2.7, 6.35, 4.07, 4.98, 3.81, 4.6, 4.54, 4.39, 2.85, 5.13, 5.11, 4.83, 4.91, 4.14, 6.19, 3.64, 4.33, 4.54, 4.55, 5.33, 4.27

23. We can take as the definition of the *conditional c.d.f.* of Y given $X = x$ the limit of

$$P[Y \leq y \mid X \in [x, x + \Delta x]]$$

as $\Delta x \longrightarrow 0$. The expression above is a well-defined conditional probability, and the limit is conventionally written as $F(y \mid x) = P[Y \leq y \mid X = x]$. Find this limit in terms of the joint and marginal densities of X and Y, and show that the derivative of $F(y \mid x)$ is the same as the conditional density function that we defined in the section $q(y \mid x) = f(x, y)/p(x)$.

Sample Problems from Actuarial Exam P

24. A device runs until either of two components fails, at which point the device stops running. The joint density function of the lifetimes of the two components, both measured in hours, is $f(x, y) = \frac{x+y}{8}$ for $0 < x < 2$ and $0 < y < 2$. What is the probability that the device fails during its first hour of operation?

25. A company offers a basic life insurance policy to its employees, as well as a supplemental life insurance policy. To purchase the supplemental policy, an employee must first purchase the basic policy. Let X denote the proportion of employees who purchase the basic policy, and Y the proportion of employees who purchase the supplemental policy.

Let X and Y have the joint density function $f(x, y) = 2(x + y)$ on the region where the density is positive. Given that 10% of the employees buy the basic policy, what is the probability that fewer than 5% buy the supplemental policy?

26. An insurance policy is written to cover a loss X where X has density function

$$f(x) = \begin{cases} \frac{3}{8}x^2 & \text{for } 0 \leq x \leq 2 \\ 0 & \text{otherwise} \end{cases}$$

The time (in hours) to process a claim of size x, where $0 \leq x \leq 2$, is uniformly distributed on the interval from x to $2x$. Calculate the probability that a randomly chosen claim on this policy is processed in three hours or more.

3.3 Continuous Expectation

To conclude this transitional chapter from discrete to continuous probability, we would like to discuss the idea of expectation. Recall that for discrete random variables, we defined

$$E[X] = \sum_x x \cdot p(x) = \sum_x x \cdot P[X = x] \tag{1}$$

The idea was to express the average of states x that could be assumed by the random variable X, using the probability mass function $p(x)$ as a system of weights to give states of higher likelihood more influence on the average value. Now if we have a continuous random variable with a state space that is a bounded interval for instance, we could think of approximating the average value by discretizing the state space into a finite number of states x_i, $i = 1, ..., n$ separated by a common spacing Δx. Because of the meaning of the density function f, the product $f(x_i)\Delta x$ is the approximate probability weight for state x_i, and therefore, the approximate expected value in the discretized state space would be

$$\sum_{x_i} x_i \cdot f(x_i)\Delta x$$

In the limit as the number of states $n \longrightarrow \infty$ and the spacing $\Delta x \to 0$, this sum converges to the integral in the definition below.

Definition 1. Let X be a continuous random variable with probability density function f. Then the *expected value* of X is the following integral, if it exists.

$$E[X] = \int_{E_x} x f(x)\, dx \tag{2}$$

(The integral is taken over the entire state space of X.) We also use the term *mean of X* and the notation μ for $E[X]$. The *expected value of a function g* of X is

$$E[g(X)] = \int_{E_x} g(x) f(x) \, dx \tag{3}$$

In formula (3), we will again be most interested in the *moments* of the distribution, i.e., expected powers of X and expected powers of $X - \mu$. Among the moments of chief interest are the mean itself, which measures the center of the probability distribution, and the *variance*:

$$\sigma^2 = \text{Var}(X) = E\left[(X - \mu)^2\right] = \int_{E_x} (x - \mu)^2 f(x) \, dx \tag{4}$$

which measures the spread of the distribution around μ. The square root σ of the variance σ^2 is the *standard deviation* of the distribution.

In this section we will discuss the meaning and computation of certain expectations, paralleling the discussion that began in Section 2.1 and continued in Section 2.6 for discrete random variables. We will feature the same basic results on means, variances, covariance and correlation. Most of the proofs of the results are entirely similar to the discrete case with sums replaced by integrals and, therefore, they will usually either be omitted or left to the exercises.

Example 1 As a first example, let us find the mean of the uniform(a, b) distribution.

The p.d.f. of the distribution is $f(x) = 1/(b-a)$ for $x \in [a, b]$ and zero otherwise. Therefore the mean is

$$\mu = E[X] = \int_a^b x \, \frac{1}{b-a} \, dx = \frac{1}{b-a} \frac{x^2}{2} \Big|_a^b = \frac{b^2 - a^2}{2(b-a)} = \frac{b+a}{2} \tag{5}$$

It is very intuitive that the mean of the uniform(a, b) distribution should come out to be the midpoint of the interval $[a, b]$, since all states in the interval contribute equally to the continuous average that is the mean. You are asked to compute the variance of X in Exercise 1. This computation will be easier if you use the formula in the next activity, which you should derive now. ∎

Activity 1 Show using basic properties of the integral that $\text{Var}(X) = E\left[X^2\right] - \mu^2$.

Example 2 Exercise 10 in Section 2.6 asked for a simulation of sample means for larger and larger sample sizes, to show the convergence of the sample mean to the theoretical mean μ. This is a good time to reconsider that investigation in the context of continuous expectation. We now write a command (essentially the same one that would be the solution of that exercise) to successively simulate sample values from a given distribution, pause every m sample values to recompute the sample mean, and update the list of sample means to date.

```
SimMeanSequence[distribution_, nummeans_, m_] :=
 Module[{nextsample, meanlist, runningsum,
   currnumobs}, currnumobs = 0; runningsum = 0;
  meanlist = {};
  While[currnumobs < nummeans,
   nextsample = RandomReal[distribution, m];
   currnumobs = currnumobs + m;
   runningsum = runningsum + Plus @@ nextsample;
                          runningsum
   AppendTo[meanlist, ─────────────]];
                          currnumobs
  ListPlot[meanlist, PlotStyle →
    {Black, PointSize[0.02]}, Joined → True]]
```

Figure 20 shows the results of executing the command on the uniform(2, 6) distribution, whose mean is the midpoint 4. We simulate 1000 sample values, and update the sample mean every 10 values. The sample mean seems to be converging to the distributional mean of 4. You should try repeating the command several times, without resetting the seed, to see this phenomenon recurring, and also see the amount of variability in the sample means early on in the sequence. This and other similar experiments provide empirical evidence for the claim that the mean of a probability distribution is that value to which the mean of a random sample converges as more and more observations are taken. This result is a pivotal theorem in probability and statistics, called the Strong Law of Large Numbers. It will be discussed more fully in Chapter 5. ■

```
SeedRandom[439 873];
SimMeanSequence[
 UniformDistribution[{2, 6}], 1000, 10]
```

Figure 3.20 - Simulated sample mean as a function of sample size/10

Example 3 Recall from Section 3.2 the Weibull distribution with parameters α and β. Its p.d.f. is the following:

$$f(x) = \alpha \beta^{-\alpha} x^{\alpha-1} e^{-(x/\beta)^{\alpha}}, \quad x \geq 0$$

Its mean is

$$E[X] = \int_{0}^{\infty} x f(x) \, dx = \int_{0}^{\infty} \alpha \beta^{-\alpha} x^{\alpha} e^{-(x/\beta)^{\alpha}} \, dx$$

It turns out that a substitution $u = (x/\beta)^{\alpha}$ expresses the integral in terms of a standard mathematical function denoted $\Gamma(r)$, which is defined by

$$\Gamma(r) = \int_{0}^{\infty} u^{r-1} e^{-u} \, du \tag{6}$$

(See Exercise 4.) The gamma function will also play an important role in later development. Among its many interesting properties is that if n is an integer, $\Gamma(n) = (n-1)!$. More generally, for all $\alpha > 1$, $\Gamma(\alpha) = (\alpha - 1)\Gamma(\alpha - 1)$, and also it can be shown that $\Gamma(1/2) = \sqrt{\pi}$. In Exercise 4 you will be asked to finish the derivation that the mean of the Weibull(α, β) distribution is

$$\mu = \beta \Gamma\left(1 + \frac{1}{\alpha}\right) \tag{7}$$

If, as in Example 3 of Section 3.2 on shock absorber lifetimes, $\alpha = 2$ and $\beta = 30$, then we can use *Mathematica* to compute the mean lifetime as

```
μ = 30 * Gamma[3 / 2]
N[μ]
```

$15 \sqrt{\pi}$

26.5868

The sketch of the Weibull density with these parameters in Figure 21 shows that 26.59 is a reasonable estimate of the center of the distribution. ■

```
f[x_] := PDF[WeibullDistribution[2, 30], x];
Plot[f[x], {x, 0, 70}, PlotStyle → Black]
```

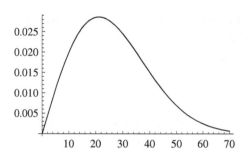

Figure 3.21 - Weibull(2,30) density function

It is worth pointing out that *Mathematica*'s kernel has commands

$$\text{Mean[distribution]} \quad \text{and} \quad \text{Variance[distribution]}$$

that return the mean and variance of any distribution in the package. For example, for the uniform distribution on $[a, b]$,

```
{Mean[UniformDistribution[{a, b}]],
 Variance[UniformDistribution[{a, b}]]}
```

$$\left\{ \frac{a+b}{2}, \frac{1}{12}(-a+b)^2 \right\}$$

Example 4   Compute the variance of the random variable X whose probability density function is $f(x) = c/x^4$ if $x > 1$, and 0 otherwise.

By the computational formula in Activity 1, we must compute the first moment $\mu = E[X]$ and the second moment $E[X^2]$. But first we should determine the constant c. In order for the density to integrate to 1, we must have

$$1 = c\int_1^\infty 1/x^4\, dx = c\left(\frac{x^{-3}}{-3}\right)_1^\infty = c\cdot\frac{1}{3} \Longrightarrow c = 3$$

Then,

$$\mu = \int_1^\infty x\cdot\frac{3}{x^4}\, dx = \int_1^\infty \frac{3}{x^3}\, dx = 3\left(\frac{x^{-2}}{-2}\right)_1^\infty = \frac{3}{2}$$

$$E[X^2] = \int_1^\infty x^2\cdot\frac{3}{x^4}\, dx = \int_1^\infty \frac{3}{x^2}\, dx = 3\left(\frac{x^{-1}}{-1}\right)_1^\infty = 3$$

Therefore by the computational formula for variance in Activity 1,

$$\sigma^2 = E[X^2] - \mu^2 = 3 - \frac{9}{4} = \frac{3}{4}$$

Notice that if we tried to do a similar computation of the variance of a random variable with density $f(x) = c/x^2$, $x > 1$, and 0 otherwise, we would run into a problem. When computing $E[X^2]$ for instance, we would try to integrate $x^2 \cdot \frac{c}{x^2} = c$ from 1 to infinity, which would be infinite. Therefore, the second moment would not exist, nor would the variance exist. ■

Activity 2 For densities of the form $f(x) = c/x^p$, $x > 1$, and 0 otherwise, for what powers p does the variance exist and for what powers does it not exist? How about the mean?

Example 5 Expected value is an important tool in decision-making. For example, suppose an investor has two alternatives to choose between: the first gives a rate of return R_1 that is a continuous random variable with the uniform distribution on $[0, .1]$, and the second gives a rate of return R_2 that has probability density $f(r_2) = 20\,e^{-20\,r_2}$, $r_2 > 0$. If we make the sensible assumption that high mean rate of return is desirable, but low risk as measured by variance is also desirable, is there a clear winner between these two investment alternatives?

We should compute the means and variances of R_1 and R_2. By the formulas above,

$$\mu_1 = E[R_1] = \frac{.1 + 0}{2} = .05; \quad \sigma_1^2 = Var(R_1) = \frac{(.1 - 0)^2}{12} = .000833$$

For the second investment, we compute $\mu_2 = E[R_2]$ and $\sigma_2^2 = Var(R_2)$ in *Mathematica* below.

$$\mu_2 = \text{N}\left[\int_0^\infty \text{r} * 20 * \text{E}^{-20\,\text{r}}\,d\text{r}\right]$$

$$\text{o2sq} = \text{N}\left[\int_0^\infty (\text{r} - \mu_2)^2 * 20 * \text{E}^{-20\,\text{r}}\,d\text{r}\right]$$

0.05

0.0025

So we see that the mean returns on the two investments are the same, but investment 1 has a smaller variance than investment 2. Since it is therefore less risky, the investor should prefer investment 1. ■

We can take our cue from the single-variable case in order to formulate a version of expectation in the context of jointly distributed random variables.

Definition 2. Let $X_1, X_2, \ldots, X_n$ be continuous random variables with joint density function $f(x_1, x_2, \ldots, x_n)$. Then the *expected value of a real-valued function g* of the random variables is

$$E[g(X_1, \ldots, X_n)] = \int \cdots \int_{E_x} g(x_1, \ldots, x_n) \cdot f(x_1, \ldots, x_n)\, dx_1 \cdots dx_n \qquad (8)$$

where the integral is taken over all possible joint states $\mathbf{x} = (x_1, x_2, \ldots, x_n)$.

Most of the time, except when the random variables are independent and the multiple integral simplifies to a product of single integrals, we will consider the bivariate case, in which formula (8) reduces to

$$E[g(X, Y)] = \int_{E_y} \int_{E_x} g(x, y) f(x, y)\, dx\, dy \qquad (9)$$

where again f is the joint density of X and Y. Perhaps the most interesting such expectation is the *covariance* of X and Y:

$$\begin{aligned} \text{Cov}(X, Y) &= E\big[(X - \mu_x)(Y - \mu_y)\big] \\ &= \int_{E_y} \int_{E_x} (x - \mu_x)(y - \mu_y) f(x, y)\, dx\, dy \end{aligned} \qquad (10)$$

which is as before a measure of the degree of association between X and Y. When X and Y tend to be large together and small together, the integrand will be large with high probability, and so the double integral that defines the covariance will be large. The definition of the *correlation* ρ between X and Y is also the same as before:

$$\rho = \text{Corr}(X, Y) = \frac{\text{Cov}(X, Y)}{\sigma_x \sigma_y} \qquad (11)$$

where the σ's are the respective standard deviations of X and Y.

Let's move to a couple of example computations, and then catalogue the general results about continuous expectation.

Example 6 Recall the random variables U and V from Example 4 of Section 3.2, which were the smaller and larger of two independent uniform$(0,1)$ random variables. We found that the joint density of U and V was

$$f(u, v) = 2, \quad \text{if } 0 \le u \le v \le 1, \quad \text{and } 0 \text{ otherwise.}$$

Then the expected distance between the smaller and the larger is

$$E[V - U] = \int_0^1 \int_u^1 2(v - u)\,dv\,du$$

$$= 2\int_0^1 \left(\frac{v^2}{2} - vu\right)\Big|_u^1\,du$$

$$= 2\int_0^1 \frac{1}{2} - u + \frac{u^2}{2}\,du = \frac{2}{6}$$

Alternatively in the right side of the first line above we could have noted that the double integral breaks into a difference of two double integrals:

$$\int_0^1 \int_u^1 2v\,dv\,du - \int_0^1 \int_u^1 2u\,dv\,du$$

which is the same as $E[V] - E[U]$, and we could have computed the two easier double integrals to finish the problem. ∎

Activity 3 Finish the computation of $E[V] - E[U]$ in the manner suggested at the end of Example 6.

Example 7 Find the covariance and correlation of the random variables X and Y from Example 5 of Section 3.2, whose joint density is

$$f(x, y) = \frac{3}{2}(x^2 + y^2), \quad x,\, y \in [0, 1]$$

There are quite a few ingredients in the computation, but we will let *Mathematica* do the tedious integrations for us. It is helpful that X and Y have the same marginal distribution, hence the same mean and standard deviation. First let us find $\mu_x = \mu_y$. We can do this by integrating the function $g(x, y) = x$ times the density, which we define below, to get $E[X]$.

```
f[x_, y_] := 3/2 * (x^2 + y^2)
```

```
mux = ∫₀¹ ∫₀¹ x * f[x, y] dy dx;   muy = mux
```

$\dfrac{5}{8}$

To get the variance of X, which equals the variance of Y, we can integrate to find $E[(X - \mu_x)^2]$:

$$\texttt{varx = }\int_0^1 \int_0^1 \texttt{(x - mux)}^2 \texttt{ * f[x, y] dy dx; vary = varx}$$

$$\frac{73}{960}$$

The covariance and correlation can be done by using formulas (10) and (11):

$$\texttt{covarxy = }\int_0^1 \int_0^1 \texttt{(x - mux) * (y - muy) * f[x, y] dy dx}$$

$$\rho = \frac{\texttt{covarxy}}{\sqrt{\texttt{varx}} \texttt{ * } \sqrt{\texttt{vary}}}$$

$$-\frac{1}{64}$$

$$-\frac{15}{73}$$

Notice that X and Y have a slight negative correlation, and that, as usual, the correlation is less than or equal to 1 in magnitude. ∎

Here are the main theorems about expectation, listed in the continuous context. You will observe that they are the same as the ones we proved for discrete random variables. First is the linearity of expectation, which extends (see Exercise 13) to many random variables.

Theorem 1. If X and Y are continuous random variables with finite means, and c and d are constants, then

$$E[c \cdot X + d \cdot Y] = c \cdot E[X] + d \cdot E[Y] \quad \blacksquare \tag{12}$$

If continuous random variables are independent, then the variance of a linear combination is easily expressed:

Theorem 2. If X and Y are independent continuous random variables with finite variances, and c and d are constants, then

$$\text{Var}(c \cdot X + d \cdot Y) = c^2 \cdot \text{Var}(X) + d^2 \cdot \text{Var}(Y) \quad \blacksquare \tag{13}$$

If you check back to the proof of the analogue of Theorem 2 for discrete random variables, you will find that it still works word-for-word here, because it relies only on the linearity of expectation. Theorem 2 also extends readily to many independent random variables; constant coefficients, if present, come out as squares and the variance of the sum breaks into the sum of the variances.

When continuous random variables are independent, expected products factor into products of expectations. We will include the proof of this theorem. It can be extended to many random variables as well (see Exercise 14).

Theorem 3. If X and Y are independent continuous random variables, then $E[g(X) h(Y)] = E[g(X)] E[h(Y)]$ provided these expectations exist.

Proof. If f is the joint density of X and Y, then by the independence assumption, $f(x, y) = p(x) q(y)$, where p and q are the marginal densities of X and Y, respectively. Therefore,

$$
\begin{aligned}
E[g(X) h(Y)] &= \int_{E_x} \int_{E_y} g(x) h(y) f(x, y)\, d\,y\, d\,x \\
&= \int_{E_x} \int_{E_y} g(x) h(y) p(x) q(y)\, d\,y\, d\,x \\
&= \int_{E_x} g(x) p(x)\, d\,x \int_{E_y} h(y) q(y)\, d\,y \\
&= E[g(X)] E[h(Y)] \quad \blacksquare
\end{aligned}
$$

In the case where the random variables are not independent, the formula for the variance of a linear combination includes covariance terms as well, as indicated in the next theorem.

Theorem 4. If $X_1, X_2, \ldots, X_n$ are jointly distributed random variables, then

$$
\operatorname{Var}\left(\sum_{i=1}^{n} c_i X_i\right) = \sum_{i=1}^{n} c_i^2 \operatorname{Var}(X_i) + 2 \sum_{i=1}^{n} \sum_{j=1}^{i-1} c_i c_j \operatorname{Cov}(X_i, X_j) \quad \blacksquare
\tag{14}
$$

A couple of familiar miscellaneous results about covariance are in the next theorem. You should read about another interesting property in Exercise 15. Formula (15) is an often useful computational formula for the covariance, and formulas (16) and (17) show that whereas the covariance is dependent on the units used to measure the random variables, the correlation is not. Therefore the correlation is a standardized index of association.

Theorem 5. Let X and Y be jointly distributed random variables with means μ_x and μ_y, and let $a, b, c,$ and d be positive constants. Then

$$
\operatorname{Cov}(X, Y) = E[X \cdot Y] - \mu_x \mu_y
\tag{15}
$$

$$
\operatorname{Cov}(a X + b, c Y + d) = a \cdot c \cdot \operatorname{Cov}(X, Y)
\tag{16}
$$

$$
\operatorname{Corr}(a X + b, c Y + d) = \operatorname{Corr}(X, Y) \quad \blacksquare
\tag{17}
$$

Lastly we have the result that gives meaning to the correlation ρ.

Theorem 6. If ρ is the correlation between random variables X and Y, then $|\rho| \le 1$. Moreover, ρ equals positive 1 or negative 1 if and only if, with certainty, Y is a linear function of X. If X and Y are independent, then $\rho = 0$. ∎

Activity 4   If X and Y are independent random variables, what is $\text{Var}(X - Y)$? (Be careful.)

We would like to conclude this section by mentioning the version of conditional expectation for continuous, jointly distributed random variables. It follows along the same lines as in the discrete case.

Definition 3. Let X and Y be continuous random variables with joint density $f(x, y)$, and let $q(y \mid x)$ be the conditional density of Y given $X = x$. Then the *conditional expectation* of a function g of Y given $X = x$ is

$$E[g(Y) \mid X = x] = \int g(y)\, q(y \mid x)\, d\, y \tag{18}$$

So, to find the average value of $g(Y)$ given an observed value of X, we take the weighted average of all possible values $g(y)$, weighted by the conditional density of y for the observed x. The properties are similar to the discrete case; for instance, it is very easy to show that conditional expectation is linear; and if X and Y are independent, then $E[Y \mid X = x] = E[Y]$. Again the law of total probability for expectation holds:

$$E[E[g(Y) \mid X]] = E[g(Y)] \tag{19}$$

where we think of $E[g(Y) \mid X]$ as the function $h(x) = E[g(Y) \mid X = x]$ composed with X, which then becomes a random variable $h(X)$ whose expectation can be taken relative to the marginal density of X. Loosely speaking, formula (19) states that we can find the expectation of $g(Y)$ by first conditioning on X, then unconditioning by averaging out over all X values.

Example 8 Returning to the random variables U and V again from Example 6, i.e., the smaller and larger of two independent, identically distributed uniform(0,1) random variables, we found in Section 3.2 that the marginal densities of U and V are:

$$p(u) = 2\,(1 - u),\ \ 0 \le u \le 1 \quad \text{and} \quad q(v) = 2\,v,\ \ 0 \le v \le 1$$

Since the joint density was $f(u, v) = 2$, if $0 \le u \le v \le 1$, the two conditional densities are

$$q(v \mid u) = \frac{f(u, v)}{p(u)} = \frac{2}{2\,(1 - u)} = \frac{1}{1 - u} \text{ for } v \in [u,\, 1]$$

$$p(u \mid v) = \frac{f(u, v)}{q(v)} = \frac{2}{2\,v} = \frac{1}{v} \text{ for } u \in [0,\, v]$$

This would mean for instance that

$$E[V \mid U = u] = \int_u^1 v \frac{1}{1-u} \, dv = \frac{1}{1-u} \frac{1-u^2}{2} = \frac{1+u}{2}$$

which is unsurprising, since V is uniform on $[u, 1]$ conditioned on $U = u$, and for the uniform distribution the mean is the midpoint of the state space. For similar reasons, the conditional mean of U given $V = v$ would be $v/2$. This says that if the larger of two uniform random numbers is known, then the smaller is expected to be half of the larger. The conditional second moment of U given $V = v$ is

$$E[U^2 \mid V = v] = \int_0^v u^2 \frac{1}{v} \, du = \frac{1}{v} \frac{v^3}{3} = \frac{v^2}{3}$$

Hence the conditional variance of U given $V = v$ is

$$E[U^2 \mid V = v] - (E[U \mid V = v])^2 = \frac{v^2}{3} - \left(\frac{v}{2}\right)^2 = \frac{v^2}{12} \quad \blacksquare$$

Exercises 3.3

1. Derive an expression for the variance of the uniform(a, b) distribution.

2. If X is a random variable with probability density function

$$f(x) = \frac{1}{\pi(1 + x^2)}, \quad x \in \mathbb{R}$$

find, if it exists, the mean of X. Does the variance of X exist?

3. Find the mean and variance of the random variable X with the triangular density function:

$$f(x) = \begin{cases} 1 + x & \text{if } -1 \leq x \leq 0 \\ 1 - x & \text{if } 0 < x \leq 1 \\ 0 & \text{otherwise} \end{cases}$$

4. By carrying out the integral by hand, verify the expression in (7) for the mean of the Weibull(α, β) distribution.

5. (*Mathematica*) Use *Mathematica* to find the first three moments $E[T]$, $E[T^2]$, and $E[T^3]$ for the random variable T of Exercise 6 of Section 3.2, whose density is

$$f(t) = \begin{cases} \frac{1}{100} t \, e^{-t/10} & \text{if } t \geq 0 \\ 0 & \text{otherwise} \end{cases}$$

6. Let μ be the mean of the uniform(0, 4) distribution, and let σ be the standard deviation. If X is a random variable with this distribution, find: (a) $P[\,|X - \mu| < \sigma]$; (b) $P[\,|X - \mu| < 2\sigma]$; (c) $P[\,|X - \mu| < 3\sigma]$

7. (*Mathematica*) Use simulation to approximate the mean of the Weibull(4, 6) distribution. Then use *Mathematica* and formula (7) to calculate the mean analytically. Graph the density function, noting the position of the mean on the x-axis.

8. (*Mathematica*) Recall the insurance policy problem of Exercise 3.2-11, in which there is a policy that pays \$40,000 to the holder's heirs if he dies within the next 25 years. The death time (in years beginning from the present) has the Weibull(1.5, 20) distribution. Now let us take into account the premiums paid by the policy holder, which for simplicity we suppose are collected at a constant rate of \$1800 per year until the time of death. Find the expected profit for the insurance company.

9. Suppose that X and Y have joint density

$$f(x, y) = c(x^3 + y^3), \quad x, y \in [0, 2]$$

Find the covariance and correlation between X and Y.

10. (*Mathematica*) Build a simulator of differences $V - U$ for the random variables of Example 6. Look at several histograms of samples of such differences, and comment on the relationship of the histograms to the expected value of $V - U$ derived in that example.

11. Find Cov(U, V) for the random variables of Example 6.

12. (*Mathematica*) Compute $E[X + 6Y]$ using *Mathematica* for the random variables X and Y of Exercise 20 of Section 3.2, whose joint density is

$$f(x, y) = \frac{c}{x + y}, \quad x, y \in [1, 2]$$

13. Prove that if $X_1, X_2, \ldots, X_n$ are continuous random variables with finite means and $c_1, c_2, \ldots, c_n$ are constants then

$$E[c_1 X_1 + c_2 X_2 + \cdots + c_n X_n] = c_1 E[X_1] + c_2 E[X_2] + \cdots + c_n E[X_n]$$

14. Prove that if $X_1, X_2, \ldots, X_n$ are continuous random variables with finite means and if the X's are independent of each other, then

$$E[X_1 X_2 \cdots X_n] = E[X_1] E[X_2] \cdots E[X_n]$$

15. Show that if X and Y are random variables and $a, b, c,$ and d are constants then

$$\text{Cov}(aX + bY, cX + dY) = ac \, \text{Var}(X) + (bc + ad) \, \text{Cov}(X, Y) + bd \, \text{Var}(Y)$$

16. Let X and Y be random variables with joint density $f(x, y) = 2x \cdot 4e^{-4y}$, $x \in [0, 1]$, $y \in [0, \infty)$. Find $E[2XY]$.

17. If random variables X and Y are uncorrelated (i.e., $\rho = 0$), are the random variables $2X - 1$ and $3Y + 1$ also uncorrelated? Explain.

18. Show that if X and Y are continuous random variables with joint density $f(x, y)$, then

$$E[g(X) | X = x] = g(x)$$

19. Find $E[Y | X = x]$ for the random variables X and Y of Example 7.

Sample Problems from Actuarial Exam P

20. The warranty on a machine specifies that it will be replaced at failure or age 4, whichever occurs first. The machine's age at failure, X, has density function $f(x) = 1/5$ for $x \in [0, 5]$ and 0 otherwise. Let Y be the age of the machine at the time of replacement. Determine the variance of Y.

21. Let T_1 be the time between a car accident and the report of a claim to the insurance company. Let T_2 be the time between the report of the claim and payment of the claim. The joint density function of T_1 and T_2, $f(t_1, t_2)$, is constant over the region $0 < t_1 < 6$, $0 < t_2 < 6$, $t_1 + t_2 < 10$, and zero otherwise. Determine $E[T_1 + T_2]$, the expected time between a car accident and payment of the claim.

22. The stock prices of two companies at the end of any given year are modeled with random variables X and Y that follow a distribution with joint density function

$$f(x, y) = \begin{cases} 2x & \text{if } 0 < x < 1, \ x < y < x + 1 \\ 0 & \text{otherwise} \end{cases}$$

What is the conditional variance of Y given that $X = x$?

CHAPTER 4

CONTINUOUS DISTRIBUTIONS

4.1 The Normal Distribution

With most of the general ideas of continuous distributions in hand, we can now move to a survey of some of the important continuous probability distributions. The continuous uniform distribution and the Weibull distribution that we met in the last chapter are certainly two of them. But in this section we study what is perhaps the most important: the normal distribution.

Recall that continuous random variables take values in subintervals of the real line (ideally at least). In many cases, those values tend to distribute across the line in a roughly symmetrical hill shape, with most of the observations near the center, and successively fewer as you move away from the center. Consider for example Figure 1(a), which is a scaled histogram of a large set of 291 National League batting averages on August 5, 2008. You may open up the first of the two closed cells above the figure to see the data that produced the plot. The center of the distribution seems to be around .270, and the great majority of the observations are between about .190 and .350, that is, about .080 on either side of the center.

```
Needs["KnoxProb7`Utilities`"]
```

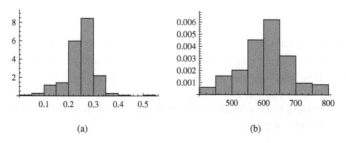

(a)   (b)

Figure 4.1 - (a) Empirical distribution of National League batting averages; (b) Distribution of SAT scores

Observed distributions like this are pervasive in the social and physical sciences. Another example is the data set charted in Figure 1(b). (The first closed cell above the figure contains the list of raw data.) This time 167 SAT math scores for entering students at my college show the same sort of distribution, with a center around 600, fewer observations near the tails, and data distributed more or less symmetrically about the center (though there is a small left skew, created mainly by a few observations under 450).

Random sampling gives other instances of data distributed in this hill shape. For instance, below we take random samples of size 40 from the uniform(0, 2) distribution, compute the sample means for 100 such random samples, and plot a histogram of the resulting sample means. Almost all of the sample means are within about .25 of the mean $\mu = 1$ of the uniform distribution that we sampled from. This histogram shape is no coincidence, because a theorem called the *Central Limit Theorem*, discussed in Chapter 5, guarantees that the probability distribution of the sample mean is approximately normal.

```
SeedRandom[98996];
sampmeans = Table[
    Mean[RandomReal[UniformDistribution[{0, 2}], 40]],
    {i, 1, 100}];
g1 = Histogram[sampmeans, {.1}, "ProbabilityDensity",
    ChartStyle → Gray, BaseStyle → 8]
```

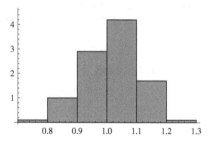

Figure 4.2 - Histogram of 100 sample means of size 40 from uniform(0, 2)

Activity 1 What is the standard deviation of the uniform(0, 2) distribution? What is the standard deviation of the sample mean $\overline{X}$ of a random sample of size 40 from this distribution? How many standard deviations of $\overline{X}$ does .25 represent?

The probability density function that provides a good fit to distributions such as the three above is the following.

Definition 1. The *normal distribution* with mean parameter μ and variance parameter σ^2 is the continuous distribution whose density function is

$$f(x) = \frac{1}{\sqrt{2\pi\sigma^2}} e^{-(x-\mu)^2/2\sigma^2}, \quad x \in \mathbb{R} \tag{1}$$

Therefore, if X is a normally distributed random variable (which we abbreviate as $X \sim N(\mu, \sigma^2)$), then the probability that X takes a value in the interval $[a, b]$ is

$$P[a \le X \le b] = \int_a^b \frac{1}{\sqrt{2\pi\sigma^2}} e^{-(x-\mu)^2/2\sigma^2} \, dx \tag{2}$$

This area is shown in Figure 3.

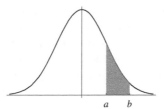

Figure 4.3 - Normal probability $P[a \le X \le b]$ is the shaded area under normal density

In *Mathematica*'s kernel is the distribution object

NormalDistribution[μ, σ]

which represents the $N(\mu, \sigma^2)$ distribution. Note that the second argument is the standard deviation σ, which is the square root of the variance σ^2. The functions PDF, CDF, and RandomReal can therefore be applied to this distribution object in the same way as before. To illustrate, suppose that a random variable X has the normal distribution with mean $\mu = 0$ and variance $\sigma^2 = 4$, hence standard deviation $\sigma = 2$. The commands below define the density function in general and the particular c.d.f., integrate the density function to find $P[0 \le X \le 2]$, then find the same probability by using the c.d.f.

```
f[x_, μ_, σ_] := PDF[NormalDistribution[μ, σ], x];
F[x_] := CDF[NormalDistribution[0, 2], x];
NIntegrate[f[x, 0, 2], {x, 0, 2}]
N[F[2] - F[0]]
```

0.341345

0.341345

(If you remove the numerical evaluation function N from the last command and reexecute, you will probably get a closed form that involves a function called Erf, which is defined in terms of a normal integral.)

By using a clever technique to change from a single to a double integral, followed by a substitution, it is possible (see Exercise 3) to show that the normal density in formula (1) integrates to 1 over the real line, as it should. Here we check this fact in *Mathematica* for the $N(0, 4)$ density we used above. You should try changing the mean and standard deviation parameters several times to make sure that this property still holds.

$$\int_{-\infty}^{\infty} f[x, 0, 2] \, dx$$

1

Figure 3, as well as the density formula itself, suggests that the normal density function is symmetric about its mean μ. Figure 4 gives you another look. We plot in three dimensions the normal density as a function of x and the mean μ for constant standard deviation $\sigma = 2$. In the foreground we see the symmetry of $f[x, -2, 2]$ about $\mu = -2$, and as μ increases to 2 we see the density curve cross section receding into the background in such a way that its center of symmetry is always at μ.

```
Plot3D[f[x, μ, 2], {x, -8, 8},
  {μ, -2, 2}, PlotPoints → 30,
  ViewPoint → {-0.012`, -3.293`, 0.779`},
  AxesLabel → {"x", "μ", None},
  PlotStyle → Gray, Lighting → "Neutral"]
```

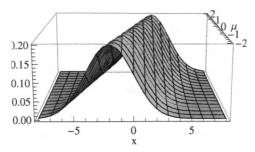

Figure 4.4 - Standard normal density as function of x and μ

Activity 2 Show that the points $\mu \pm \sigma$ are inflection points of the normal density with mean μ and standard deviation σ. (Use either hand computation or *Mathematica*.)

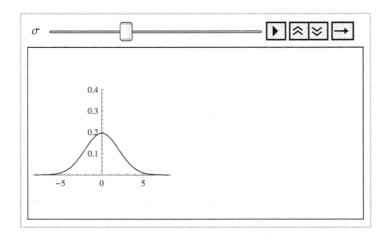

If you think about what Activity 2 says, you will reach the conclusion that the normal density curve spreads out as the standard deviation σ increases. You may run the animation above to confirm this visually. We fix the mean at 0 and vary σ from 1 to 4 to observe that the larger is σ, the more probability weight is distributed to the tails of the distribution.

Example 1 In Activity 1 we asked some questions about the scenario of taking repeated random samples of size 40 from the uniform(0, 2) distribution and finding their sample means. Figure 2 was a histogram of 100 such sample means. Let us fill in the details now and see how an appropriate normal distribution fits the data.

The mean of the uniform(0, 2) distribution is 1, and the variance is $(2-0)^2/12$, i.e. 1/3. Recall that if $\overline{X}$ is the mean of a random sample of size n from a distribution whose mean is μ and whose variance is σ^2, then the expected value of $\overline{X}$ is also μ and the variance of $\overline{X}$ is σ^2/n. So in our situation, $\overline{X}$ has mean 1 and variance (1/3)/40 = 1/120, hence, its standard deviation is $\sqrt{1/120} \approx .09$. The observed maximum distance of data points from $\mu = 1$ of about .25 that we noted in Figure 3 translates into a distance of around 3 standard deviations of $\overline{X}$. Now we superimpose the graph of the $N(1, 1/120)$ density function on the histogram of Figure 3. Figure 5 shows a fairly good fit, even with our small sample size of 40, and only 100 repetitions of the experiment. ∎

```
g2 = Plot[f[x, 1, √(1/120)], {x, 0.75, 1.25},

    PlotStyle → Black, BaseStyle → Gray];
Show[g1, g2, Ticks → {{0, 0.75, 1, 1.25}, Automatic}]
```

Figure 4.5 - $N(1, 1/120)$ density and histogram of 100 sample means of size 40 from uniform(0, 2)

Activity 3   Use the RandomReal command to simulate 1000 observations from the $N(0, 1)$ distribution, and plot a scaled histogram of this sample together with the $N(0, 1)$ density.

Example 2   Assume that demands for a manufactured item are normally distributed with unknown mean and variance. Suppose that there are fixed costs of $10,000 plus a cost of $5 per item to manufacture them. If the plan is to sell the items for $10 apiece, estimate the net profit and the probability of breaking even if a historical random sample of demands is as follows:

```
demands = {1844, 1845, 2299, 2229, 2143,
    2133, 2476, 1539, 2152, 2344, 2017, 2016};
```

The total cost of manufacture for x items is $10\,000 + 5\,x$, and the revenue for selling these x items is $10\,x$. So the net profit is $10\,x - (5\,x + 10\,000) = 5\,x - 10\,000$. The breakeven point is where net profit is zero, which happens at the point $x = 2000$. But demands are random, so if X is the random variable that models the number of items demanded, we can assume that X is normally distributed, but we are not given the mean and standard deviation parameters of the distribution. Let us use the sample mean $\overline{X}$ of the twelve observed random sample values to estimate μ.

```
N[Mean[demands]]
```

2086.42

The random profit is $5\,X - 10\,000$, and by linearity of expectation we see that the expected profit $E[5\,X - 10\,000]$ is approximately:

```
5 * 2086 - 10 000
```

430

With a selling price of $10 per item, this particular venture is not a great money-maker. To find the break-even probability we will also have to estimate the variance σ^2 and corresponding standard deviation σ. We can estimate σ^2 by the *sample variance*, which is an average squared deviation of the data values from the sample mean. The exact formula for the sample variance is

$$S^2 = \frac{1}{n-1} \sum_{i=1}^{n} (X_i - \overline{X})^2 \tag{3}$$

The so-called *sample standard deviation* S is the square root $S = \sqrt{S^2}$ of the sample variance, and serves as an estimate of σ. *Mathematica* can compute these statistics by the functions

Variance[datalist]   and   StandardDeviation[datalist]

which are contained in the kernel. Here are the sample variance and standard deviation for our list of demands.

```
N[{Variance[demands], StandardDeviation[demands]}]
```

{65 437.2, 255.807}

Rounding to the nearest integer, we estimate σ by 256. The event of breaking even (or better) is the event that $X \geq 2000$. With our estimates of μ and σ we can compute:

```
(*P[X ≥ 2000]=*)
N[(1 - CDF[NormalDistribution[2086, 256], 2000])]
```

0.631541

As we have seen before, a large portion of the time a random variable will take on a value within 2 standard deviations of its mean, and so we expect the demand for the item to be within 2(256) = 512 of the true mean demand, estimated by the sample mean of 2086. Here are the profits for the demands that are 2 standard deviations below and 2 standard deviations above the estimated mean:

```
{5 * (2086 - 2 * 256) - 10 000, 5 * (2086 + 2 * 256) - 10 000}
```

{-2130, 2990}

With high probability, the profit will be between -\$2130 and \$2990. ∎

We have been calling the two parameters of the normal distribution by the names "mean" and "variance," but we have not yet justified that this is what they are. Here are *Mathematica* computations of the integrals that would define the mean and variance of the $N(0, 1)$ distribution. The second integral is actually $E[X^2]$, but this is the same as the variance in light of the fact that the first integral shows that $E[X] = 0$.

```
μ = ∫_{-∞}^{∞} x * ─────── * E^{-x²/2} dx
                    √(2 π)

sigsq = ∫_{-∞}^{∞} x² * ─────── * E^{-x²/2} dx
                         √(2 π)
```

0

1

As expected, the mean is the same as the parameter μ, that is 0, and the variance is $\sigma^2 = 1$. You should try to integrate by hand to check these results; it will help to note that the first integrand is an odd function and the second one is an even function. We now state and prove the general result.

Theorem 1. If X is a random variable with the $N(\mu, \sigma^2)$ distribution, then

$$E[X] = \mu \text{ and } \text{Var}(X) = \sigma^2 \tag{4}$$

Proof. To show that $E[X] = \mu$, it is enough to show that $E[X - \mu] = 0$. This expectation is

$$\int_{-\infty}^{\infty} (x - \mu) \cdot \frac{1}{\sqrt{2\pi\sigma^2}} e^{-(x-\mu)^2/2\sigma^2} \, dx$$

The substitution $z = (x - \mu)/\sigma$ changes the integral to

$$\sigma \cdot \int_{-\infty}^{\infty} z \frac{1}{\sqrt{2\pi}} e^{-z^2/2} \, dz$$

which equals zero from above, because it is a constant times the expected value of a $N(0, 1)$ random variable. For the second part, to show that $\text{Var}(X) = E[(X - \mu)^2] = \sigma^2$ it is enough to show that $E[(X - \mu)^2/\sigma^2] = 1$. The expectation on the left is

$$\int_{-\infty}^{\infty} \frac{(x - \mu)^2}{\sigma^2} \frac{1}{\sqrt{2\pi\sigma^2}} e^{-(x-\mu)^2/2\sigma^2} \, dx$$

Again we make the substitution $z = (x - \mu)/\sigma$ to get

$$\int_{-\infty}^{\infty} z^2 \frac{1}{\sqrt{2\pi}} e^{-z^2/2} \, dz$$

As mentioned above, it can be verified that this integral equals 1, which completes the proof. ∎

The next theorem, usually called the *standardization theorem* for normal random variables, shows that by subtracting the mean from a general $N(\mu, \sigma^2)$ random variable and dividing by the standard deviation, we manage to center and rescale the random variable in such a way that it is still normally distributed, but the new mean is 0 and the new variance is 1. The $N(0, 1)$ distribution therefore occupies an important role in probability theory and applications, and it is given the special name of the *standard normal distribution*.

Theorem 2. If the random variable X has the $N(\mu, \sigma^2)$ distribution, then the random variable defined by

$$Z = \frac{X - \mu}{\sigma} \tag{5}$$

has the $N(0, 1)$ distribution.

Proof. We will use an important technique that will be discussed systematically in Section 4.3, called the *cumulative distribution function* method. In it we write an expression for the c.d.f. of the transformed random variable, take advantage of the known distribution of the original random variable to express that c.d.f. conveniently, and then differentiate it to get the density of the transformed variable.

The c.d.f. of Z can be written as

$$F_Z(z) = P[Z \leq z] = P\left[\frac{X - \mu}{\sigma} \leq z\right] = P[X \leq \mu + \sigma\, z] \tag{6}$$

But we know that X has the $N(\mu, \sigma^2)$ distribution. If $G(x)$ is the c.d.f. of that distribution, then formula (6) shows that $F_Z(z) = G(\mu + \sigma\, z)$. Differentiating, the density of Z is

$$f_Z(z) = F_Z{}'(z) = G'(\mu + \sigma\, z)\, \sigma = g(\mu + \sigma\, z)\, \sigma$$

where g is the $N(\mu, \sigma^2)$ density function. Substituting into formula (1) we obtain

$$f_Z(z) = g(\mu + \sigma\, z)\, \sigma = \frac{1}{\sqrt{2\pi\sigma^2}}\, e^{-(\mu + \sigma z - \mu)^2/2\sigma^2}\, \sigma = \frac{1}{\sqrt{2\pi}}\, e^{-z^2/2}$$

which is the $N(0, 1)$ density. Thus, Z is standard normal. ∎

Activity 4   Use the c.d.f. technique of the proof of Theorem 2 to try to show a converse result: If $Z \sim N(0, 1)$ then the random variable $X = \sigma\, Z + \mu$ has the $N(\mu, \sigma^2)$ distribution.

Example 3   It has been a tradition in probability and statistics to go on at some length about standardizing normal random variables, and then to use tables of the standard normal c.d.f. to find numerical answers to probability questions. The availability of technology that can quickly give such numerical answers without recourse to standardization has reduced the necessity for doing this, but I cannot resist the temptation to show a couple of quick example computations.

Suppose that the National League batting averages in the introductory discussion are indeed normally distributed. Let us estimate their mean and variance by the sample mean and sample variance of the list of averages.

```
μ = Mean[averages]
sigsq = Variance[averages]
σ = √sigsq
```

0.248585

0.00338447

0.0581762

Rounding to three significant digits, we are assuming that a generic randomly selected batting average X has the $N(.249, .00338)$ distribution, so the values of μ and σ will be taken to be .249 and .0582, respectively. What is the probability that this random batting average exceeds .300?

Standardizing by subtracting the mean and dividing by the standard deviation yields:

$$P[X > .300] = P\left[\frac{X - \mu}{\sigma} > \frac{.300 - \mu}{\sigma}\right] = P\left[Z > \frac{.300 - .249}{.0582}\right] = P[Z > .876]$$

Thus we have converted a non-standard normal probability involving X to a standard normal probability involving Z. Now many statistical tables of the standard normal c.d.f. are set up to give, for various values of z, the area under the standard normal density that is shaded in Figure 6(a), which is the probability $P[0 \leqslant Z \leqslant z]$.

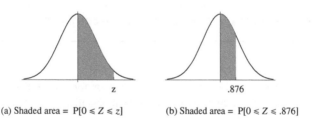

(a) Shaded area = $P[0 \leqslant Z \leqslant z]$ (b) Shaded area = $P[0 \leqslant Z \leqslant .876]$

Figure 4.6 - Standard normal probabilities

Such a table would tell us that the area in Figure 6(b), which is $P[0 \leqslant Z \leqslant .876]$, is .3095. How do we find $P[Z > .876]$ from this? Because of the symmetry of the standard normal density about 0, the area entirely to the right of 0 must be exactly .5. This breaks up into the shaded area in Figure 6(b) plus the unshaded area to the right of .876. Thus, .5 = .3095 + $P[Z > .876]$; hence, $P[Z > .876] = .5 - .3095 = .1905$. To obtain this result from *Mathematica* is easier. We can either call for the complement of the standard normal c.d.f. at .876:

```
1 - CDF[NormalDistribution[0, 1], .876]
```

0.190515

or before standardizing we can ask for the complement of the normal c.d.f. with $\mu = .249$ and $\sigma = .0582$ at the point .300:

```
1 - CDF[NormalDistribution[.249, .0582], .300]
```

0.190437

(The slight difference in the numerical answers is due to the fact that the argument .876 in the first call to the standard normal c.d.f. was rounded.)

As a second example computation, what is the probability that a randomly selected batting average is between .244 and .249? Standardizing, we obtain

$$P[.244 \leq X \leq .249] = P\left[\frac{.244 - .249}{.0582} \leq Z \leq \frac{.249 - .249}{.0582}\right] = P[-.086 \leq Z \leq 0]$$

Most standard normal tables do not give probabilities for intervals that lie to the left of 0, so you would usually not be able to read this off directly. But again by symmetry (refer to Figure 6(a)) the area between $-.086$ and 0 equals the area between 0 and .086, which can be read from the standard normal table, or computed in *Mathematica* as follows. (Why are we subtracting .5 from this c.d.f. value to obtain the probability?)

```
CDF[NormalDistribution[0, 1], 0.086] - .5
```

0.0342668

Again, with *Mathematica* we have no need for tables or for standardizing in this kind of computation, because we can compute the original probability by subtracting c.d.f. values directly:

```
CDF[NormalDistribution[.249, .0582], .249] -
  CDF[NormalDistribution[.249, .0582], .244]
```

0.0342313

Again the small difference between the two answers is due to the rounding of .086. ∎

Activity 5   Suppose that a standard normal table gives the probabilities
$P[0 \leqslant Z \leqslant 1.20] = .3849$ and $P[0 \leqslant Z \leqslant .42] = .1628$. Use the symmetry of the normal
density to find both $P[-1.20 \leqslant Z \leqslant .42]$ and $P[Z > -.42]$.

Example 4 By no means is it irrelevant to know about standardization in the era of widely
available technology. Standardization gives us the ability to make very general statements
about normal densities based on properties of the standard normal density. For instance,
there is a rule of thumb called the "68-95 rule" which says that roughly 68% of the area
under a normal curve lies within one standard deviation of the mean, and 95% of the area
lies within two standard deviations of the mean. Let us see where this rule comes from.

Let X be a general $N(\mu, \sigma^2)$ random variable. Now the area under the normal curve
with these parameters that is within one standard deviation of the mean is exactly the
probability

$$P[\mu - \sigma \leqslant X \leqslant \mu + \sigma] = P\left[\frac{\mu - \sigma - \mu}{\sigma} \leqslant \frac{X - \mu}{\sigma} \leqslant \frac{\mu + \sigma - \mu}{\sigma}\right] \tag{7}$$
$$= P[-1 \leqslant Z \leqslant 1]$$

and the standard normal probability on the far right does not depend on the particular values
of μ and σ. *Mathematica* computes this probability as:

```
N[CDF[NormalDistribution[0, 1], 1] -
   CDF[NormalDistribution[0, 1], -1]]
```

0.682689

By a standardization entirely similar to the one in formula (7),

$$P[\mu - 2\sigma \leqslant X \leqslant \mu + 2\sigma] = P[-2 \leqslant Z \leqslant 2] \tag{8}$$

and the probability on the right is

```
N[CDF[NormalDistribution[0, 1], 2] -
   CDF[NormalDistribution[0, 1], -2]]
```

0.9545

Thus, the "68-95" rule. ∎

Our last example involves an alternative graphical check to see whether a data set
may reasonably be considered to be a random sample from a normal distribution.

Example 5. The pth *percentile* of the probability distribution of a random variable X is the point x_p such that $P[X \leq x_p] = p$. (a) Find a general relationship between the pth percentile x_p of the $N(\mu, \sigma^2)$ distribution and the pth percentile z_p of the standard normal distribution. Then (b), defining the ith smallest member $X_{(i)}$ of a random sample X_1, X_2, ..., X_n as the pth *percentile of the sample*, where $p = i/(n + 1)$, discuss what a plot of $X_{(i)}$ vs. z_p would look like if the random sample came from some normal distribution. (A plot of the pairs $(z_p, X_{(i)})$ described in part (b) is called a *normal scores*, or *normal quantile* plot.)

(a) By standardizing in the defining equation for the pth percentile of a distribution, we get

$$p = P[X \leq x_p] = P\left[Z \leq \frac{x_p - \mu}{\sigma}\right] = P[Z \leq z_p]$$

Thus, since the pth percentile of the standard normal distribution is clearly a unique value, we must have that

$$z_p = \frac{x_p - \mu}{\sigma}. \tag{9}$$

(b) We leave unproved the intuitively reasonable fact that the $p = i/(n + 1)^{th}$ percentile of the random sample should estimate the corresponding percentile x_p of the distribution being sampled from. That is, $X_{(i)} \approx x_p$. But by formula (9) above, if the sample was taken from a normal distribution, then x_p is in turn related to the standard normal percentile by $x_p = \mu + \sigma z_p$. Hence we have $X_{(i)} \approx \mu + \sigma z_p$. A normal quantile plot should show an approximately linear relationship between the standard normal percentiles and the ordered sample values, whose slope is roughly σ, and whose y-intercept is roughly μ.

Let us illustrate this conclusion in three ways: (1) for a simulated sample that does come from a normal distribution; (2) for a simulated sample that does not; and (3) for a real data set, specifically the SAT scores from the start of this section. In the *Mathematica* input cell below, we first initialize the random seed, and then we create random samples of size 30 from the $N(0, 2^2)$ and the uniform(−4, 4) distribution and sort them into increasing order.

```
SeedRandom[17 351];
normsample =
   Table[RandomReal[NormalDistribution[0, 2]], {30}];
sortednormsample = Sort[normsample];
unifsample = Table[
     RandomReal[UniformDistribution[{-4, 4}]], {30}];
sortedunifsample = Sort[unifsample];
```

Next we use *Mathematica*'s Quantile[distribution, p] command to create a list of the standard normal percentiles for probabilities $p = i/(30 + 1)$, $i = 1, 2, \ldots, 30$ to pair up with the sorted values. This command, as suggested by its name, returns the pth percentile of the given named distribution. The Transpose command forms the ordered pairs, which we then ListPlot for both the normal and uniform case.

```
normquants = Table[Quantile[
    NormalDistribution[0, 1], i / 31], {i, 1, 30}];
normpairs = Transpose[{normquants,
    sortednormsample}];
unifpairs = Transpose[{normquants,
    sortedunifsample}];
normplotnorm = ListPlot[normpairs,
    AxesLabel → {z_p, "X(i)"}, PlotStyle → Black];
normplotunif = ListPlot[unifpairs,
    AxesLabel → {z_p, "U(i)"}, PlotStyle → Black];
GraphicsRow[{normplotnorm, normplotunif}]
```

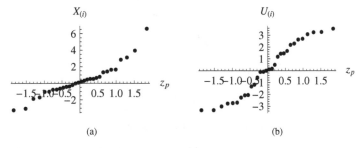

(a)   (b)

Figure 4.7 - (a) Normal scores plot for simulated $N(0, 2^2)$ sample; (b) Normal scores plot for simulated unif $(-4, 4)$ sample

The difference is rather subtle, and is further clouded by the occurrence of an outlying value of more than 6 in the simulated normal random sample. In part (a) of Figure 7, note that except for this extreme value, the points are relatively linear with a slope of around 2, and an intercept of around 0, as predicted by our analysis. Part (b) shows the normal scores plot for a sample from the uniform distribution with a roughly equivalent range of values. Note that there is curvature reminiscent of the cube root graph, or the inverse tangent graph. This can be explained by the fact that the standard normal percentiles increase quickly away from the origin, and slowly near the origin, owing to the heavy distribution of probability weight near 0. But the uniform percentiles will increase at a constant rate throughout the interval $[-4, 4]$. So the rate at which the uniform percentiles change with respect to the normal percentiles will be greatest near zero, and will be smaller on the tails, as the simulated sample bears out.

The main benefit of normal scores plots, however, is to diagnose departures from normality in unknown data sets. Figure 1(b) was a histogram of a set of 167 SAT scores that were entered as the *Mathematica* list SATscores. The histogram seemed to suggest that a normal distribution of scores was reasonable to assume. Let us construct a normal scores plot and check this. Below we use the same approach to construct the plot. First we sort the scores, next we table the 167 percentiles with $p = i/(167 + 1)$ of the standard normal distribution, then we transpose the quantiles and sorted scores to a list of pairs, and finally we ListPlot that data.

```
sortedSAT = Sort[SATscores];
normquants167 = Table[Quantile[
    NormalDistribution[0, 1], i / 168], {i, 1, 167}];
datapoints = Transpose[{normquants167, sortedSAT}];
ListPlot[datapoints,
 AxesLabel → {zp, "X(i)"}, PlotStyle → Black]
```

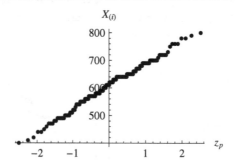

Figure 4.8 - Normal scores plot for SAT scores

The linearity here is quite good, so we should have very little discomfort about the assumption of normality. ∎

Activity 6 Use the graph in Figure 8 to visually estimate the mean and standard deviation of the distribution of SAT scores, and then calculate the sample mean and standard deviation in *Mathematica*.

Exercises 4.1

1. (*Mathematica*) The list below consists of 110 Nielsen television ratings of primetime shows for the CBS, NBC, ABC, and Fox networks selected from the time period of July 7–July 11, 2008. Does the distribution of ratings seem approximately normal? Regardless of the apparent normality or lack of it, estimate the normal parameters, and superimpose the corresponding density function on a scaled histogram of the ratings.

```
ratings = {5.8, 6.4, 7.0, 7.9, 7.2, 6.9, 3.5, 3.5, 5.9,
    4.5, 5.6, 5.7, 2.9, 3.4, 3.3, 3.6, 3.7, 4.2, 4.1, 4.5,
    4.7, 5.4, 5.7, 6.4, 4.2, 3.6, 4.5, 4.7, 5.6, 6.2,
    5.0, 5.6, 4.4, 4.3, 4.9, 4.9, 7.5, 8.2, 5.0, 4.7,
    3.0, 3.6, 5.2, 5.5, 2.7, 2.6, 2.6, 2.9, 4.0, 4.5,
    4.7, 5.2, 4.9, 5.1, 5.7, 6.1, 3.1, 3.0, 4.3, 4.5,
    3.4, 3.0, 5.1, 5.6, 5.7, 5.8, 2.2, 2.1, 2.1, 2.3,
    3.7, 3.9, 6.1, 6.2, 6.5, 6.8, 3.9, 3.7, 2.4, 2.3,
    3.1, 3.0, 3.2, 3.5, 4.0, 4.7, 5.2, 5.3, 2.1, 2.0,
    2.2, 2.4, 3.3, 3.8, 3.6, 3.8, 4.3, 4.5, 5.3, 5.4,
    2.9, 3.3, 4.2, 4.7, 4.5, 4.3, 1.7, 1.9, 2.1, 2.3};
```

2. (*Mathematica*) The data below are unemployment rates (in %) in a sample of 65 Illinois counties in June, 2008. Does the distribution of unemployment rates seem approximately normal? Estimate the normal parameters, and superimpose the corresponding density function on a scaled histogram of the unemployment rates.

```
unemployment =
    {5.4, 7.1, 8.5, 9.0, 6.6, 6.6, 7.0, 8.0, 7.1, 7.1, 6.0,
    7.6, 7.4, 7.6, 7.6, 7.2, 5.7, 6.5, 6.0, 8.2, 8.4, 8.2,
    7.1, 6.2, 6.7, 6.2, 7.0, 6.4, 5.5, 6.2, 6.7, 6.4, 5.5,
    5.5, 7.1, 6.6, 6.4, 5.2, 8.7, 7.5, 6.4, 5.4, 6.5, 8.4,
    6.3, 5.7, 6.5, 6.2, 7.6, 5.6, 6.5, 6.9, 7.0, 6.5, 8.3,
    7.6, 6.5, 7.4, 6.5, 8.1, 7.4, 7.0, 5.7, 7.8, 6.9};
```

3. Show that the $N(0, 1)$ density integrates to 1 over all of $\mathbb{R}$; therefore, it is a good density.

(Hint: To show that $\int_{-\infty}^{\infty} f(x)\, d\,x = 1$ it is enough to show that the square of the integral is one.

Then express the square of the integral as an iterated two-fold integral, and change to polar coordinates.)

4. Show that the $N(\mu, \sigma^2)$ density function is symmetric about μ.

5. Use calculus to show that the $N(\mu, \sigma^2)$ density function reaches its maximum value at μ, and compute that maximum value.

6. (*Mathematica*) Recall that if $X_1, X_2, \ldots, X_n$ is a random sample of size n, then the sample variance is $S^2 = \frac{1}{n-1} \sum_{i=1}^{n} \left(X_i - \overline{X}\right)^2$. Simulate 100 samples of size 20 from the $N(0, 1)$ distribution, the $N(2, 1)$ distribution, and the $N(4, 1)$ distribution and, in each case, sketch a histogram of the 100 observed values of S^2. Discuss the dependence of the empirical

distribution of S^2 on the value of μ. Conduct a similar experiment and discuss the dependence of the empirical distribution of S^2 on the value of σ^2.

7. (*Mathematica* or tables) If X is a $N(\mu, \sigma^2)$ random variable, compute (a) $P[X - \mu < \sigma]$; (b) $P[X - \mu < 2\sigma]$; (c) $P[X - \mu < 3\sigma]$.

8. (*Mathematica* or tables) Suppose that compression strengths of human tibia bones are normally distributed with mean $\mu = 15.3$ (units of $kg/mm^2 \times 1000$) and standard deviation $\sigma = 2.0$. Among four independently sampled tibia bones, what is the probability that at least three have strengths of 17.0 or better?

9. (*Mathematica* or tables) Assume that in a certain area the distribution of household incomes can be approximated by a normal distribution with mean \$30,000 and standard deviation \$6000. If five households are sampled at random, what is the probability that all will have incomes of \$24,000 or less?

10. (*Mathematica*) Simulate 500 values of $Z = (X - \mu)/\sigma$ where $X \sim N(\mu, \sigma^2)$ for several cases of the parameters μ and σ. For each simulation, superimpose a scaled histogram of observed z values and a graph of the standard normal density curve. What do you see, and why should you have expected to see it?

11.   Suppose that a table of the standard normal distribution tells you that $P[0 \le Z \le 1.15] = .3749$, and $P[0 \le Z \le 2.46] = .4931$.   Find (a) $P[1.15 \le Z \le 2.46]$;   (b) $P[-1.15 \le Z \le 2.46]$; (c) $P[Z > 2.46]$; (d) $P[Z < -1.15]$.

12. Assume that X has the $N(1.8, 0.7^2)$ distribution. Without using *Mathematica* or tables, find $P[1.65 \le X \le 1.85]$ if a table of the standard normal distribution tells you that $P[0 \le Z \le .07] = .0279$ and $P[0 \le Z \le .21] = .0832$.

13. (*Mathematica*) For each of 200 simulated random samples of size 50 from the uniform(4, 6) distribution, compute the sample mean. Then plot a scaled histogram of the sample mean together with an appropriate normal density that you expect to be the best fitting one to the distribution of $\overline{X}$. Repeat the simulation for the Weibull(2, 30) distribution that we encountered in Example 3 of Section 3.3.

14. Suppose that the times people spend being served at a post office window are approximately normal with mean 1.5 minutes and variance .16 minutes. What are the probabilities that:
(a) a service requires either more than 1.9 minutes or less than 1.1 minutes?
(b) a service requires between .7 minutes and 2.3 minutes?

15. (*Mathematica*) The *deciles* of the probability distribution of a random variable X are the points $x_{.1}, x_{.2}, x_{.3}, \ldots, x_{.9}$ such that

$$P[X \le x_{.1}] = .1, \; P[X \le x_{.2}] = .2, \; P[X \le x_{.3}] = .3,$$

etc. Find the deciles of the $N(0, 1)$ distribution.

16. (a) Produce a normal scores plot for 40 simulated values from the Weibull(2,3) distribution. Does it appear to be about linear? (b) Produce a normal scores plot for the unemployment data in Exercise 2. Are you confident in an assumption of normality?

17. (*Mathematica*) Suppose that an insurance company determines that the distribution of claim sizes on their auto insurance policies is normal with mean 1200 and standard deviation 300. When a claim is submitted, the company pays the difference between the amount claimed and the "deductible" amount listed on the particular policy. Half of the company's policies have deductible amounts of $250, 1/4 have $500 deductibles, and the remaining 1/4 have $1000 deductibles. What is the expected payout of a random claim?

18. (*Mathematica* or tables) The rate of return on investment is defined as (final value − initial value)/(initial value). Suppose that the rate of return on a particular investment is normal with mean .10 and standard deviation .2. Find: (a) the probability that the investment loses money; (b) the probability that an investment of $1000 has a final value of at least $1050.

4.2 Bivariate Normal Distribution

There is a two-dimensional version of the normal distribution that is of importance in the study of how continuous random variables depend on one another. This so-called *bivariate normal distribution* is the topic of this section.

```
statsdata = {hwk, quiz, final} =
   {{47.1, 27.7, 73.1, 41.5, 85.6, 65.2, 58.5, 43.1,
     87.5, 44.6, 54.0, 41.0, 86.0, 68.5, 67.1, 84.2,
     75.2, 61.5, 54.2, 86.5, 46.3, 86,5, 79.6},
    {65.8, 72.5, 76.7, 50.8, 86.7, 67.5, 75.0, 55.0,
     98.3, 51.7, 75.0, 65.0, 88.3, 75.8, 81.7, 61.7,
     66.7, 46.7, 68.3, 91.7, 72.5, 78.3, 80.0},
    {111, 110, 141, 89, 162, 137, 120, 109,
     191, 93, 152, 104, 165, 147, 171, 145,
     140, 117, 105, 169, 104, 165, 142}};
```

Example 1 I recently taught an elementary statistics class and became interested in the accuracy with which the final exam score could be predicted by other classwork such as homework and quizzes. Above is the data set of homework percentages, quiz percentages, and final exam scores (out of 200 points) for the 23 students in the class, in the form of three *Mathematica* lists called hwk, quiz, and final. The dependence of the final exam score on both of the predictor variables of homework and quiz percentage is illustrated in the plots in Figure 9. There is a strong tendency for the final score to increase, following a roughly linear function, with both variables. The points exhibit variability, falling in a cloud rather than perfectly along a line.

```
g1 = ListPlot[Transpose[{hwk, final}],
    AxesLabel → {"hwk %", "final"}, PlotStyle → Black];
g2 = ListPlot[Transpose[{quiz, final}], AxesLabel →
        {"quiz %", "final"}, PlotStyle → Black];
GraphicsRow[{g1, g2}]
```

Figure 4.9 - Dependence of final score on homework and quiz percentage

In many (but not all) cases of bivariate data in which the pairs plot as they do in Figure 9, the individual variables also display a roughly normal shape. This is a rather small data set, but the dot plots in Figure 10 for the quiz and final variables are reasonably normal. (A *dot plot* of a single variable data set simply graphs each data point on an axis, stacking dots for values that are the same or nearly the same. The DotPlot command in Knox-Prob7`Utilities` sketches such plots. It takes only one argument, the list of data that it is plotting, but has several options including VariableName to put a text name on the axis.) ∎

```
Needs["KnoxProb7`Utilities`"]
g3 = DotPlot[quiz, VariableName → "quiz"];
g4 = DotPlot[final, VariableName → "final"];
GraphicsRow[{g3, g4}]
```

Figure 4.10 - Empirical distributions of quiz percentage and final exam score

---

Activity 1 Produce a dot plot of the homework percentage data. Does it look normally distributed? Can you speculate on the reason for the shape that you see?

Example 2 As another example, consider the data set below, obtained by the Centers for Disease Control (CDC). For the year 2007, they estimated for each state plus the District of Columbia the percentage of children in the state aged 19-35 months who had been immunized with three different types of common vaccine: polio, MMR, and hepatitis. Each triple corresponds to one state, and we have used *Mathematica*'s Transpose function to obtain three lists of data, one for each vaccine.

```
{polio, MMR, hep} =
  Transpose[{{94.9, 95.0, 95.1}, {92.9, 89.7, 93.9},
      {90.6, 89.0, 92.0}, {90.2, 92.5, 92.4},
      {92.3, 94.6, 91.6}, {95.0, 91.2, 94.1},
      {97.9, 95.3, 97.5}, {93.7, 94.8, 95.7},
      {93.3, 95.2, 95.2}, {93.2, 92.3, 94.9},
      {93.9, 91.4, 93.9}, {96.0, 93.8, 95.2},
      {87.6, 86.1, 88.8}, {91.0, 93.1, 92.3},
      {91.9, 90.4, 93.6}, {93.5, 93.0, 92.7},
      {95.7, 93.1, 95.0}, {92.5, 90.8, 94.0},
      {94.0, 92.9, 94.2}, {91.2, 90.2, 88.3},
      {97.3, 97.1, 96.6}, {92.8, 93.3, 96.1},
      {87.8, 89.5, 89.9}, {94.8, 94.9, 93.4},
      {90.9, 87.2, 94.2}, {94.5, 89.0, 91.8},
      {90.2, 89.6, 92.4}, {95.9, 94.0, 96.9},
      {84.8, 86.3, 84.4}, {97.6, 96.6, 98.6},
      {92.6, 91.2, 92.4}, {89.7, 90.6, 90.8},
      {93.6, 93.6, 93.5}, {93.2, 96.9, 91.7},
      {97.8, 95.2, 97.8}, {93.6, 90.7, 95.3},
      {92.3, 89.9, 92.0}, {87.7, 88.9, 87.0},
      {93.7, 93.8, 92.5}, {98.0, 94.7, 97.7},
      {91.1, 92.5, 92.1}, {97.7, 95.0, 97.4},
      {94.4, 94.5, 94.4}, {92.6, 90.4, 91.7},
      {91.8, 90.9, 88.6}, {95.1, 93.6, 93.7},
      {92.8, 90.9, 91.5}, {85.1, 90.5, 85.0}, {93.8, 96.2,
       94.6}, {93.8, 91.4, 94.5}, {90.8, 87.5, 90.6}}];
```

As Figures 11 and 12 show, the individual distributions of the MMR and hepatitis variables look reasonably normal, and the joint distribution of the two again shows a roughly linear increasing pattern, with variability inherent in the way the points form a cloud around a line. A couple of outlying data points around (90.5,85) and (87,94) seem to fit the elliptically shaped cloud of points less well. ■

```
g5 = DotPlot[MMR, VariableName → "MMR"];
g6 = DotPlot[hep, VariableName → "hepatitis"];
GraphicsRow[{g5, g6}]
```

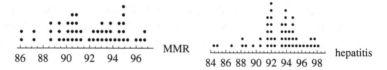

Figure 4.11 - Empirical distributions of immunization coverage

```
ListPlot[Transpose[{MMR, hep}],
   AxesLabel → {"MMR", "hepatitis"}, PlotStyle → Black]
```

Figure 4.12 - Empirical joint distribution of immunization coverage for MMR and hepatitis

Activity 2 Produce a dot plot of the polio data in Example 2 to see whether it also appears normally distributed. Construct scatter plots like Figure 12 to see whether the joint behavior is similar for the other pairs of variables: MMR and polio, and hepatitis and polio.

Bivariate Normal Density

Examples 1 and 2 suggest to us that there may frequently be bivariate probabilistic models in which the individual random variables in the pair have a normal distribution, but instead of the two random variables being independent, they are correlated. This correlation shows up in sample data as a tendency for observed pairs to lie in an elliptical cloud around some line. Because of this dependence, we may be able to make predictions about one of the variables given the value of the other.

One way of forming and using a probabilistic model for this situation is to postulate a particular two-dimensional distribution called the *bivariate normal distribution* for the pair (X, Y), which gives rise to the distribution of samples of pairs as in the examples. It will be proved that both X and Y have marginal normal distributions. We will also have the capability of modeling the linear relationship between X and Y, and using it for predictive purposes.

Definition 1. The pair of random variables (X, Y) is said to have the *bivariate normal distribution* with parameters μ_x, μ_y, σ_x^2, σ_y^2, and ρ if its joint density function is

$$f(x, y) = \frac{1}{2\pi\sigma_x\sigma_y\sqrt{1-\rho^2}}$$
$$\times \exp\left[\frac{-1}{2(1-\rho^2)}\left(\frac{(x-\mu_x)^2}{\sigma_x^2} - 2\rho\frac{(x-\mu_x)(y-\mu_y)}{\sigma_x\sigma_y} + \frac{(y-\mu_y)^2}{\sigma_y^2}\right)\right] \quad (1)$$

Activity 3 Let the parameter $\rho = 0$ in formula (1). What do you notice about the joint density $f(x, y)$?

As usual with joint densities, the probability that the pair (X, Y) falls into a set B is

$$P[(X, Y) \in B] = \iint_B f(x, y)\, dx\, dy \quad (2)$$

which is the volume bounded below by the set B in the x-y plane and above by the surface $z = f(x, y)$.

Mathematica knows about this distribution. In the standard package MultivariateStatistics` is the object

MultinormalDistribution[meanvector, covariancematrix]

where the argument *meanvector* is the list $\{\mu_x, \mu_y\}$ and the argument *covariancematrix* is the 2×2 matrix

$$\begin{pmatrix} \sigma_x^2 & \rho\,\sigma_x\,\sigma_y \\ \rho\,\sigma_x\,\sigma_y & \sigma_y^2 \end{pmatrix}$$

(written in *Mathematica* list form as $\{\ \{\sigma_x^2,\ \rho\,\sigma_x\,\sigma_y\},\{\rho\,\sigma_x\,\sigma_y,\ \sigma_y^2\}\ \}$). We will see shortly that these parameters are aptly named. The parameters μ_x and μ_y are the means of X and Y, σ_x^2 and σ_y^2 are the variances of X and Y, and ρ is the correlation between X and Y, which means that the matrix entry $\rho\,\sigma_x\,\sigma_y$ is the covariance. (You will show that $\mathrm{Corr}(X,\ Y) = \rho$ in Exercise 12.)

Here for example is a definition of the bivariate normal density with $\mu_x = 0$, $\mu_y = 2$, $\sigma_x^2 = 1$, $\sigma_y^2 = 4$, and $\rho = .6$. Then $\rho\sigma_x\,\sigma_y = (.6)\,(1)\,(2) = 1.2$. Note the use of the list $\{x,\ y\}$ as the argument for the two variable form of the PDF function.

```
Needs["MultivariateStatistics`"]
meanvector = {0, 2};
covariancematrix = {{1, 1.2}, {1.2, 4}};
f[x_, y_] := PDF[MultinormalDistribution[
    meanvector, covariancematrix], {x, y}]
```

We can understand the graph of f in two ways: by just plotting the density surface or by looking at its contour graph, i.e., the curves of equal probability density. (The closed cell shows the code for generating Figure 13.)

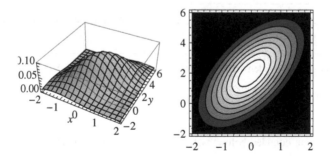

Figure 4.13 - Bivariate normal density surface and contour plot

The probability mass is concentrated most densely in elliptical regions around the point $(0,\ 2)$, which is the mean. The major and minor axes are tilted, and there is more spread in the y-direction than in the x-direction. (Look at the tick marks carefully.) Simulated data from this distribution show the same tendency. Here are 200 points randomly sampled from the same distribution and plotted.

```
simlist = RandomReal[MultinormalDistribution[
    meanvector, covariancematrix], {200}];
ListPlot[simlist, PlotStyle →
    {Black, PointSize[0.015]}]
```

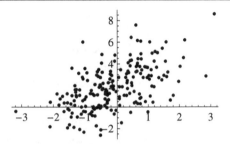

Figure 4.14 - Simulated bivariate normal data

Note the similarity between the simulated data in Figure 14 and the real data in Figures 9 and 12. It appears that the bivariate normal distribution could be a good model for such real data. The role of the mean vector (μ_x, μ_y) in determining the center of the distribution and the point cloud is clear. Try the next activity to investigate the exact role of the variance parameters and the correlation parameter.

Activity 4 Sketch contour graphs of the bivariate normal density, holding μ_x and μ_y at 0, σ_x and σ_y at 1, and varying ρ between $-.9$ and $.9$. Try it again with $\sigma_x = \sigma_y = 2$ and $\sigma_x = \sigma_y = 3$. What seems to be the effect of changing ρ when the standard deviations are held fixed and equal? Next fix $\rho = .7$, and plot contours for the cases: $\sigma_x = 1$, $\sigma_y = 2$ and $\sigma_x = 1$, $\sigma_y = 3$ and $\sigma_x = 1$, $\sigma_y = 4$. For fixed ρ, what is the effect of changing the ratio of standard deviations?

Exercise 10 asks you to show from the density formula (1) that the contours of the bivariate normal density are indeed ellipses centered at (μ_x, μ_y), rotated at an angle α with the x-axis satisfying

$$\cot(2\,\alpha) = \frac{\sigma_x^2 - \sigma_y^2}{2\,\rho\sigma_x\,\sigma_y} \tag{3}$$

Therefore, when the variances are equal, the cotangent is 0, hence $2\alpha = \pi/2$, and so the angle of rotation is $\alpha = \pi/4$, as you probably noticed in Activity 4. You may have also noticed that the variance parameters affected the lengths of the two axes of symmetry of the ellipses: as σ_x^2 increases, the length of the axis parallel to the rotated x-axis grows, and as σ_y^2 increases the rotated y-axis grows. As in the one-dimensional case, these parameters control the spread of the distribution. In the electronic book, you can execute the cell below which enables you to fix the means μ_x and μ_y, the variance $\sigma_x{}^2$, and the correlation ρ, and manipulate the variance σ_y^2 to see the changes in the elliptical contours.

```
Clear[g];
g[x_, y_, μx_, μy_, σx_, σy_, ρ_] :=
    PDF[MultinormalDistribution[{μx, μy},
        {{σx², ρ * σx * σy}, {ρ * σx * σy, σy²}}], {x, y}];
Manipulate[ContourPlot[g[x, y, 0, 0, 1, σy, .8],
    {x, -3, 3}, {y, -3, 3}, PlotRange → All], {σy, 1, 4}]
```

The role of the parameters σ_x^2 and σ_y^2 is easiest to see in a very special case. When you did Activity 3 you should have deduced that when $\rho = 0$ the joint density $f(x, y)$ simplifies to the product $p(x) q(y)$ of a $N(\mu_x, \sigma_x^2)$ density for X and a $N(\mu_y, \sigma_y^2)$ density for Y. It is clear in this case that X and Y are independent, and the μ and σ^2 parameters determine the centers and spreads of their marginal distributions. It is not a general rule that lack of correlation implies independence, but it is true when the two random variables have the bivariate normal distribution.

Activity 5 In the closed cell below is a data list of pairs of atmospheric ozone levels collected by the Economic Research Service of the United States Department of Agriculture for a number of counties in the East Coast in the year 1987. The first member of each pair is an average spring 1987 ozone level, and the second is the summer ozone for the same county. Plot the pairs as points in the plane to observe the relationship between the two variables: spring and summer ozone level. Also, plot a histogram of the individual season levels. Do the underlying assumptions of the model for random bivariate normal pairs seem to hold for this data? What anomalies do you see, and can you guess the reasons for them?

Example 3 Let us estimate for the statistics data in Example 1 the probability that a student receives at least 80% on both quizzes and final exam (i.e., a final exam score of at least 160 points).

Have faith for a moment in the earlier claims that for a bivariate normal pair (X, Y), $X \sim N(\mu_x, \sigma_x^2)$ and $Y \sim N(\mu_y, \sigma_y^2)$ and $\rho = \text{Corr}(X, Y)$. Assume that the quiz percentage X and the final exam score Y of a randomly selected student form a bivariate normal pair. We can estimate the means of X and Y by the component sample means below.

```
μ_x = N[Mean[quiz]];
μ_y = N[Mean[final]];
meanvector = {μ_x, μ_y}
```

{71.813, 134.304}

Here are the sample variances and standard deviations, which estimate the variances and standard deviations of X and Y.

```
varx = N[Variance[quiz]]; vary = N[Variance[final]];
σ_x = N[StandardDeviation[quiz]];
σ_y = N[StandardDeviation[final]];
{varx, vary}
{σ_x, σ_y}
```

{175.443, 812.767}

{13.2455, 28.5091}

The correlation ρ must also be estimated. Recall that its defining formula is

$$\rho = \frac{\mathrm{E}\left[(X - \mu_x)(Y - \mu_y)\right]}{\sigma_x \sigma_y}$$

The expectation in the numerator is an average product of differences between X and its mean with differences between Y and its mean. Given n randomly sampled pairs (X_i, Y_i) a sensible estimate of the correlation is the *sample correlation coefficient R*, defined by

$$R = \frac{\sum_{i=1}^{n}(X_i - \overline{X})(Y_i - \overline{Y})/(n-1)}{S_x S_y} = \frac{\sum_{i=1}^{n}(X_i - \overline{X})(Y_i - \overline{Y})}{\sqrt{\sum_{i=1}^{n}(X_i - \overline{X})^2}\sqrt{\sum_{i=1}^{n}(Y_i - \overline{Y})^2}} \tag{4}$$

So the probabilistic weighted average of products of differences from means is replaced by the average among the sample pairs of products of differences from sample means, and the standard deviations are estimated by sample standard deviations. (Make sure you see why the simplified formula on the right of (4) holds.)

Mathematica has a Correlation function in its kernel, whose syntax is

Correlation[variable1,variable2]

where the arguments are each lists of observations to be correlated. Here we estimate the covariance matrix for the statistics data set.

```
R = Correlation[quiz, final]
covariancematrix =
    {{varx, R * σ_x * σ_y}, {R * σ_x * σ_y, vary}};
MatrixForm[covariancematrix]
```

0.810708

$$\begin{pmatrix} 175.443 & 306.137 \\ 306.137 & 812.767 \end{pmatrix}$$

Actually, we need not have gone to this trouble of writing the estimated covariance matrix out longhand, because there is another command

Covariance[datalist]

which does the whole job of computing the estimated covariance matrix, as you can see below. Its argument is the full data list of pairs, which we create using Transpose.

```
covmatrix2 = Covariance[Transpose[{quiz, final}]];
MatrixForm[covmatrix2]
```

$$\begin{pmatrix} 175.443 & 306.137 \\ 306.137 & 812.767 \end{pmatrix}$$

Now we can define the bivariate normal density function using the parameters we have estimated.

```
f[x_, y_] := PDF[MultinormalDistribution[
    meanvector, covariancematrix], {x, y}]
```

The question asks for $P[X > 80, Y > 160]$, which would be the following integral.

```
NIntegrate[f[x, y], {x, 80, ∞}, {y, 160, ∞}]
```

0.144642

In this particular statistics course, it was very difficult to do well on both items. The estimated joint density of quiz and final scores is in Figure 15. ∎

```
Plot3D[f[x, y], {x, 50, 100}, {y, 90, 200},
  PlotRange → All, PlotPoints → 25,
  AxesLabel → {"quiz", "final", None},
  PlotStyle → Gray, Lighting → "Neutral"]
```

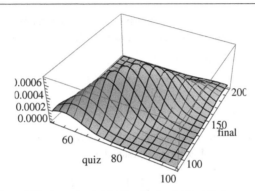

Figure 4.15 - Approximate joint density of quiz % and final exam score

Marginal and Conditional Distributions

We will now show that if the pair of random variables (X, Y) has the bivariate normal distribution with the usual parameters, then $X \sim N(\mu_x, \sigma_x^2)$ and $Y \sim N(\mu_y, \sigma_y^2)$. As an incidental but important by-product, we will obtain the conditional distributions of each random variable given the other.

The argument rests on the following algebraic identity, which is tedious but straightforward to check. (It comes from completing the square on y in the exponent of the bivariate normal density formula (1).)

$$\frac{1}{2(1-\rho^2)}\left(\frac{(x-\mu_x)^2}{\sigma_x^2} - 2\rho(x-\mu_x)\frac{(y-\mu_y)}{\sigma_x\sigma_y} + \frac{(y-\mu_y)^2}{\sigma_y^2} \right)$$
$$= \frac{(x-\mu_x)^2}{2\sigma_x^2} + \frac{\left(y - \left[\mu_y + \frac{\rho\sigma_y}{\sigma_x}(x-\mu_x)\right]\right)^2}{2\sigma_y^2(1-\rho^2)}$$

(5)

The negative of this expression is just the exponent in the density formula. To shorten notation write

$$\mu_{y|x} = \mu_y + \frac{\rho\sigma_y}{\sigma_x}(x-\mu_x) \quad \text{and} \quad \sigma_{y|x}^2 = \sigma_y^2(1-\rho^2)$$

(6)

Substituting the right side of formula (5) into formula (1), we see that the bivariate normal joint density can be written in factored form as

$$f(x,\,y) = \frac{1}{\sqrt{2\pi}\,\sigma_x}\exp\!\left(-\frac{(x-\mu_x)^2}{2\sigma_x^{\,2}}\right)\cdot\frac{1}{\sqrt{2\pi}\,\sigma_{y|x}}\exp\!\left(-\frac{(y-\mu_{y|x})^2}{2\sigma_{y|x}^{\,2}}\right) \qquad (7)$$

The marginal density of X can be found by computing the integral $\int_{-\infty}^{\infty} f(x,\,y)\,d\,y$. But the two leading factors in (7) come out of this integral, and the integral of the product of the two rightmost factors is 1, because they constitute a normal density with mean $\mu_{y|x}$ and variance $\sigma_{y|x}^{\,2}$. Hence the marginal density of X is

$$p(x) = \frac{1}{\sqrt{2\pi}\,\sigma_x}\exp\!\left(-\frac{(x-\mu_x)^2}{2\sigma_x^{\,2}}\right)$$

which is the normal density with mean μ_x and variance σ_x^2.

But in fact more is true. Equation (7) states that

$$f(x,\,y) = p(x)\cdot\frac{1}{\sqrt{2\pi}\,\sigma_{y|x}}\exp\!\left(-\frac{(y-\mu_{y|x})^2}{2\sigma_{y|x}^{\,2}}\right)$$

where $p(x)$ is the marginal density of X. Therefore, by the definition of conditional density,

$$q(y\,|\,x) = \frac{f(x,\,y)}{p(x)} = \frac{1}{\sqrt{2\pi}\,\sigma_{y|x}}\exp\!\left(-\frac{(y-\mu_{y|x})^2}{2\sigma_{y|x}^{\,2}}\right) \qquad (8)$$

Hence, the conditional density of Y given $X = x$ is also normal, with parameters $\mu_{y|x}$ and $\sigma_{y|x}^{\,2}$ given by formula (6).

The formula for $\mu_{y|x}$ in (6) is particularly interesting, because it gives a predictive linear relationship between the variables; if $X = x$ is known to have occurred, then the conditional expected value of Y is $\mu_y + \dfrac{\rho\sigma_y}{\sigma_x}(x - \mu_x)$. It is rather surprising however that the conditional variance $\sigma_y^2(1 - \rho^2)$ of Y does not depend on the particular value that X takes on. But it is intuitively reasonable that the largest value of the conditional variance is σ_y^2 in the case $\rho = 0$ where the variables are uncorrelated, and the conditional variance decreases to 0 as ρ increases to 1. The more strongly that X and Y are correlated, the less is the variability of Y when the value of X is known.

Symmetric arguments yield the marginal distribution of Y and the conditional p.d.f. of X given $Y = y$, in the theorem below, which summarizes all of our results.

Theorem 1. Let (X, Y) have the bivariate normal density in (1) with parameters $\mu_x, \mu_y, \sigma_x^2, \sigma_y^2$, and ρ. Then,

    (a) the marginal distribution of X is $N(\mu_x, \sigma_x^2)$;

    (b) the marginal distribution of Y is $N(\mu_y, \sigma_y^2)$;

    (c) ρ is the correlation between X and Y;

    (d) the curves of constant probability density are ellipses centered at (μ_x, μ_y) and rotated at an angle α with the x-axis satisfying $\cot(2\,\alpha) = (\sigma_x^2 - \sigma_y^2)/(2\,\rho\sigma_x\sigma_y)$;

    (e) if $\rho = 0$, then X and Y are independent;

    (f) the conditional density of Y given $X = x$ is normal with conditional mean and variance

$$\mu_{y|x} = \mu_y + \frac{\rho\sigma_y}{\sigma_x}(x - \mu_x) \quad \text{and} \quad \sigma_{y|x}^{\;2} = \sigma_y^2\left(1 - \rho^2\right)$$

    (g) the conditional density of X given $Y = y$ is normal with conditional mean and variance

$$\mu_{x|y} = \mu_x + \frac{\rho\sigma_x}{\sigma_y}\left(y - \mu_y\right) \quad \text{and} \quad \sigma_{x|y}^{\;2} = \sigma_x^2\left(1 - \rho^2\right) \quad \blacksquare$$

Activity 6 Finish the details of the proof that Y has a normal distribution, and that the conditional distribution of X given $Y = y$ is normal.

Example 4 You may have heard of the statistical problem of *linear regression*, which involves fitting the best linear model to a data set of pairs (x_i, y_i). The presumption is that the data are instances of a model

$$Y = a + bX + \epsilon \tag{9}$$

where X is a random variable that has the $N(\mu_x, \sigma_x^2)$ distribution, and ϵ is another normally distributed random variable with mean 0 and some variance σ^2, whose correlation with X is zero. The game is to estimate the coefficients a and b and to use them to predict new y values for given x values. Let us see that the bivariate normal model implies a linear regression model, and also gives information about the coefficients.

Suppose that X and Y are bivariate normal. Define a random variable ϵ by:

$$\epsilon = \left(Y - \mu_y\right) - \frac{\rho\sigma_y}{\sigma_x}(X - \mu_x)$$

Since X and Y are normally distributed, and ϵ is a linear combination of them, it can be shown that ϵ is normally distributed. (We will show this fact in the independent case in the next section, though we omit the proof in the correlated case.) It is also easy to see that $E[\epsilon]$ = 0 (why?).

Finally, ϵ is uncorrelated with X because

$$
\begin{aligned}
\text{Cov}\,(\epsilon,\,X) &= \text{E}[\epsilon\,(X-\mu_x)] \\
&= \text{E}\!\left[(Y-\mu_y)(X-\mu_x) - \frac{\rho\sigma_y}{\sigma_x}(X-\mu_x)^2\right] \\
&= \text{Cov}\,(X,\,Y) - \frac{\rho\sigma_y}{\sigma_x}\sigma_x^{\,2} = 0
\end{aligned}
$$

The last line follows because $\text{Cov}(X,\,Y) = \rho\sigma_x\sigma_y$. Therefore, Y and X are related linearly by

$$
(Y-\mu_y) = \frac{\rho\sigma_y}{\sigma_x}(X-\mu_x) + \epsilon \Longrightarrow Y = \mu_y + \frac{\rho\sigma_y}{\sigma_x}(X-\mu_x) + \epsilon \tag{10}
$$

and X and ϵ satisfy the distributional assumptions of linear regression. In particular, the slope coefficient $b = \rho\sigma_y/\sigma_x$, and the predicted value of Y given $X = x$ is the conditional mean $\mu_{y|x} = \mu_y + \dfrac{\rho\sigma_y}{\sigma_x}(x-\mu_x)$. Notice that the farther x is from μ_x, the more the predicted y differs from μ_y. ∎

Activity 7   Simulate 100 pairs $(X,\,Y)$ using model (9) with $a = 1$, $b = 2$, $\mu_x = 0$, $\sigma_x^{\,2} = 4$, $\sigma^2 = 1$. You should do this by simulating independent X and ϵ observations, using each pair $(X,\,\epsilon)$ to compute the corresponding Y, and repeating 100 times. Does a graph of the resulting pairs $(X_i,\,Y_i)$ provide convincing evidence that indeed X and Y are bivariate normal?

Let us close this section with an example of a different set of data on which Theorem 1 sheds light.

```
hwydata = {{1, 14.5, 11.3}, {2 , 14.1, 9.8},
   {3, 12.6 , 8.8}, {4 , 12.7, 8.1}, {5 , 12.9, 7.5},
   {6, 12.6 , 7.6}, {7, 13.0 , 7.5}, {8 , 11.1 , 7.4},
   {9 , 11.7 , 7.0}, {10 , 10.3,  6.3},
   {11 , 9.4 , 6.4}, {12, 9.6 , 6.4}, {13, 9.7, 6.0},
   {14, 8.9 , 5.6}, {15 , 9.3 , 5.4}, {16, 8.0, 5.3},
   {17, 6.9, 5.2}, {18 , 8.2 , 5.3}, {19, 6.8 , 5.5},
   {20, 6.9, 5.7}, {21, 8.3, 5.5}, {22, 7.2 , 5.7},
   {23 , 7.3 , 5.5}, {24, 7.4, 5.4}, {25, 8.0 , 5.3},
   {26, 7.6, 4.9}, {27, 6.7, 4.7}, {28, 6.6, 4.5},
   {29, 6.7, 4.3}, {30, 5.7, 3.6}, {31, 5.8, 3.4},
   {32, 5.3, 3.3}, {33, 6.1, 3.4}, {34, 5.6, 3.4},
   {35, 5.7, 3.5}, {36, 5.5, 3.5}, {37, 4.7, 3.3},
   {38, 4.9, 2.9}, {39, 4.4, 2.7}, {40, 3.8, 2.7}};
```

Example 5 The data above have to do with highway death rates in fatalities per million vehicle miles over a forty-year span beginning in 1945. The first component is an index to the year, the second component is the fatality rate in the state of New Mexico in that year, and the third is the death rate in the entire U.S. in that year. We wonder how highway death rates in New Mexico relate to those in the U.S. as a whole. Is a bivariate normal model appropriate? Can we make predictions about highway fatalities in New Mexico (or in the U.S.)?

```
{year, newmexico, us} = Transpose[hwydata];
```

The scatter plot in Figure 16 shows a very strong linear dependence between the two variables, which supports a bivariate normal model. But the histograms following in Figure 17 show right skewed marginal distributions.

```
ListPlot[Transpose[{newmexico, us}],
  AxesLabel → {"New Mex", "U.S."},
  PlotStyle → {Black, PointSize[0.02`]}]
```

Figure 4.16 - U.S. vs. New Mexico highway fatality rates, 1945-1984

```
GraphicsRow[{Histogram[newmexico,
    5, ChartStyle → Gray, BaseStyle → 8],
  Histogram[us, 5, ChartStyle → Gray, BaseStyle → 8]}]
```

Figure 4.17 - Marginal distributions of New Mexico (left) and U.S. (right) fatality rates

At times, data analysts can apply stabilizing transformations such as x^p, $p < 1$, and $\log(x)$ to knock down large data values and make the transformed variables look a bit more symmetric. Let us work from here onward with the logarithms of our two variables. You can check that the scatterplot still looks good after the log transformation. The individual histograms for the logged variables show somewhat more symmetry than those for the original variables, although they are by no means perfect.

```
{lognewmex, logus} = {Log[newmexico], Log[us]};
GraphicsRow[
  {Histogram[lognewmex, 5, ChartStyle → Gray,
    BaseStyle → 8], Histogram[logus, 5,
    ChartStyle → Gray, BaseStyle → 8]}]
```

Figure 4.18 - Histograms of logs of variables in Figure 17

Now let us estimate the bivariate normal parameters. We use the Mean, StandardDeviation, and Correlation commands as before.

```
{μ₁, μ₂, σ₁, σ₂, ρ} = {Mean[lognewmex],
  Mean[logus], StandardDeviation[lognewmex],
  StandardDeviation[logus],
  Correlation[lognewmex, logus]}
```

$\{2.05947, 1.64157, 0.346194, 0.356417, 0.960863\}$

Notice the very high value $\rho = .96$ which indicates the strength of the relationship. By Theorem 1(a), since the logged New Mexico death rate X has approximately the $N(2.05947, .346194^2)$ distribution, we can find the probability $P[1.7 \le X \le 2.3]$ for example as

```
CDF[NormalDistribution[μ₁, σ₁], 2.3] -
  CDF[NormalDistribution[μ₁, σ₁], 1.7]
```

0.606852

Since X is the log of the actual death rate, say $X = \log(R)$, .606852 is also the probability that $R \in \left[e^{1.7}, e^{2.3}\right] \approx [5.47, 9.97]$.

We can also estimate the linear regression relationship between the logged variables. Let Y be the log of the U.S. death rate and write using Theorem 1(f) the conditional mean of Y given $X = x$ as:

```
Y[x_] := μ₂ + ρ σ₂
               ―――  (x - μ₁);
               σ₁
Y[x]
```

$$1.64157 + 0.989239 \, (-2.05947 + x)$$

Then the predicted value of the logged U.S. death rate in a year in which the logged New Mexico death rate was 2.2 is

```
Y[2.2]
```

1.78059

Notice how the occurrence of an x value somewhat above the mean of X led to a predicted Y value above the mean of Y.   This predicted Y is actually the mean of the conditional distribution of Y given $X = 2.2$. The variance of that distribution is

```
σ₂² (1 - ρ²)
```

0.00974876

Therefore, given $X = 2.2$ we have for instance that the conditional probability that Y is at least as large as μ_y is

```
1 - CDF[NormalDistribution[1.78059, √.00974876 ], μ₂]
```

0.92043

Knowledge of this X value has drastically altered the distribution of Y. ∎

Activity 8 How would you predict the actual (not logged) U.S. death rate in the last example given the actual New Mexico death rate?

Exercises 4.2

1. (*Mathematica*) Consider the set of pairs below, which are sale prices of homes (in dollars), and square footage of the home for single family homes up for sale in Pigeon Forge, Tennessee in 2008. The first cell is actually a dummy cell to give you an idea of the structure of the data set; the full data set is in the closed cell beneath it, and it consists of 115 such pairs. Investigate whether this data set seems to obey the assumptions of the bivariate normal model. Does a log transformation on one or both variables help?

```
homedata = {{769 000, 4768},
    {699 000, 6814}, {689 000, 1456}, {659 000, 2964},
    {649 900, 2925}, {599 900, 2568}, {599 000, 2812},
    {535 000, 3642}, ..., {229 900, 1044}};
```

2. (*Mathematica*) Two company stocks have weekly rates of return X and Y which are bivariate normal with means .001 and .003, variances .00002 each, and correlation $-.2$. Simulate 500 observations of the pair (X, Y) and make a scatterplot of your simulated data. Compute the probability that simultaneously the rate of return X exceeds .0015 and the rate of return Y is less than .002.

3. (*Mathematica*) The Bureau of Transportation Statistics reports the pairs below, which are percentages of late arrivals and departures of U.S. airline flights from 1995–2008. Check that the assumptions of the bivariate normal model are reasonable. Estimate the parameters of that model. Using the estimated parameters, compute (a) the probability that a year has no more than 20% late arrivals and no more that 18% late departures; (b) the probability, given that a year has 21% late arrivals that it has at least 19% late departures; (c) the predicted percentage of late departures in a year that has 21% late arrivals.

```
lateflights = {{19.65, 15.20}, {22.89, 18.49},
    {20.93, 16.47}, {21.80, 17.35}, {21.98, 17.77},
    {22.72, 18.72}, {20.23, 17.05}, {17.22, 14.06},
    {15.37, 12.20}, {20.26, 16.39}, {19.86, 17.12},
    {21.66, 19.39}, {24.61, 21.45}, {24.11, 20.73}};
{latearrivals, latedepartures} =
    Transpose[lateflights];
```

4. (*Mathematica*) The data below are final 2007–8 team statistics for the 30 NBA teams for the variables: field goal percentage, free throw percentage, assists per game, and points per game (all for home games only). Which of the individual variables seem to be about normally distributed? Which of the first three variables seems to predict points per game best? Assuming bivariate normal models, predict (a) the points per game for a team that shoots a 44% field goal percentage; (b) the points per game for a team that shoots a 75% free throw percentage; and (c) the points per game for a team that averages 22 assists per game.

```
fgpct = {0.459, 0.47, 0.5, 0.476, 0.497, 0.474,
    0.444, 0.464, 0.466, 0.454, 0.475, 0.464,
    0.468, 0.446, 0.454, 0.444, 0.458, 0.435,
    0.452, 0.449, 0.439, 0.448, 0.46, 0.439,
    0.443, 0.451, 0.448, 0.457, 0.438, 0.443};
ftpct = {0.752, 0.751, 0.783, 0.769, 0.759, 0.721,
    0.768, 0.798, 0.769, 0.723, 0.771, 0.814,
    0.812, 0.782, 0.772, 0.77, 0.767, 0.756,
    0.714, 0.733, 0.727, 0.726, 0.706, 0.717,
    0.736, 0.736, 0.767, 0.761, 0.781, 0.727};
assists = {22.35, 24.68, 26.68, 24.42, 26.4, 20.8,
    22.73, 19.11, 21.76, 19.19, 22.35, 20.97,
    23.81, 19.59, 22.00, 21.31, 22.32, 22.08,
    21.31, 21.46, 18.65, 21.42, 20.4, 20.00,
    23.47, 19.87, 21.13, 20.95, 21.12, 20.04};
ppg = {111.03, 110.65, 110.07, 108.58, 106.22,
    104.47, 103.98, 102.53, 100.85, 100.69, 100.54,
    100.41, 100.18, 98.82, 98.22, 97.54, 97.46,
    97.32, 97.06, 97.02, 96.92, 96.72, 96.62,
    96.37, 95.81, 95.59, 95.36, 95.36, 93.8, 91.35};
```

5. (*Mathematica*) The following data from the Bureau of Labor Statistics are incidence rates of work-related non-fatal injuries and illnesses per 100 workers per year for a sample of business and industry categories in 2005. The first element of the pair X is the rate of such incidents in which work time was lost and the second element Y is the rate of incidents that did not result in lost work time. Do you find a bivariate normal model to be reasonable? Estimate the functional form of the relationship of X to Y, and use it to predict the work loss rate X for an industry in which the rate of non-work loss incidents Y is 3.0. Compare your answer to the sample mean of X and comment.

```
injuries =
  {{2.8, 2.5} , {3.5, 3.1}, {3.3, 2.5}, {1.6, 3.1},
   {3.1, 3.9}, {7.7, 4.3}, {5.6, 3.5}, {4.8, 2.5},
   {3.4, 3.0}, {2.6, 1.6}, {3.5, 1.6}, {1.4, 1.4},
   {2.3, 1.4}, {2.2, 1.4}, {2.1, 1.5}, {2.5, 1.3},
   {2.1, 1.0}, {2.8, 2.5}, {3.1, 2.5}, {3.6, 3.2},
   {3.1, 3.6}, {3.6, 2.6}, {3.1, 2.2}, {4.7, 3.1},
   {4.4, 3.0}, {6.1, 3.2}, {6.1, 3.0}, {3.6, 3.1},
   {4.3, 2.0}, {3.6, 2.2}, {7.2, 3.0}, {2.3, 2.1},
   {2.8, 2.0}, {1.5, 1.8}, {3.8, 2.8}, {5.2, 4.2},
   {2.5, 1.9}, {2.3, 1.8}, {1.8, 1.2}, {1.8, 1.4},
   {2.2, 1.5}, {4.1, 3.0}, {4.9, 3.2}, {4.7, 4.4},
   {3.6, 2.8}, {3.8, 4.2}, {4.6, 4.8}, {3.8, 4.0},
   {3.0, 3.5}, {2.9, 4.0}, {2.6, 2.8}, {1.0, 1.0},
   {1.2, 1.0}, {2.6, 2.7}, {4.6, 3.7}, {2.4, 1.9},
   {4.0, 3.3}, {1.6, 1.4}, {2.2, 1.9}, {2.1, 2.7}};
{offday, nooffday} = Transpose[injuries];
```

6. (*Mathematica*) The pairs below are, respectively, the age-adjusted mortality rate and sulfur dioxide pollution potential in 60 cities (data from the U.S. Department of Labor Statistics). Do you find evidence that these variables can be modeled by a bivariate normal distribution? Try a log transformation and check again. Estimate the parameters for the log transformed data. Does it seem as if mortality is related to the amount of sulfur dioxide present? What is the conditional variance and standard deviation of log mortality given the log sulfur dioxide level?

```
pollution = {{922, 59}, {998, 39}, {962, 33}, {982, 24},
   {1071, 206}, {1030, 72}, {935, 62}, {900, 4},
   {1002, 37}, {912, 20}, {1018, 27}, {1025, 278},
   {970, 146}, {986, 64}, {959, 15}, {860, 1}, {936, 16},
   {872, 28}, {959, 124}, {942, 11}, {892, 1}, {871, 10},
   {971, 5}, {887, 10}, {953, 1}, {969, 33}, {920, 4},
   {844, 32}, {861, 130}, {989, 193}, {1006, 34},
   {861, 1}, {929, 125}, {858, 26}, {961, 78},
   {923, 8}, {1113, 1}, {995, 108}, {1015, 161},
   {991, 263}, {894, 44}, {939, 18}, {946, 89},
   {1026, 48}, {874, 18}, {954, 68}, {840, 20},
   {912, 86}, {791, 3}, {899, 20}, {904, 20}, {951, 25},
   {972, 25}, {912, 11}, {968, 102}, {824, 1},
   {1004, 42}, {896, 8}, {912, 49}, {954, 39}};
```

7. (*Mathematica*) Simulate 200 data pairs from the bivariate normal distribution with $\mu_x = \mu_y = 0$, $\sigma_x^2 = \sigma_y^2 = 1$, and each of $\rho = 0, .1, .3, .5, .7, .9$. For each value of ρ, produce individual histograms of the two variables, and a scatterplot of the pairs. Report on the key features of your graphs and how they compare for changing ρ.

8. (*Mathematica*) Consider again the vaccination coverage data of Example 2. (a) Find the approximate probability that a state with 90% coverage on polio has at least 90% coverage on MMR; (b) Find the approximate probability that a state with 95% coverage on MMR has at least 95% coverage on polio.

9. (*Mathematica*) As in Exercise 2, compute the conditional probability that the second stock has a rate of return of at least .005 given that the first earns a rate of return of .002.

10. Verify that the contours of the bivariate normal density are ellipses centered at (μ_x, μ_y), rotated at an angle α with the x-axis satisfying formula (3).

11. (*Mathematica*) In the statistics data of Example 1, estimate (a) the unconditional probability that a student scores at least 120 on the final; and (b) the conditional probability that a student scores at least 120 on the final given that he averages at least 80% on the quizzes.

12. Show that the parameter ρ in the bivariate normal density is indeed the correlation between X and Y. (Hint: simplify the computation by noting that

$$\rho = E\left[\frac{X - \mu_x}{\sigma_x} \frac{Y - \mu_y}{\sigma_y}\right]$$

and making the two-variable substitution $z_1 = (x - \mu_x)/\sigma_x$, $z_2 = (y - \mu_y)/\sigma_y$. Then complete the square in the exponent.)

13. What do you expect to see if you take cross-sections of the bivariate normal density surface by planes perpendicular to the x-axis? Are the resulting curves probability density functions? If not, what transformation can you apply to make them into density functions?

14. Suppose that Y and X satisfy a linear regression model given by $Y = \mu_y + \frac{\rho\sigma_y}{\sigma_x}(X - \mu_x) + \epsilon$. Compute the correlation between them.

15. *Mathematica* is capable of using its RandomReal command to simulate from the bivariate normal distribution with given $\mu_x, \mu_y, \sigma_x, \sigma_y$, and ρ. Explain how you could simulate such data yourself using the linear regression model in equation (9).

16. If X and Y are bivariate normal, write the integral for $P[X \le x, Y \le y]$, then change variables in the integral as in the hint for Exercise 12. What density function is now in the integrand? Use this to generalize an important result from Section 4.1.

17. A generalization of the bivariate normal distribution to three or more jointly distributed normal random variables is possible. Consider three random variables X_1, X_2, and X_3, and let

$$\mu = \begin{pmatrix} \mu_1 \\ \mu_2 \\ \mu_3 \end{pmatrix} \text{ and } \sum = \begin{pmatrix} \sigma_1^2 & \rho_{12}\,\sigma_1\,\sigma_2 & \rho_{13}\,\sigma_1\,\sigma_3 \\ \rho_{12}\,\sigma_1\,\sigma_2 & \sigma_2^2 & \rho_{23}\,\sigma_2\,\sigma_3 \\ \rho_{13}\,\sigma_1\,\sigma_3 & \rho_{23}\,\sigma_2\,\sigma_3 & \sigma_3^2 \end{pmatrix}$$

be, respectively, the column vector of means of the X's and the matrix of variances and paired covariances of the X's. Here, ρ_{ij} is the correlation between X_i and X_j. Consider the *trivariate normal density in matrix form*:

$$f(x) = \frac{1}{(2\pi)^{3/2}\sqrt{\det(\Sigma)}}\, e^{-1/2\,(x-\mu)^t\,\Sigma^{-1}(x-\mu)}$$

in which x is a column vector of variables x_1, x_2, and x_3, the notation "det" stands for determinant, Σ^{-1} is the inverse of the covariance matrix, and the notation y^t means the transpose of the vector y. Show that if all paired correlations are 0 then the three random variables X_1, X_2, and X_3 are mutually independent.

18. Referring to the formula in Exercise 17 for the multivariate normal density function $f(x)$ in matrix form, show that in the two-variable case $f(x)$ agrees with the formula for the bivariate normal density function in (1). (In place of $(2\pi)^{3/2}$ in the denominator, put $(2\pi)^{2/2}$ for the two-variable case. In general, it would be $(2\pi)^{n/2}$ for n variables.)

4.3 New Random Variables from Old

From the very beginning of this book we have set the stage for the subject of this section: new random variables created from others by a transformation. Recall the following examples, among others:

■ In Section 1.1 we looked at the empirical distribution of the minimum among 5 randomly sampled values from $\{1, 2, \ldots, 50\}$. The transformation is $Y = \min\{X_1, X_2, X_3, X_4, X_5\}$.

■ In Example 3 of Section 1.4 we simulated the distribution of the range of a sample of five integers from $\{1, 2, \ldots, 50\}$. The transformation is $Y = \max\{X_1, X_2, X_3, X_4, X_5\} - \min\{X_1, X_2, X_3, X_4, X_5\}$.

■ In Example 3 of Section 2.1 we simulated random numerical grades that had a target distribution. The grade G was a piecewise-defined function of a uniform random variable U.

■ In Example 4 of Section 3.2 we used simulation to support an intuitive argument that random variables U and V, which were the smaller and larger of two uniform random variables, had a uniform density over a triangular region. In this case we have a joint transformation taking the two uniform random variables onto two others, U and V.

■ In the proof of Theorem 2 of Section 4.1 we used cumulative distribution functions to show that if $X \sim N(\mu, \sigma^2)$, then the standardized $Z = (X - \mu)/\sigma$ has the $N(0, 1)$ distribution.

In this section we will build on your earlier exposure to develop some of the basic ideas and techniques of *transformation theory* in probability. The problem is to find the probability distribution of a random variable that is a function of other random variables. In the first subsection we look at the *cumulative distribution function technique*, which was first introduced in the proof of the normal standardization theorem. We will highlight its implications to simulating continuous random variables. Then in the second subsection we define *moment-generating functions* and explore their use in finding distributions of sums of independent random variables.

Activity 1 Look up and reread the bulleted references above. Try to find other examples from earlier in the book of transformed random variables.

C.D.F. Technique and Simulation

Consider a random variable X whose distribution is known, and a transformed random variable

$$Y = g(X) \tag{1}$$

The idea of the cumulative distribution function technique is to write down the c.d.f. $F_Y(y) = P[Y \le y]$ of Y, then substitute $Y = g(X)$ in, and express the probability $P[g(X) \le y]$ as a function of y by using the known distribution of X. We can do this in either the discrete or continuous case, but we will focus mostly on continuous random variables in this section. In the case that X and Y are continuous, once we have the c.d.f. of Y, differentiation produces the density function. We illustrate in the next example.

Example 1 Let X be a continuous random variable with the p.d.f. $f(x) = 2x$, $x \in [0, 1]$, $f(x) = 0$ otherwise. Find the probability density function of $Y = X^2$.

Since X takes on only non-negative values here, we have

$$F_Y(y) = P[Y \le y] = P[X^2 \le y] = P\left[0 \le X \le \sqrt{y}\right] \tag{2}$$

The probability on the far right is

$$P\left[0 \le X \le \sqrt{y}\right] = \int_{0}^{\sqrt{y}} 2x\,dx = x^2 \,\Big|_{0}^{\sqrt{y}} = y \tag{3}$$

and this computation is valid for all $y \in [0, 1]$. So, the c.d.f. of Y is $F_Y(y) = y$ for these y's, and clearly F_Y vanishes for $y < 0$ and $F_Y = 1$ for $y > 1$. (Why?) Differentiation of the c.d.f. of Y gives the density of Y as $f_Y(y) = 1$ for $y \in [0, 1]$. Our conclusion is that if X has the distribution given in the problem, then $Y = X^2$ has the uniform(0,1) distribution. ■

Equations (2) and (3) in the last example illustrate beautifully the heart of the c.d.f. method; make sure you understand how they are obtained and why we are looking at them. The next activity should serve as a good check.

Activity 2 This is a converse to the result in Example 1. Let $X \sim$ uniform(0,1) and consider the random variable $Y = \sqrt{X}$. Show that Y has the p.d.f. $f_Y(y) = 2y$, $y \in [0, 1]$, $f_Y(y) = 0$ otherwise.

Example 2 Let X and Y be independent uniform(0,1) random variables, and let U be the smaller and V the larger of X and Y. We will show that the pair (U, V) has the constant joint density function $f_{U,V}(u, v) = 2$, $0 \le u \le v \le 1$. (This fills in the analytical gap in Example 4 in Section 3.2.)

Because U is the smaller and V is the larger of X and Y, the state space is clearly the set of pairs (u, v) indicated in the statement: $0 \le u \le v \le 1$. Now just as it is true in the single variable case that a density $f(x)$ is the derivative $F'(x)$ of the c.d.f., it is also true in the two variable case that the joint density is the second order mixed partial of the joint c.d.f., that is:

$$f(x, y) = \frac{\partial^2 F}{\partial x \, \partial y} \tag{4}$$

You may see this by applying the Fundamental Theorem of Calculus twice to the double integral that defines the joint c.d.f. In the case at hand we have that the joint c.d.f. of U and V is

$$\begin{aligned}
F_{U,V}(u, v) &= \mathrm{P}[U \le u, V \le v] \\
&= \mathrm{P}[X \le u, Y \le v, X \le Y] + \mathrm{P}[X \le v, Y \le u, X > Y],
\end{aligned}$$

by separately considering the two cases where X is the smaller and where Y is the smaller. Now to find the total probability, we would integrate the joint density $f(x, y) = 1$ over the shaded region in Figure 19, but that is just the area of the shaded region. This area, by adding two rectangle areas, is

$$F_{U,V}(u, v) = uv + u(v - u) = 2uv - u^2$$

The partial derivative of this with respect to v is $2u$, and the second order mixed partial is

$$f_{U,V}(u, v) = \frac{\partial^2 F_{U,V}}{\partial u \, \partial v} = \frac{\partial}{\partial u}(2u) = 2$$

as desired. ∎

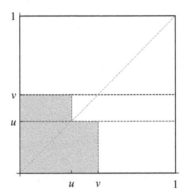

Figure 4.19 - Region of integration for $U \le u, V \le v$, when $u \le v$

Activity 3 In Example 2, though there is no probability density when $u > v$, the c.d.f. $P[U \le u, V \le v]$ is actually defined and non-zero even when $u > v$. Draw a graph similar to Figure 19 for this case, and find an expression for $P[U \le u, V \le v]$. Then take the second order mixed partial and see what happens.

The c.d.f. technique is particularly useful in the proof of the next important theorem on simulation.

Theorem 1. Let $U \sim$ uniform$(0,1)$, and let F be a continuous, strictly increasing c.d.f. Then $X = F^{-1}(U)$ is a random variable with the distribution associated with F.

Proof. Since F is strictly increasing and continuous, F^{-1} exists and is strictly increasing and continuous. The c.d.f. of $X = F^{-1}(U)$ is

$$P[X \le x] = P\big[F^{-1}(U) \le x\big] = P\big[F\big(F^{-1}(U)\big) \le F(x)\big] = P[U \le F(x)] = F(x)$$

since the c.d.f. of U is $F_U(u) = u$, $u \in [0, 1]$. Thus, X has the distribution function F. ∎

Theorem 1 completes the story of simulation of continuous random variables. By the random number generating methods discussed in Chapter 1, one can simulate a pseudo-random observation U from uniform$(0,1)$. To obtain a simulated X that has a given distribution characterized by F, just apply the transformation $X = F^{-1}(U)$.

Example 3 An important continuous distribution that we will look at later in this chapter is the *exponential*(λ) *distribution*, whose p.d.f. is

$$f(t) = \lambda e^{-\lambda t}, \ t \geq 0 \tag{5}$$

This turns out to be the distribution of the amount of time $T_n - T_{n-1}$ that elapses between successive arrivals of a Poisson process. The Poisson process itself can be simulated if we can simulate an exponential random variable S. By Theorem 1, $S = F^{-1}(U)$ has this distribution, where $U \sim$ unif (0, 1) and F is the c.d.f. associated with f. It remains only to find F^{-1}. First, the c.d.f. associated with the density in formula (5) is

$$F(t) = \int_0^t \lambda e^{-\lambda u} \, du = 1 - e^{-\lambda t}, \ t \geq 0 \tag{6}$$

Hence,

$$t = F\big(F^{-1}(t)\big) \Longrightarrow t = 1 - e^{-\lambda F^{-1}(t)} \Longrightarrow F^{-1}(t) = \frac{-1}{\lambda} \log(1 - t)$$

So, we can simulate an exponential random variable by

$$S = \frac{-1}{\lambda} \log(1 - U) \tag{7}$$

Below is a *Mathematica* command that uses what we have done to simulate a list of n exponential(λ) observations. Observe that the data histogram of the simulated values has the exponentially decreasing shape that one would expect, given the form of the density function. ∎

```
SimulateExp[n_, λ_] :=
    Table[- Log[1 - RandomReal[]] / λ, {n}]
```

```
Needs["KnoxProb7`Utilities`"]
```

```
SeedRandom[13 645]
```

```
datalist = SimulateExp[200, 0.5];
g1 = Histogram[datalist, 8,
    "ProbabilityDensity", ChartStyle → Gray];
g2 = Plot[0.5 e^-0.5 t, {t, 0, 10}, PlotStyle → Black];
Show[g1, g2, BaseStyle → 8]
```

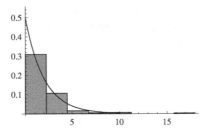

Figure 4.20 - Histogram of 200 simulated exp(1/2) observations with exp(1/2) density function

Example 4 We have used *Mathematica*'s RandomReal[dist] command and NormalDistribution[μ,σ] object before to simulate $N(\mu, \sigma^2)$ observations. We now know how to simulate normal observations ourselves using uniform(0,1) pseudo-random numbers. But we must compute the inverse c.d.f. numerically rather than in closed form. This is because the normal c.d.f. is

$$F(x) = \int_{-\infty}^{x} \frac{1}{\sqrt{2\pi\sigma^2}} e^{-(t-\mu)^2/2\sigma^2} \, dt$$

which is not expressible in closed form, nor is its inverse. But recall that *Mathematica* has the function

Quantile[distribution, p]

in the kernel, which returns the inverse c.d.f. evaluated at p for the given distribution. For example,

```
CDF[NormalDistribution[0, 2], 1.5]
Quantile[NormalDistribution[0, 2], .773373]
```

0.773373

1.5

These computations are telling us that if F is the $N(0, 2^2)$ c.d.f., then $F(1.5) = .773373$, and correspondingly $F^{-1}(.773373) = 1.5$. The following command simulates a list of n normal observations, using Theorem 1, i.e., by computing $F^{-1}(U)$ for the desired number of pseudo-random uniform random variables U. The dotplot of the observations in Figure 21 illustrates that the shape of the empirical distribution is appropriate. ∎

```
SimNormal[n_, μ_, σ_] := Table[Quantile[
    NormalDistribution[μ, σ], RandomReal[]], {n}]
```

```
datalist = SimNormal[200, 0, 2];
DotPlot[datalist]
```

Figure 4.21 - 200 simulated $N(0, 2^2)$ observations

Moment-Generating Functions

In many interesting applications of probability we need to find the distribution of the sum of independent random variables:

$$Y = X_1 + X_2 + \cdots + X_n$$

or the sample mean of independent, identically distributed random variables:

$$\overline{X} = \frac{X_1 + X_2 + \cdots + X_n}{n}$$

In Example 2 of Section 2.6 we simulated such sample means, and also observed by properties of expectation that $E[\overline{X}] = \mu$ and $\text{Var}(\overline{X}) = \sigma^2/n$. But can we draw conclusions analytically about the entire distribution of $\overline{X}$? Remember that at least for large sample sizes, it appeared from the simulation experiment that this distribution was approximately normal.

A new probabilistic device, defined below, is very useful for problems involving the distributions of sums of random variables.

Definition 1. The *moment-generating function* (m.g.f.) of a probability distribution is the function

$$M(t) = E\left[e^{tX}\right] \tag{8}$$

valid for all real numbers t such that the expectation is finite.

Notice that M is just a real-valued function of a real variable t. It would be calculated as

$$M(t) = \sum_{k \in E} e^{tk} q(k) \tag{9}$$

for a discrete distribution with p.m.f. q and state space E, and

$$M(t) = \int_E e^{tx} f(x)\, dx \tag{10}$$

for a continuous distribution with density function f.

There is a one-to-one correspondence between discrete p.m.f.'s and their m.g.f.'s that is most easily seen in a small example. Consider the discrete distribution with weights 1/3, 1/6, and 1/2 respectively on the points $x = 1, 2, 3$. Then the expected value of e^{tX} is just:

$$\frac{1}{3} e^{1 \cdot t} + \frac{1}{6} e^{2 \cdot t} + \frac{1}{2} e^{3 \cdot t}$$

Notice that the states can be read off as the coefficients of t in the exponents, and the probabilities attached to those states are the coefficients of the exponential terms in the sum.

Let us compute m.g.f.'s for two important distributions from earlier in the book: binomial and normal.

Example 5 Find the moment-generating function of the $b(n, p)$ distribution.

By formula (9) and the binomial theorem,

$$
\begin{aligned}
M(t) = E\left[e^{tX}\right] &= \sum_{k=0}^{n} e^{tk} \binom{n}{k} p^k (1-p)^{n-k} \\
&= \sum_{k=0}^{n} \binom{n}{k} (p\, e^t)^k (1-p)^{n-k} \\
&= (p\, e^t + (1-p))^n
\end{aligned}
\tag{11}
$$

The formula in (11) is valid for all $t \in \mathbb{R}$. ∎

Example 6 Find the moment-generating function of the $N(\mu, \sigma^2)$ distribution.

It suffices to find the moment-generating function $M_Z(u)$ of the standard normal distribution, because the general normal m.g.f. may be written as

$$
\begin{aligned}
M(t) = \mathrm{E}\left[e^{tX}\right] &= \mathrm{E}\left[e^{\sigma t(X-\mu)/\sigma}\, e^{\mu t}\right] \\
&= e^{\mu t}\, \mathrm{E}\left[e^{\sigma t(X-\mu)/\sigma}\right] \\
&= e^{\mu t}\, \mathrm{E}\left[e^{\sigma t Z}\right] \\
&= e^{\mu t}\, M_Z(\sigma t)
\end{aligned}
$$

The standard normal m.g.f. is

$$
\begin{aligned}
M_Z(u) = \mathrm{E}\left[e^{uZ}\right] &= \int_{-\infty}^{\infty} \frac{e^{uz}}{\sqrt{2\pi}}\, e^{-z^2/2}\, dz \\
&= \int_{-\infty}^{\infty} \frac{1}{\sqrt{2\pi}}\, e^{-(z^2 - 2uz + u^2)/2} \cdot e^{u^2/2}\, dz \\
&= e^{u^2/2} \int_{-\infty}^{\infty} \frac{1}{\sqrt{2\pi}}\, e^{-(z-u)^2/2}\, dz \\
&= e^{u^2/2}, \quad u \in \mathbb{R}
\end{aligned}
$$

The second line follows by completing the square on z in the exponent, and the fourth line is true since the last integral on the right side of the third line is that of a $N(u, 1)$ density, which equals 1. Therefore the general $N(\mu, \sigma^2)$ m.g.f. is

$$
M(t) = e^{\mu t}\, M_Z(\sigma t) = e^{\mu t}\, e^{(\sigma t)^2/2} = e^{\mu t + \sigma^2 t^2/2}, \quad t \in \mathbb{R} \ ∎ \tag{12}
$$

Activity 4 Derive the m.g.f. of the Bernoulli(p) distribution.

The phrase "moment-generating function" comes from the fact that

$$
M'(0) = \mathrm{E}[X], \quad M''(0) = \mathrm{E}\left[X^2\right], \tag{13}
$$

etc.; in general, the nth derivative of the m.g.f. evaluated at 0 is $M^{(n)}(0) = \mathrm{E}[X^n]$. So, the function $M(t)$ generates all of the moments of the distribution of X. You will prove this in Exercise 10.

As interesting as the previous fact might be, it takes second place to the following theorem, whose proof is beyond the level of this book: the m.g.f. is unique to the distribution, meaning that no two distributions share the same m.g.f. This implies that if we can recognize the m.g.f. of a transformed random variable Y by knowing the m.g.f.'s of those random variables that Y depends on, then we have the distribution of Y. This line of reasoning is known as the *moment-generating function technique*. We use it in two examples below to prove important results about the normal distribution.

Example 7 Let $X \sim N(\mu, \sigma^2)$. Use the moment-generating function technique to show that $Y = a X + b$ is normally distributed.

From formula (12), the m.g.f. of X is $M(t) = \exp\left(\mu t + \frac{1}{2}\sigma^2 t^2\right)$. Therefore, the m.g.f. of Y is

$$
\begin{aligned}
M_Y(t) = \mathrm{E}\left[e^{t Y}\right] &= \mathrm{E}\left[e^{t(a X + b)}\right] \\
&= e^{t b}\, \mathrm{E}\left[e^{(t a) X}\right] \\
&= e^{t b}\, M(t a) \\
&= e^{t b}\, e^{\mu t a + 1/2\, \sigma^2 (t a)^2} \\
&= e^{(a \mu + b) t + 1/2\left(\sigma^2 a^2\right) t^2}
\end{aligned}
$$

The last expression is the m.g.f. of the $N\left(a \mu + b, a^2 \sigma^2\right)$ distribution; hence by the uniqueness of m.g.f's, Y must have this distribution. ∎

Activity 5 Use the moment-generating function technique to show the normal standardization theorem: if $X \sim N(\mu, \sigma^2)$, then $Z = \frac{X - \mu}{\sigma} \sim N(0, 1)$.

So by Example 7, a linear transformation of a normal random variable results in another normal random variable. In fact, linear combinations of many independent normal random variables are also normal, as we show in the next example. (The hypothesis of independence can be dropped, but we will not prove that here.)

Example 8 Let $X_i \sim N(\mu_i, \sigma_i^2)$, $i = 1, 2, \ldots, n$, and suppose that the X_i's are mutually independent. Prove that the random variable

$$Y = a_1 X_1 + a_2 X_2 + \cdots + a_n X_n$$

is normal with mean and variance

$$\mu_Y = \sum_{i=1}^{n} a_i \mu_i, \quad \sigma_Y^2 = \sum_{i=1}^{n} a_i^2 \sigma_i^2$$

First recall that if $X_1, X_2, \ldots, X_n$ are mutually independent, then

$$\mathrm{E}[g_1(X_1)\, g_2(X_2) \cdot g_n(X_n)] = \mathrm{E}[g_1(X_1)]\, \mathrm{E}[g_2(X_2)] \cdots \mathrm{E}[g_n(X_n)] \tag{14}$$

Therefore, the m.g.f. of Y is

$$M_Y(t) = \mathrm{E}[e^{tY}] = \mathrm{E}[e^{t(a_1 X_1 + a_2 X_2 + \cdots + a_n X_n)}]$$
$$= \mathrm{E}[\textstyle\prod_{i=1}^{n} e^{t\,a_i\,X_i}]$$
$$= \textstyle\prod_{i=1}^{n} \mathrm{E}[e^{(t\,a_i\,X_i)}]$$
$$= \textstyle\prod_{i=1}^{n} e^{t\,a_i\,\mu_i + \frac{1}{2}\sigma_i^2(a_i\,t)^2}$$
$$= e^{t\,\Sigma a_i\,\mu_i + \frac{1}{2}\left(\Sigma \sigma_i^2\,a_i^2\right)t^2}$$

The latter is the m.g.f. of $N\!\left(\Sigma a_i\,\mu_i,\ \Sigma \sigma_i^2\,a_i^2\right)$, which completes the proof. ∎

A special case of the last example deserves our attention. Let $X_1, X_2, \ldots, X_n$ be a random sample from the $N(\mu, \sigma^2)$ distribution. In particular, the X_i's are independent. Consider the sample mean $\overline{X}$. It can be rewritten as

$$\overline{X} = \frac{\Sigma X_i}{n} = \frac{1}{n} X_1 + \frac{1}{n} X_2 + \cdots + \frac{1}{n} X_n$$

In Example 8, let $a_i = 1/n$, $\mu_i = \mu$, and $\sigma_i^2 = \sigma^2$ for all $i = 1, 2, \ldots, n$. Then the conclusion of Example 8 implies that

$$\overline{X} \sim N\!\left(\sum_{i=1}^{n} \frac{1}{n}\mu,\ \sum_{i=1}^{n} \frac{1}{n^2}\sigma^2\right) = N\!\left(\mu,\ \frac{\sigma^2}{n}\right) \tag{15}$$

This is one of the most important results in probability and statistics. The sample mean of a normal random sample is a random variable with a normal distribution also, whose mean is the same as the mean μ of the distribution being sampled from, and whose variance σ^2/n is $1/n$ times the underlying variance. We have seen these results about the mean and variance before, but we have just now proved the normality of $\overline{X}$. Let us close this section with an illustration of the power of this observation.

Example 9 A balance is suspected of being in need of calibration. Several weight measurements of a known 1 gram standard yield the data below. Give a probability estimate of the likelihood that the balance needs recalibration. (Assume that the standard deviation of measurements is $\sigma = .0009$ grams.)

```
datalist =
  {1.0010, .9989, 1.0024, 1.0008, .9992, 1.0015,
   1.0020, 1.0004, 1.0018, 1.0005, 1.0013, 1.0002};
```

Apparently a correctly calibrated balance should yield observations with a mean of 1 gram on the standard weight. There are 12 observations, which we will assume constitute a random sample from a normal distribution with $\sigma = .0009$. It is helpful to draw a legal analogy in order to respond to the question. Assume that the balance is innocent until proven guilty beyond a reasonable doubt. If the 12 observations together are unlikely to have occurred if the balance was correctly calibrated, then we convict the balance of the crime of miscalibration. We can summarize the information contained in the raw data by the sample mean $\overline{X}$; then if the particular $\overline{x}$, a possible value of the random variable $\overline{X}$, that we observe is a very unlikely one when $\mu = 1$, we would conclude that μ is most probably not 1.

First, we let *Mathematica* give us the observed value $\overline{x}$ of the sample mean.

```
Mean[datalist]
```

1.00083

So the observed sample mean is a bit bigger than 1. But is it big enough to constitute "beyond reasonable doubt?" As a random variable, under the "assumed innocence" of the balance, $\overline{X}$ would have the $N\left(1, (.0009)^2/12\right)$ distribution by formula (15). Now the likelihood that $\overline{X}$ comes out to be identically equal to 1.00083 is 0 of course, since $\overline{X}$ is continuously distributed. But that is not a fair use of the evidence. It is more sensible to ask how large is the probability that $\overline{X} \geq 1.00083$ if the true $\mu = 1$. If that is very small then we have a lot of probabilistic evidence against $\mu = 1$. We can compute:

```
1 - CDF[NormalDistribution[1, .0009/√12], 1.00083]
```

0.00069995

We now see that if $\mu = 1$ were true, it would only be about .0007 likely for us to have observed a sample mean as large or larger than we did. This is very strong evidence that the true μ is greater than 1, and the balance should be recalibrated. We might say that in deciding this, our risk of an erroneous decision is only about .0007.

It would be a good idea for you to reread this example a couple of times, paying particular attention to the use of statistical evidence to draw a conclusion. It illustrates the kind of thinking that you will see frequently if you take a course in statistics. ∎

Activity 6 How sensitive is the decision we made in Example 9 to the largest observation 1.0024? Delete that and rerun the problem. Do we still prove the balance guilty beyond reasonable doubt? What is the new error probability?

Exercises 4.3

1. Use the c.d.f. technique to find the density function of (a) $Y = X^2$; (b) $Y = \sqrt[3]{X}$ if X is a continuous random variable with density function $f(x) = 3x^2$, $x \in [0, 1]$.

2. Use the c.d.f. technique to find the density function of $U = Z^2$ if $Z \sim N(0, 1)$. This density is called the *chi-square density*, and will be of interest to us in a later section.

3. Explain, with justification, how to simulate a uniform(a, b) observation using a uniform$(0, 1)$ observation.

4. Prove this converse of the simulation theorem. If F is the strictly increasing, continuous c.d.f of the random variable X, then the random variable $U = F(X)$ has the uniform$(0,1)$ distribution.

5. (*Mathematica*) Write a *Mathematica* command to simulate a list of n observations from the distribution whose density function is $f(x) = 3\,x^2$, $x \in [0, 1]$. Run the command several times, and compare histograms of data lists to the density function.

6. If a random temperature observation in Fahrenheit is normally distributed with mean 70 and standard deviation 5, find the probability that the temperature recorded in degrees Celsius exceeds 20.

7. Derive the moment-generating function of the uniform(a, b) distribution.

8. Derive the moment-generating function of the geometric(p) distribution.

9. A residence has three smoke alarms. The owners test them every three months to be sure that they are functioning; let random variables X_1, X_2, and X_3 be 1 if alarm 1, 2, or 3 respectively is still functioning and 0 if not. Assume that the alarms operate independently and each functions with probability 2/3. Find the moment-generating function of the product $X_1 X_2 X_3$. What does this product random variable represent, and what p.m.f. corresponds to the m.g.f. of the product?

10. Prove that if X is a random variable with moment-generating function $M(t)$, then the nth derivative of the m.g.f. at 0 is the nth moment, i.e., $M^{(n)}(0) = E[X^n]$.

11. Derive the moment-generating function of the Bernoulli(p) distribution (see Activity 4), and use it to show that the sum of n i.i.d. Bernoulli(p) random variables has the binomial(n, p) distribution.

12. Derive the moment-generating function of the Poisson(μ) distribution, and use it to show that if $X_1 \sim \text{Poisson}(\mu_1)$, $X_2 \sim \text{Poisson}(\mu_2)$, ..., $X_n \sim \text{Poisson}(\mu_n)$ are independent random variables, then $X_1 + X_2 + \cdots + X_n$ has a Poisson distribution with parameter $\mu_1 + \mu_2 + \cdots + \mu_n$.

13. Recall from Section 4.1 the example data set on National League batting averages. Accepting as correct the estimates of the mean batting average (.249) and the standard deviation (.0582), what is the probability that a sample of 20 such batting averages will have a sample mean of less than .250? Greater than .280?

14. As in Exercise 4.1.14, suppose that the times people spend being served at a post office window are normal with mean 1.5 minutes and variance .16 minutes. What are the probabilities that

(a) the average service time among ten people requires more than 1.7 minutes;

(b) the total service time among ten people requires less than 16 minutes?

Sample Problems from Actuarial Exam P

15. An investment account earns an annual interest rate R that follows a uniform distribution on the interval (0.04, 0.08). The value of a 10,000 initial investment in this account after one year is given by $V = 10,000\, e^R$. Determine the cumulative distribution function $F(v)$ of V for values of v that satisfy $0 < F(v) < 1$.

16. Let T denote the time in minutes for a customer service representative to respond to 10 telephone inquiries. T is uniformly distributed on the interval with endpoints 8 minutes and 12 minutes. Let R denote the average rate, in customers per minute, at which the representative responds to inquiries. What is the density function of the random variable R on the interval $(10/12, 10/8)$?

17. A company insures homes in three cities, J, K, and L . Since sufficient distance separates the cities, it is reasonable to assume that the losses occurring in these cities are independent. The moment-generating functions for the loss distributions of the cities are:

$$M_J(t) = (1 - 2\,t)^{-3}; \quad M_K(t) = (1 - 2\,t)^{-2.5}; \quad M_L(t) = (1 - 2\,t)^{-4.5}$$

Let X represent the combined losses from the three cities. Calculate $E[X^3]$.

4.4 Order Statistics

Earlier in the text we have seen examples in which a random sample X_1, X_2, ..., X_n was taken from a population, and we wanted to gain information about the population using a summary measure or *statistic*, which is a function of the sample values. The sample mean $\overline{X} = \sum_{i=1}^{n} X_i/n$ and the sample variance $S^2 = \sum_{i=1}^{n}(X_i - \overline{X})^2/(n-1)$ are examples of statistics. But in this section we would like to consider statistics of a different sort, called *order statistics*, and other statistics that are based on them. In short, the order statistics of a random sample are just the sample values arranged in increasing order. The simplest examples of order statistics are the minimum value:

$$X_{(1)} = \min\{X_i;\ i = 1, 2, \ldots, n\} \tag{1}$$

and the maximum value:

$$X_{(n)} = \max\{X_i;\ i = 1, 2, \ldots, n\} \tag{2}$$

of the sample. In general:

Definition 1. The ith order statistic $X_{(i)}$ of a random sample $X_1, X_2, \ldots, X_n$ from a population that is characterized by a p.d.f. (or p.m.f.) $f(x)$ is the ith smallest member of the sample.

The smallest and largest order statistics $X_{(1)}$ and $X_{(n)}$ estimate the smallest and largest possible states of the random variable X whose density function (or mass function) is $f(x)$. The *sample range*, which is the difference:

$$X_{(n)} - X_{(1)} \tag{3}$$

estimates the maximum spread of values of X. One of the most important statistics based on the order statistics is the *sample median M* of the random sample, defined by

$$M = \begin{cases} X_{\left(\frac{n+1}{2}\right)} & \text{if } n \text{ is odd} \\ \frac{X_{(n/2)} + X_{(n/2+1)}}{2} & \text{if } n \text{ is even} \end{cases} \tag{4}$$

i.e., the middle value of the ordered sample in the odd case, and the average of the two middle values in the even case. The sample median estimates the median of the probability distribution of X, which is the point to the left of which is half of the total probability. So order statistics and functions of them play an important role in statistical estimation, and we need to know about their probability distributions.

Activity 1 If the observed values of a random sample are 23, 4, 16, 7, and 8, what are the sample median and sample range? If an additional sample value of 9 is included, what are the new median and range?

Our goal in this section is to derive formulas for the joint density, single variable marginal densities, and two-variable joint marginal densities of order statistics. We can then use these to make inferences about populations based on observed values of the order statistics; for instance, we can answer questions such as: if a population has the characteristics that it is claimed to have, how likely is it that the order statistics would come out as they did? If the answer is: not very likely, then the underlying assumptions about the population are probably wrong. We will see other applications of order statistics in examples and exercises, especially involving the minimum and maximum.

To avoid the sticky problem of sample values possibly being equal, meaning that there is not a unique way of mapping sample values $X_1, X_2, \ldots, X_n$ to order statistics

$X_{(1)}$, $X_{(2)}$, ..., $X_{(n)}$, we make the simplifying assumption that the sample is taken from a continuous distribution. Then with probability 1, strict inequality prevails:

$$X_{(1)} < X_{(2)} < ... < X_{(n)} \tag{5}$$

We will not give a rigorous derivation of the joint distribution of order statistics, which requires knowledge of how joint substitutions in multiple integrals work for non-invertible transformations. An intuitive argument in the next paragraph actually sheds more light on why the formula for which we will argue is true. To set it up, recall that the meaning of a probability density is that the approximate probability that the random variable takes a value in the infinitesimally small interval $[x, x + d\,x]$ around the point x is $f(x)\,d\,x$, where f is the density of the random variable.

Consider the case of a random sample of size 3, X_1, X_2, X_3 from a continuous distribution with density function $f(x)$. The joint density of the three sample variables is

$$f(x_1, x_2, x_3) = f(x_1)\,f(x_2)\,f(x_3)$$

Now except for a set of outcomes of probability 0 (where X values may be tied), the sample space Ω can be partitioned into $3! = 6$ disjoint events:

$$A_1 = \{\omega \in \Omega : X_1(\omega) < X_2(\omega) < X_3(\omega)\}$$
$$A_2 = \{\omega \in \Omega : X_1(\omega) < X_3(\omega) < X_2(\omega)\}$$
$$A_3 = \{\omega \in \Omega : X_2(\omega) < X_1(\omega) < X_3(\omega)\}$$
$$A_4 = \{\omega \in \Omega : X_2(\omega) < X_3(\omega) < X_1(\omega)\}$$
$$A_5 = \{\omega \in \Omega : X_3(\omega) < X_1(\omega) < X_2(\omega)\}$$
$$A_6 = \{\omega \in \Omega : X_3(\omega) < X_2(\omega) < X_1(\omega)\}$$

On each individual event A_i, the values of the order statistics relate directly to the sample values; for instance for outcomes ω in A_3, the values of the order statistics are $X_{(1)}(\omega) = X_2(\omega)$, $X_{(2)}(\omega) = X_1(\omega)$, $X_{(3)}(\omega) = X_3(\omega)$. Consequently, if we try to write an expression for the joint probability that the three order statistics take values in infinitesimally small intervals $[y_i, y_i + d\,y_i]$, where $y_1 < y_2 < y_3$, we have for outcomes in A_3:

$$P[X_{(1)} \in [y_1, y_1 + d\,y_1], X_{(2)} \in [y_2, y_2 + d\,y_2],$$
$$X_{(3)} \in [y_3, y_3 + d\,y_3] \cap A_3]$$
$$= P[X_2 \in [y_1, y_1 + d\,y_1], X_1 \in [y_2, y_2 + d\,y_2], X_3 \in [y_3, y_3 + d\,y_3]] \tag{6}$$
$$\approx f(y_1)\,d\,y_1 \cdot f(y_2)\,d\,y_2 \cdot f(y_3)\,d\,y_3$$
$$= f(y_1)\,f(y_2)\,f(y_3)\,d\,y_1\,d\,y_2\,d\,y_3$$

A derivation similar to formula (6) can be done for each of the 6 disjoint events A_i, and the right side would be the same in each case (try at least one other case yourself). Thus the total probability that the order statistics jointly take values in the respective intervals $[y_i, y_i + d\,y_i]$ is:

$$P[X_{(1)} \in [y_1, y_1 + d\, y_1], X_{(2)} \in [y_2, y_2 + d\, y_2], X_{(3)} \in [y_3, y_3 + d\, y_3]]$$
$$= 6 \cdot f(y_1)\, f(y_2)\, f(y_3)\, d\, y_1\, d\, y_2\, d\, y_3; \quad y_1 < y_2 < y_3 \tag{7}$$

This suggests that the joint density function of $X_{(1)}$, $X_{(2)}$, and $X_{(3)}$ is $6 \cdot f(y_1)\, f(y_2)\, f(y_3)$, non-zero on the set of states for which $y_1 < y_2 < y_3$. Generalizing to a random sample of size n motivates the following theorem.

Theorem 1. The joint density of the order statistics of a random sample of size n from a continuous distribution with p.d.f. $f(x)$ is

$$f(y_1, y_2, \ldots, y_n) = n!\, f(y_1)\, f(y_2) \cdots f(y_n); \quad y_1 < y_2 < \cdots < y_n \quad \blacksquare \tag{8}$$

Example 1 Suppose that X_1, X_2, X_3, X_4 is a random sample of size 4 from a distribution with density $f(x) = 2\,x$; $x \in [0, 1]$. Find the probability that simultaneously the smallest order statistic is less than .1 and the largest exceeds .9.

      By Theorem 1 the joint density of the four order statistics is $f(y_1, y_2, y_3, y_4) = 4! \cdot 2\, y_1 \cdot 2\, y_2 \cdot 2\, y_3 \cdot 2\, y_4$ for values of the y_i's such that $0 < y_1 < y_2 < y_3 < y_4 < 1$. Since no extra restrictions have been placed on y_2 and y_3 we integrate them out first and then integrate with respect to the other variables:

$$P[X_{(1)} < .1, X_{(4)} > .9] = 4! \cdot \int_0^{.1} \int_{.9}^1 \int_{y_1}^{y_4} \int_{y_2}^{y_4} 2\, y_1 \cdot 2\, y_2 \cdot 2\, y_3 \cdot 2\, y_4\, d\, y_3\, d\, y_2\, d\, y_4\, d\, y_1$$

$$= 4! \cdot \int_0^{.1} 2\, y_1 \left(\int_{.9}^1 2\, y_4 \left(\int_{y_1}^{y_4} 2\, y_2 \left(\int_{y_2}^{y_4} 2\, y_3\, d\, y_3 \right) d\, y_2 \right) d\, y_4 \right) d\, y_1$$

$$= 4! \cdot \int_0^{.1} 2\, y_1 \left(\int_{.9}^1 2\, y_4 \left(\int_{y_1}^{y_4} 2\, y_2 \left(y_4{}^2 - y_2{}^2 \right) d\, y_2 \right) d\, y_4 \right) d\, y_1$$

The rest of the integration is rather tedious, so we will give it to *Mathematica*:

```
24 * ∫₀^.1 2 y1 ( ∫.9^1 2 y4 ( ∫y1^y4 2 y2 ( ∫y2^y4 2 y3 dy3 ) dy2 ) dy4 ) dy1
```

0.0185368

The probability that the smallest order statistic is less than .1 and the largest exceeds .9 is about .0185. ∎

Activity 2 In the example above, use the c.d.f. technique to find $P[X_{(1)} < .1]$ and $P[X_{(4)} > .9]$ separately. Do you expect the product of these to equal $P[X_{(1)} < .1, X_{(4)} > .9]$?

Next we show the formulas for the marginal densities of single order statistics and pairs of order statistics. We could derive these formally by setting up multiple integrals of the joint density with respect to all of the other order statistics, but again it is instructive to take the more intuitive approach to motivating the formulas.

Theorem 2. (a) The marginal density of the ith order statistic in a random sample of size n from a continuous distribution with p.d.f. $f(x)$ and c.d.f. $F(x)$ on state space (a, b) is:

$$f_i(y_i) = \frac{n!}{(i-1)!\,(n-i)!} (F(y_i))^{i-1} (1 - F(y_i))^{n-i} f(y_i); \quad a < y_i < b \tag{9}$$

(b) The joint marginal density of the jth and k^{th} order statistic $(j < k)$ in a random sample of size n from a continuous distribution with p.d.f. $f(x)$ and c.d.f. $F(x)$ on state space (a, b) is:

$$f_{jk}(y_j, y_k) =$$
$$\frac{n!}{(j-1)!\,(k-j-1)!\,(n-k)!} (F(y_j))^{j-1} (F(y_k) - F(y_j))^{k-j-1} \tag{10}$$
$$\cdot (1 - F(y_k))^{n-k} f(y_j) f(y_k); \quad a < y_j < y_k < b$$

Proof. (a) (Informal) In order for the ith smallest order statistic $X_{(i)}$ to be in the infinitesimal interval $[y_i, y_i + d\,y_i]$, one of the sample values X_j must be in that interval, exactly $i - 1$ of them must be less than y_i, and the remaining $n - i$ of them must exceed $y_i + d\,y_i$. A multinomial experiment is generated with n trials corresponding to the sample values, with three success categories (sample value $< y_i$), (sample value in $[y_i, y_i + d\,y_i]$), and (sample value $> y_i + d\,y_i$). The success probabilities for the categories are approximately $F(y_i)$, $f(y_i)\,d\,y_i$, and $1 - F(y_i)$, respectively. The appropriate multinomial probability of $i - 1$ type 1 successes, 1 type 2 success, and $n - i$ type 3 successes is therefore:

$$P[X_{(i)} \in [y_i, y_i + d\,y_i]] \approx \binom{n}{i-1,\,1,\,n-i} (F(y_i))^{i-1} (f(y_i)\,d\,y_i)^1 (1 - F(y_i))^{n-i}$$

On slight rearrangement, this equation implies the density in formula (9) for the random variable $X_{(i)}$.

(b) Do this argument yourself in the next Activity. ∎

Activity 3 Intuitively justify the form of the joint density of $X_{(j)}$ and $X_{(k)}$ in formula (10). (Hint: In order for both $X_{(j)} \in [y_j, y_j + d\,y_j]$ and $X_{(k)} \in [y_k, y_k + d\,y_k]$ to occur, how many sample values must fall into what intervals, and with what probabilities?)

Example 2 Let X_1, X_2, X_3, X_4, X_5 be a random sample from the continuous uniform distribution on $(0, 1)$. Find the probability distribution of the median $X_{(3)}$. Use this result to compute $P[X_{(3)} < 1/2]$ and $E[X_{(3)}]$. Write a *Mathematica* program to simulate values of the sample median and superimpose a scaled histogram of simulated medians on a graph of the p.d.f. of $X_{(3)}$. Lastly, if the observed values of the random sample were .45, .76, .82, .92, and .68, does the value of the sample median give you cause to seriously doubt the hypothesis that the sample was taken from the uniform(0,1) distribution?

At the heart of all of the questions posed in this example is the p.d.f. of $X_{(3)}$. Since we have $n = 5$, $i = 3$, $f(x) = 1$, and $F(x) = x$ for $x \in (0, 1)$, formula (9) of Theorem 1 yields:

$$f(y_3) = \frac{5!}{2! \cdot 2!} y_3^2 (1 - y_3)^2 \cdot 1 = 30 \left(y_3^2 - 2 y_3^3 + y_3^4 \right); \quad y_3 \in (0, 1)$$

Notice that 1/2 is the median of the distribution of each sample value, so it is interesting to know if the distribution of the sample median $X_{(3)}$ puts half of its weight to the left of the distributional median $m = 1/2$ and half to the right. We compute:

```
(* P{X(3) <1/2] = *)
⌠.5
⌡0  30 (y3² - 2 y3³ + y3⁴) dy3
```

0.5

So we do observe this intuitively reasonable phenomenon.

The expected value of $X_{(3)}$ is:

```
(* E{X(3)] = *)
⌠1
⌡0  30 * y3 (y3² - 2 y3³ + y3⁴) dy3
```

$\dfrac{1}{2}$

Therefore, considered as an estimator of the distributional median $m = .5$, the sample median has the desirable property that its expected value is m.

The program below simulates a desired number of sample medians of a desired sample size n from a given population distribution. The sample itself is taken and sorted, then the median of that sample is computed and added to a list. In computing the sample median, the Mod function is used to tell the difference between even and odd sample size, and the appropriate case in formula (4) is employed.

```
SimulateMedians[nummeds_, n_, dist_] := Module[
  {medlist, medindex, sample, sortedsample, med},
  medlist = {};
  If[Mod[n, 2] == 1 (* odd case *),
   medindex = (n + 1) / 2;
   Do[sample = RandomReal[dist, n];
    sortedsample = Sort[sample];
    med = sortedsample[[medindex]];
    AppendTo[medlist, med], {nummeds}],
   (* even case *)
   medindex = n / 2;
   Do[sample = RandomReal[dist, n];
    sortedsample = Sort[sample];
    med = (sortedsample[[medindex]] +
        sortedsample[[medindex + 1]]) / 2;
    AppendTo[medlist, med], {nummeds}]];
  medlist]
```

Now we apply the function to our question, simulating 200 random samples, each of size 5. The histogram of sample medians and the plot of the density function of the sample median follow.

```
Needs["KnoxProb7`Utilities`"]
SeedRandom[451 623]
```

```
medianlist = SimulateMedians[
    200, 5, UniformDistribution[{0, 1}]];
g1 = Histogram[medianlist, 8,
    "ProbabilityDensity", ChartStyle → Gray];
g2 = Plot[30 (y^2 - 2 y^3 + y^4), {y, 0, 1},
    PlotStyle → Black];
Show[{g1, g2}, BaseStyle → 8]
```

Figure 4.22 - Histogram of 200 simulated medians of samples of size 5 from unif(0,1)

Visually it does appear that the distribution of the sample median is symmetric about the distributional median of 1/2 in this case, and the histogram of the simulated medians follows the density function fairly well. (Try to check the symmetry mathematically.)

Finally, if the observed sample values are as given in the problem, then the sample median is the middle value of the sorted sample, which is .76. We ask the question: if the uniform(0,1) distribution is the correct one, what is the probability that the sample median is .76 or more? This probability is computed in the integral below.

```
(* P{X(3) >.76]  =  *)
⌠¹  30 (y₃² - 2 y₃³ + y₃⁴) dy₃
⌡.76
```

$$ (*\ P\{X_{(3)} > .76] \ = \ *) $$
$$ \int_{.76}^{1} 30 \left(y_3{}^2 - 2\, y_3{}^3 + y_3{}^4 \right)\, dy_3 $$

0.0932512

So it is only about 9.3% likely that a sample of size 5 would produce such a large median. This is suspicious, but it does not provide conclusive evidence against the hypothesis that the underlying distribution is uniform(0,1). ■

Example 3 Using the same assumptions as Example 2, find the p.d.f. of the sample range $X_{(5)} - X_{(1)}$ and find $P[X_{(5)} - X_{(1)} < .5]$.

By formula (10) of Theorem 2, the joint p.d.f. of $X_{(1)}$ and $X_{(5)}$ is

$$ f_{15}(y_1,\, y_5) \ = \ \frac{5!}{0!\cdot 3!\cdot 0!}\, y_1{}^0 (y_5 - y_1)^3 (1 - y_5)^0 \cdot 1 \cdot 1 $$
$$ = \ 20\, (y_5 - y_1)^3, \ 0 < y_1 < y_5 < 1 $$

Then if we denote the sample range by $Y = X_{(5)} - X_{(1)}$, the graph of the region of $(y_1,\, y_5)$ pairs satisfying $Y \le y$ in Figure 23 indicates that the c.d.f. of Y can be computed as:

$$F(y) = P[Y \le y] = P[X_{(5)} - X_{(1)} \le y]$$

$$= \int_0^y \int_0^{y_5} f_{15}(y_1, y_5)\, d\,y_1\, d\,y_5 + \int_y^1 \int_{y_5-y}^{y_5} f_{15}(y_1, y_5)\, d\,y_1\, d\,y_5$$

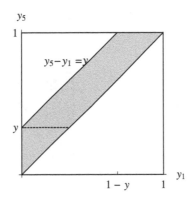

Figure 4.23 - Region of integration for $Y = X_{(5)} - X_{(1)} \le y$

Using our formula for f_{15}, the c.d.f. of the range is:

```
(*  F(y)  =  *)
  ⌠y ⎛⌠y5                   ⎞
  ⎮  ⎜⎮   20 * (y5 - y1)³ dy1⎟ dy5 +
  ⌡0 ⎝⌡0                    ⎠
  ⌠1 ⎛⌠y5                   ⎞
  ⎮  ⎜⎮    20 * (y5 - y1)³ dy1⎟ dy5
  ⌡y ⎝⌡y5-y                 ⎠
```

$$5\,(1 - y)\, y^4 + y^5$$

Simplifying a bit, the c.d.f. of Y is $F(y) = 5\,y^4 - 4\,y^5$, and the p.d.f. of the range is its derivative, $f(y) = 20\,y^3 - 20\,y^4$, $y \in (0, 1)$. Substituting directly into the c.d.f. formula, $P[X_{(5)} - X_{(1)} < .5] = F(.5) = 5\left(.5^4\right) - 4\left(.5^5\right) = .1875.$ ∎

Example 4 Find the probability distribution of the sample median M of a sample of size 4 from the distribution with p.d.f. $f(x) = 3\,x^2$; $x \in [0, 1]$. Find also the expected value of M.

Since we are in the case of even sample size, $M = \frac{1}{2}\,(X_{(2)} + X_{(3)})$. Integrating f from 0 to x gives easily that the underlying c.d.f. is $F(x) = x^3$; $x \in [0, 1]$. The joint density of $X_{(2)}$ and $X_{(3)}$ is, by formula (10),

$$f_{23}(y_2, y_3) = \frac{4!}{1! \cdot 0! \cdot 1!}\left(y_2{}^3\right)^1 \cdot \left(y_3{}^3 - y_2{}^3\right)^0 \cdot \left(1 - y_3{}^3\right) \cdot 3\,y_2{}^2 \cdot 3\,y_3{}^2$$

$$= 216 \cdot y_2{}^5 \cdot \left(y_3{}^2 - y_3{}^5\right);\ 0 < y_2 < y_3 < 1$$

The c.d.f. of the sample median can be written:

$$G(m) = P[M \leq m] \;=\; P\left[\tfrac{1}{2}(X_{(2)} + X_{(3)}) \leq m\right]$$
$$= P[X_{(2)} + X_{(3)} \leq 2m] \tag{11}$$

(a) case $m < 1/2$;   (b) case $m \geq 1/2$

Figure 4.24 - Regions of integration for $M \leq m$

The region of integration for the last probability in formula (11) has a different shape depending on whether or not $2m < 1$, that is, $m < 1/2$. Part (a) of Figure 24 shows the case where $m < 1/2$, in which the boundary line has slope -1 and intercept $2m < 1$. In that case we can just integrate out y_3 first between the two slanted boundaries, and let y_2 range from 0 to m. The c.d.f. of M in this case is:

```
(* Case m < 1/2 *)
G1[m_] := Integrate[ (Integrate from y2 to 2m-y2 of 216 * y2^5 * (y3^2 - y3^5) dy3) dy2;
G1[m]
```

$$\frac{74\, m^9}{7} - \frac{1024\, m^{12}}{77}$$

And for $m < 1/2$ the density is the derivative of the above:

```
gdensity1[m_] := D[G1[m], m];
gdensity1[m]
```

$$\frac{666\, m^8}{7} - \frac{12\,288\, m^{11}}{77}$$

In the case where $m \geq 1/2$ illustrated by part (b) of Figure 24, the region of integration is the more complicated polygonal region that is shaded. But it is easier to complement and integrate over the small triangle in the upper right corner as follows:

```
(* Case m ≥ 1/2 *)
G2[m_] := 1 - ∫ₘ¹ (∫₂ₘ₋ᵧ₃^y₃ 216 * y₂⁵ * (y₃² - y₃⁵) dy₂) dy₃;
G2[m]
```

$$1 - \frac{162\,m}{11} + \frac{648\,m^2}{7} - 320\,m^3 + 648\,m^4 -$$
$$\frac{5184\,m^5}{7} + 384\,m^6 - \frac{438\,m^9}{7} + \frac{1024\,m^{12}}{77}$$

Thus for $m \geq 1/2$, the density is:

```
gdensity2[m_] := D[G2[m], m]
gdensity2[m]
```

$$-\frac{162}{11} + \frac{1296\,m}{7} - 960\,m^2 + 2592\,m^3 -$$
$$\frac{25\,920\,m^4}{7} + 2304\,m^5 - \frac{3942\,m^8}{7} + \frac{12\,288\,m^{11}}{77}$$

To compute the expected value of M we must integrate m times the density separately over the two regions and add. The command below shows that $E[M] = .771429$. ∎

```
(* E[M] = *)
∫₀^.5 m * gdensity1[m] dm + ∫.₅¹ m * gdensity2[m] dm
```

0.771429

Activity 4 Compute the median of the underlying distribution of X in Example 4 to see whether $E[M]$ agrees with it.

Example 5 The computer system controlling the electronics of an automobile is divided into 3 control modules. Each fails at a time X (measured in auto mileage, thousands) whose density function is of the form $f(x) = c\,x(40 - x)$, $x \in [0, 40]$, and we may assume that module failures occur independently. When a failure of any module occurs, the whole system must be inspected and repaired or replaced. Find the expected time T at which this happens.

We may consider the three failure times of modules, X_1, X_2, X_3 as a random sample from the given distribution. The time T is then just the first order statistic $X_{(1)}$. For the sake of variety and review, this time let us use first principles to find the c.d.f. and density function of $X_{(1)}$, and then check to see that the result agrees with formula (9).

First we need to determine the value of the constant c, by forcing the integral of the density function over the given state space to equal 1:

$$\texttt{Solve}\left[\texttt{c} * \int_0^{40} \texttt{x} * (40 - \texttt{x})\ \texttt{dx} == 1,\ \texttt{c}\right]$$

$$\left\{\left\{\texttt{c} \to \frac{3}{32\,000}\right\}\right\}$$

The c.d.f. of each module failure time is

$$\texttt{F[t\_] :=} \int_0^{\texttt{t}} \frac{3}{32\,000} * \texttt{x} * (40 - \texttt{x})\ \texttt{dx};$$

$$\texttt{F[t]}$$

$$\frac{3\,t^2}{1600} - \frac{t^3}{32\,000}$$

The c.d.f. of the system failure time $T = X_{(1)}$ is, by independence of the modules,

$$
\begin{aligned}
G(t) = P[T \le t] &= 1 - P[T > t] \\
&= 1 - P[X_1 > t]\,P[X_2 > t]\,P[X_3 > t] \\
&= 1 - (1 - F(t))^3
\end{aligned}
$$

Therefore, the p.d.f. of T is

$$g(t) = G'(t) = -3 \cdot (1 - F(t))^2 \cdot (-f(t)) = 3\,(1 - F(t))^2 \cdot f(t)$$

You can check that this is the very same formula as in (9), with $n = 3$ and $i = 1$. *Mathematica* computes the expectation as follows, which comes out to be about 12.3 thousand miles. This particular system is not behaving very satisfactorily. The control modules should be reengineered to be reliable for a longer time. ∎

```
            3
f[t_] := ─────── * t * (40 - t)
          32 000
(*  E[T]  =  *)
      ⌠40
    N │   t * 3 * (1 - F[t])² * f[t] dt
      ⌡0
```

12.2857

Exercises 4.4

1. The sample median is less sensitive to outlying data values than the sample mean is, and for this reason some statisticians prefer the median as an estimator of the center of a distribution. If a random sample consists of the observed values 6.1, 3.5, 4.8, 5.5, and 4.1 explain how one data value can be added to this list that changes the median by less than .5 but changes the mean by 10.

2. Let X_1, X_2, X_3 be a random sample from the distribution with density function $f(x) = |x|$; $x \in [-1, 1]$. Find (a) $P[X_{(1)} < -.5]$; (b) $P[X_{(3)} > .5]$; (c) $P[X_{(1)} < -.5, X_{(3)} > .5]$.

3. Suppose that X_1, X_2 is a random sample of size 2 from the distribution with density function $f(x) = 1/2$; $x \in [-1, 1]$. Find by direct probabilistic methods the joint c.d.f. $F(y_1, y_2) = P[X_{(1)} \le y_1, X_{(2)} \le y_2]$ of the order statistics of the sample, and take the partial derivatives with respect to both y_1 and y_2 to obtain the joint density. Check that the result is the same as in Theorem 1.

4. Find the p.d.f. of the sample range of a random sample of size 4 from the uniform distribution on $[-1, 1]$.

5. (*Mathematica*) Write a *Mathematica* command to simulate a desired number of sample medians of a random sample of n observations from the distribution whose density function is $f(x) = 2x$, $x \in [0, 1]$. What is the median of the underlying distribution? Use the command to produce a scaled histogram of 200 sample medians in the case $n = 11$, and superimpose on that graph the graph of the p.d.f. of the sample median. Finally estimate $E[M]$ and Var (M) using simulation in the case of a sample of size 11.

6. On a particularly uninspiring day in Mathematical Statistics class, students get up and leave independently of one another at random, uniformly distributed times in the interval $[0, 50]$ minutes. The class begins with 8 students. The dogged professor will continue class until the last student leaves. Find the probability that the class lasts no longer than 40 minutes.

7. (*Mathematica*) Approximate the probability that the minimum value in a random sample of size 5 from the standard normal distribution is positive.

8. Let X_1, X_2, X_3, X_4 be a random sample from the distribution with density $f(x) = 1/x^2$; $x > 1$. Find the probability that the sample range exceeds 3.

9. A general contractor will accept the next job offer that comes in. He has submitted bids on 3 such jobs. Suppose that he will win all of them, but the offers arrive at independent, random times each with density function $f(x) = \frac{1}{4} e^{-x/4}$; $x > 0$. Find the probability that he accepts an offer by time 2.

10. Repeat Exercise 9, except that suppose that each offer time comes at a time with its own probability density: $f_1(x) = \frac{1}{2} e^{-x/2}$; $x > 0$ for job 1, $f_2(x) = \frac{1}{3} e^{-x/3}$; $x > 0$ for job 2, and $f_3(x) = \frac{1}{4} e^{-x/4}$; $x > 0$ for job 3. Notice that the results on order statistics in this section no longer apply, since the three times are not identically distributed.

11. It is hypothesized that the distribution of rounding errors in bank interest calculations has a p.d.f. of the form $f(x) = c\left(\frac{1}{4} - x^2\right)$; $-\frac{1}{2} < x < \frac{1}{2}$. What do you think of the claim if a random sample of 5 such calculations yielded a median of .4?

12. Suppose that the distribution of waiting times at a driver's license facility is thought to be uniform on an interval $(a, 6)$ (in units of minutes). A random sample of size 6 is taken in an attempt to estimate a. Find a linear function $h(x)$ so that the statistic $h(X_{(1)})$ is a good estimator of a in the sense that $E[h(X_{(1)})] = a$.

Sample Problems from Actuarial Exam P

13. (*Mathematica*) A company agrees to accept the highest of four sealed bids on a property. The four bids are regarded as four independent random variables with common cumulative distribution function

$$F(x) = \begin{cases} 0 & \text{if } x < 3/2 \\ \frac{1}{2}(1 + \sin(\pi x)) & \text{if } 3/2 \le x \le 5/2 \\ 1 & \text{if } x > 5/2 \end{cases}$$

What is the expected value of the accepted bid?

14. Claim amounts for wind damage to insured homes are independent random variables with common density function

$$f(x) = \begin{cases} 3/x^4 & \text{if } x > 1 \\ 0 & \text{otherwise} \end{cases}$$

where x is the amount of a claim in thousands. Suppose that three such claims will be made. What is the expected value of the largest of the three claims?

4.5 Gamma Distributions

Main Properties

The family of *gamma distributions* is one of the most important of all in probability and statistics. In this section we will learn about the distributional properties of the general gamma distribution, then we will discuss some of the special features and applications of an important instance: the *exponential distribution*. Another instance of the gamma family, the *chi-square distribution*, will be introduced in the next section.

Definition 1. A continuous random variable X has the *gamma distribution* if its probability density function is:

$$f(x;\ \alpha,\ \beta) = \frac{1}{\beta^\alpha\ \Gamma(\alpha)}\ x^{\alpha-1}\ e^{-x/\beta},\ \ x > 0 \tag{1}$$

where $\alpha,\ \beta > 0$ are constant parameters. We abbreviate $X \sim \Gamma(\alpha,\ \beta)$.

As we mentioned earlier in the book, the gamma function in the denominator of (1) is defined by the integral

$$\Gamma(\alpha) = \int_0^\infty t^{\alpha-1}\ e^{-t}\ dt \tag{2}$$

which can be obtained by integration by parts for many values of α. We will consider its properties in more detail in a moment.

Activity 1 Use the substitution $u = x/\beta$ to show that the integral of the gamma density is 1.

The $\Gamma(n,\ 1/\lambda)$ distribution, where n is a positive integer, is called by some authors the *Erlang distribution* with parameters n and λ, but we will not use that terminology here.

Let's take a look at the graph of the gamma density function. In the kernel there is a GammaDistribution[α, β] object, to which you can apply the PDF function as below. In Figure 25 is the gamma p.d.f. for parameters $\alpha = 2$ and $\beta = 3$. Note the asymmetrical, right-skewed shape of the density graph.

```
Needs["KnoxProb7`Utilities`"];
f[x_] := PDF[GammaDistribution[2, 3], x];
PlotContsProb[f[x], {x, 0, 18}, {1, 5},
  Ticks → {{1, 3, 5, 7, 9, 11, 13, 15}, Automatic},
  ShadingStyle → Gray]
```

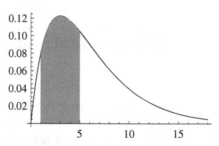

Figure 4.25 - The $\Gamma(2,3)$ density function

The area shaded is the probability $P[1 \le X \le 5]$ which can be found by integrating the density function over the interval $[1, 5]$.

$$\mathbf{N}\!\left[\int_{1}^{5} \mathbf{f}[\mathbf{x}]\,d\mathbf{x}\right]$$

0.451707

Activity 2 Try to approximate the median m of the $\Gamma(2,3)$ distribution by using the CDF function. (Remember that the median of a distribution is such that exactly half of the total probability lies to its right.) Then check your guess by using the Quantile command.

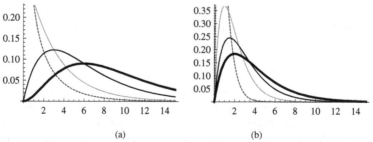

(a)   (b)

Figure 4.26 - (a) Gamma densities, $\alpha = .5$ (dashed), 1 (gray), 2 (thin), 3 (thick) and $\beta = 3$; (b) Gamma densities, $\beta = .5$ (dashed), 1 (gray), 1.5 (thin), 2 (thick) and $\alpha = 2$

The graphs in Figure 26(a) illustrate the dependence of the gamma density on the α parameter for $\alpha = .5$, 1, 2, and 3. In all cases we fix β at 3. The two cases where $\alpha \le 1$ show a markedly different behavior: for $\alpha = .5$ the density is asymptotic to the y-axis, and for $\alpha = 1$ there is a non-zero y-intercept. (What is it?) You can check that in fact $\alpha = 1$ is a break point, beyond which the density graph takes on the humpbacked shape of the other two curves. Note that the probability weight shifts to the right as α increases.

In Figure 26(b) we examine the dependence of the shape of the gamma density on the β parameter for $\beta = .5$, 1, 2, and 3. In each case α is set at 2. As β increases, notice the spreading out of the density accompanied by a shift to the right. In the electronic text you can execute the Manipulate command below to see the effect of changes to the parameters on the density graph.

```
f[x_, α_, β_] := PDF[GammaDistribution[α, β], x];
Manipulate[Plot[f[x, α, β], {x, 0, 15},
    PlotRange → {0, 1}], {α, .5, 3}, {β, .5, 3}]
```

The gamma function upon which the gamma density depends has some interesting properties that permit the density to be written more explicitly in some cases. First,

$$\Gamma(1) = \int_0^\infty e^{-t} \, dt = 1. \tag{3}$$

Second, by using integration by parts, we can derive the following recursive relationship:

$$\begin{aligned}
\Gamma(n) &= \int_0^\infty t^{n-1} e^{-t} \, dt \\
&= -t^{n-1} e^{-t} \Big|_0^\infty + \int_0^\infty (n-1) t^{n-2} e^{-t} \, dt \\
&= 0 + (n-1) \int_0^\infty t^{n-2} e^{-t} \, dt \\
&= (n-1) \Gamma(n-1)
\end{aligned} \tag{4}$$

This relationship is true regardless of whether n is an integer; however, if n is an integer then we can also iterate backward to obtain:

$$\Gamma(n) = (n-1)(n-2) \cdots 2 \cdot 1 \cdot \Gamma(1) = (n-1)! \tag{5}$$

Defining 0! to be 1 as is the custom, this formula gives values to the gamma function for all positive integers n. It also turns out that $\Gamma(1/2) = \sqrt{\pi}$, as *Mathematica* shows below.

$$\int_0^\infty t^{-1/2} E^{-t} \, dt$$

$\sqrt{\pi}$

The first two moments of the gamma distribution depend on the parameters α and β in an easy way.

Theorem 1. If $X \sim \Gamma(\alpha, \beta)$ then
    (a) $E[X] = \alpha \beta$;
    (b) $\text{Var}(X) = \alpha \beta^2$.

Proof: (a) We have

$$E[X] = \int_0^\infty t \cdot \frac{1}{\beta^\alpha \, \Gamma(\alpha)} \, t^{\alpha-1} \, e^{-t/\beta} \, dt.$$

But by multiplying and dividing by constants this integral can be rearranged into

$$E[X] = \beta \cdot \frac{\Gamma(\alpha+1)}{\Gamma(\alpha)} \cdot \int_0^\infty \frac{1}{\beta^{\alpha+1} \, \Gamma(\alpha+1)} \, t^\alpha \, e^{-x/\beta} \, dt.$$

The integral on the right is that of a $\Gamma(\alpha+1, \beta)$ density; hence it reduces to 1. Therefore,

$$E[X] = \beta \cdot \frac{\Gamma(\alpha+1)}{\Gamma(\alpha)} = \beta \cdot \frac{\alpha\Gamma(\alpha)}{\Gamma(\alpha)} = \alpha\beta.$$

(b) We leave the proof of this part as Exercise 3. ∎

The next result on sums of independent gamma random variables is crucial to much of statistical inference and applied probability. It says that if the random variables share the same β parameter, then the sum also has the gamma distribution. This will yield a very fundamental property of the Poisson process in Chapter 6, and it will loom large in the study of inference on the normal variance parameter in Section 4.6.

Theorem 2. (a) The moment-generating function of the $\Gamma(\alpha, \beta)$ distribution is

$$M(t) = E\left[e^{tX}\right] = \frac{1}{(1 - \beta t)^\alpha} \tag{6}$$

(b) If $X_1, X_2, \ldots, X_n$ are independent random variables with, respectively, $\Gamma(\alpha_1, \beta), \Gamma(\alpha_2, \beta), \ldots, \Gamma(\alpha_n, \beta)$ distributions, then

$$Y = \sum_{i=1}^n X_i \sim \Gamma(\alpha_1 + \alpha_2 + \cdots + \alpha_n, \beta)$$

Proof. (a) The m.g.f. is

$$M(t) = \int_0^\infty e^{tx} \cdot \frac{1}{\beta^\alpha \, \Gamma(\alpha)} \, x^{\alpha-1} \, e^{-x/\beta} \, dx = \int_0^\infty \frac{1}{\beta^\alpha \, \Gamma(\alpha)} \, x^{\alpha-1} \, e^{-x(1/\beta - t)} \, dx$$

In the integral on the right, for fixed t we make the change in parameter $\delta = \beta/(1 - \beta t)$. Notice that $\delta/\beta = 1/(1 - \beta t)$ and $1/\delta = 1/\beta - t$. Then,

$$
\begin{aligned}
M(t) &= \frac{1}{\beta^\alpha} \cdot \int_0^\infty \frac{1}{\Gamma(\alpha)} \, x^{\alpha-1} \, e^{-x/\delta} \, dx \\
&= \frac{\delta^\alpha}{\beta^\alpha} \cdot \int_0^\infty \frac{1}{\delta^\alpha \, \Gamma(\alpha)} \, x^{\alpha-1} \, e^{-x/\delta} \, dx \\
&= \left(\frac{\delta}{\beta}\right)^\alpha = \frac{1}{(1 - \beta t)^\alpha}.
\end{aligned}
$$

(b) The m.g.f. of the sum Y is

$$M_Y(t) = \mathrm{E}\left[e^{t(X_1 + \cdots + X_n)}\right] = \mathrm{E}\left[e^{t\,X_1} \cdot e^{t\,X_2} \cdots e^{t\,X_n}\right]$$

By independence of the X_i's and part (a), this becomes

$$M_Y(t) = \mathrm{E}\left[e^{t\,X_1}\right] \cdot \mathrm{E}\left[e^{t\,X_2}\right] \cdots \mathrm{E}\left[e^{t\,X_n}\right] = 1/(1 - \beta t)^{\Sigma \alpha_i}$$

which is the m.g.f. of the $\Gamma(\alpha_1 + \alpha_2 + \cdots + \alpha_n, \beta)$ distribution. ∎

Example 1 At a lunch counter, the customers seat themselves, then wait for an amount of time T_1 that we assume has the $\Gamma(1, 2)$ distribution for their order to be taken. Then an amount of time T_2, assumed to be independent of T_1 and $\Gamma(3, 2)$ distributed, passes until their food arrives. Find the mean and variance of the total time taken from initial arrival to delivery of the food order, and find the probability that this total time is at least 9 minutes.

The mean time until the food arrives is $\mathrm{E}[T_1 + T_2] = \mathrm{E}[T_1] + \mathrm{E}[T_2] = 1 \cdot 2 + 3 \cdot 2 = 8$, by linearity and Theorem 1(a). Because the two times are assumed to be independent, Theorem 1(b) implies that

$$\mathrm{Var}(T_1 + T_2) = \mathrm{Var}(T_1) + \mathrm{Var}(T_2) = 1 \cdot 2^2 + 3 \cdot 2^2 = 16.$$

Since the β parameters for the two gamma distributions agree, Theorem 2(b) implies that $T_1 + T_2 \sim \Gamma(1 + 3, 2)$. Thus, the probability $P[T_1 + T_2 \geq 9]$ is about .34, as computed below in *Mathematica*. ∎

```
N[1 - CDF[GammaDistribution[4, 2], 9]]
```

    0.342296

Activity 3 How would you cope with the problem in the last example if T_2 had the $\Gamma(2, 3)$ distribution?

The Exponential Distribution

The simplest member of the gamma family is also one of the most useful and interesting. Taking $\alpha = 1$ and $\lambda = 1/\beta$ in the gamma density formula, we get the following.

Definition 2. A random variable X has the *exponential distribution* if its probability density function is

$$f(x; \lambda) = \lambda e^{-\lambda x}, \quad x > 0 \tag{7}$$

where $\lambda > 0$ is a constant parameter. We abbreviate $X \sim \exp(\lambda)$.

The related *Mathematica* object is called ExponentialDistribution[λ]. For example, the following defines the exp(λ) p.d.f. in *Mathematica*.

```
f[x_, λ_] := PDF[ExponentialDistribution[λ], x]
```

Figure 4.27 - Proportional area property of the exponential distribution

The exponential distribution has a property that is a bit peculiar. Consider the picture in Figure 27 of the exp(1.5) density, and two typical times $s = .7$ and $t = 1.1$ hence $t + s = 1.8$. In the first graph, the axis origin is set at (0,0) as usual, and we shade the area to the right of $s = .7$, and on the second graph we move the axis origin to $(t, 0) = (1.1, 0)$ and shade the area to the right of $t + s = 1.8$. Though the vertical scales are different, the graphs look identical. A more careful way of saying this is that the share that the first shaded region has of the total area of 1 appears to be the same as the share that the second shaded region has of the total area to the right of $t = 1.1$. That is,

$$\frac{P[X > s]}{1} = \frac{P[X > t+s]}{P[X > t]}.$$

The right side of the equation above is also a conditional probability, which implies that

$$P[X > s] = P[X > t + s \mid X > t]. \tag{8}$$

Activity 4 Verify formula (8) algebraically for an arbitrary exp(λ) distribution.

The intuitive interpretation of equation (8) is as follows. Suppose that we are waiting for our friend Godot, who will arrive at an exponentially distributed time X. We wait for t units of time and Godot does not come. Someone else comes by, and asks how likely it is that it will be s more time units (total time $t + s$) until Godot gets here. Thus, we are asked about the conditional probability on the right. We must answer the unconditional probability $P[X > s]$, which is the same as if we had just begun waiting for Godot at time instant t. Our waiting has done no good, probabilistically. This characteristic of the exponential distribution is called *memorylessness*. It is a great help in analyzing customer service situations in which times between successive arrivals are exponentially distributed. We will have more to say about this in Chapter 6.

Some of the basic properties of the exponential(λ) distribution are below. Three of them, the mean, variance, and moment-generating function follow directly from its representation as the $\Gamma(1, 1/\lambda)$ distribution. The c.d.f. in (10) is a direct integration of the exponential density. You should be sure you know how to prove all of them.

$$\text{If } X \sim \exp(\lambda), \text{ then } E[X] = 1/\lambda \text{ and Var}(X) = 1/\lambda^2 \tag{9}$$

$$\text{If } X \sim \exp(\lambda), \text{ then } X \text{ has c.d.f. } F(x) = P[X \leq x] = 1 - e^{-\lambda x}, \ x > 0 \tag{10}$$

$$\text{If } X \sim \exp(\lambda), \text{ then the m.g.f. of } X \text{ is } M(t) = E[e^{tX}] = \frac{1}{(1 - t/\lambda)} \tag{11}$$

Example 2 If the waiting time for service at a restaurant is exponentially distributed with parameter $\lambda = .5/\text{min}$. then the expected waiting time is $1/.5 = 2$ minutes, and the variance of the waiting time is $1/(.5)^2 = 4$ so that the standard deviation is 2 minutes also. Using the *Mathematica* c.d.f. function, the probability that the waiting time will be between 1 and 4 minutes is computed as:

```
F[x_, λ_] := CDF[ExponentialDistribution[λ], x];
F[4, .5] - F[1, .5]
```

0.471195

Of course, we would get the same result by integrating the density between 1 and 4, as below. ■

```
∫₁⁴ f[x, .5] dx
```

0.471195

Example 3 A cellular phone has a 2 year warranty so that if the phone breaks prior to the end of the warranty period the owner will be paid a prorated proportion of the original purchase price of $120 based upon the fraction of time remaining until the 2 year mark. After 2 years, nothing will be paid. The phone breaks at an exponentially distributed time with mean 4 years. Find the expected amount that the warranty will pay.

The warranty pays an amount $X = 120(2 - T)/2$ if the breaking time $T \leq 2$ and zero otherwise. Since the mean of 4 years is the reciprocal of the exponential parameter λ, $T \sim \exp(1/4)$. Then the expected value of X is:

$$N\left[\int_0^2 (120\ (2-t)\ /\ 2) * (1\ /\ 4)\ E^{-(1/4)*t}\ dt\right]$$

25.5674

The company offering the warranty will have to build a margin of at least \$25.57 into the purchase price in order to expect to cover their costs. ∎

Exercises 4.4

1. (*Mathematica*) Suppose that X is a random variable with the $\Gamma(1.5,\ 1.8)$ distribution. Evaluate: (a) $P[X \le 3]$; (b) $P[X > 5]$; (c) $P[3 \le X \le 4]$.

2. Use the moment-generating function of the gamma distribution in Theorem 2 to derive the formulas for the mean and variance of the distribution.

3. Derive the formula $\text{Var}(X) = \alpha\beta^2$ for the variance of a gamma random variable by direct integration.

4. A project requires two independent phases to be completed, and the phases must be performed back-to-back. Each phase completion time has the exp(4.5) distribution. Find the probability that the project will not be finished by time 0.5.

5. Redo Exercise 4 under the assumption that the two phases may be performed concurrently.

6. Use calculus to show that if $X \sim \Gamma(2,\ \beta)$ then $P[X \ge 3]$ is an increasing function of β.

7. If we are waiting for a bus which will arrive at an exponentially distributed time with mean 10 minutes, and we have already waited for 5 minutes, what is the probability that we will wait at least 10 more minutes? What is the expected value of the remaining time (beyond the 5 minutes we have already waited) that we will wait?

8. (*Mathematica*) If $X \sim \Gamma(4,\ 1/6)$, find: (a) a number b such that $P[1 \le X \le b] = .15$; (b) a number c such that $P[X > c] = .2$.

9. Find the c.d.f. of the $\Gamma(2,\ 1/\lambda)$ distribution analytically.

10. (*Mathematica*) Gary operates his own landscape business. It is now 12:00 noon, and he has three lawns scheduled to mow. Suppose that the amounts of time T_1, T_2, and T_3 (in hours) that it takes to complete the mowing jobs have respectively the $\Gamma(1.5,\ 1)$, $\Gamma(2.5,\ 1)$, and $\Gamma(1,\ 1)$ distributions. If he has a 7:00 p.m. date and needs an hour to get ready for it, what is the probability that he will be on time? Suppose that before he leaves he also receives a phone call from his sister that lasts an exponentially distributed amount of time with mean 10 minutes. What now is the probability of being on time (if he accepts the call)?

11. (*Mathematica*) For what value of β does the $\Gamma(3,\ \beta)$ distribution have median equal to 2?

12. Find a general expression for the pth percentile of the $\exp(\lambda)$ distribution, i.e., a value x_p such that $P[X \le x_p] = p$.

One of the most important areas of application for the exponential distribution is the *Poisson process*. Exercises 13–15 preview some of the ideas that are fleshed out in Section 6.2. A constructive definition of the Poisson process is that it is a collection of random variables $(N_t)_{t \ge 0}$ such that for each outcome ω the function mapping time t to state $N_t(\omega)$ is a step function starting at 0 and increasing by jumps of size 1 only that occur at a sequence of times $T_1, T_2, T_3, \ldots$ such that the times between jumps $S_i = T_i - T_{i-1}$ are independent, identically distributed $\exp(\lambda)$ random variables. Here we take $T_0 = 0$ so that $S_1 = T_1$. The constant λ is called the *rate parameter* of the process.

13. If the first four times between jumps are 1.1, 2.3, 1.5, and 4.2, what are the first four jump times T_1, T_2, T_3, and T_4? What are the values of N at times .8, 2.6, 4.1, and 5.9? What would be a reasonable estimate of the rate parameter of the process?

14. (*Mathematica*) What is the probability distribution of the nth jump time T_n? For a Poisson process with rate $\lambda = .56$, find the probability that $T_3 > 6.1$.

15. (*Mathematica*) The event $\{N_t \ge n\}$ is equivalent to what event involving T_n? For a Poisson process with rate $\lambda = 2$, compute $P[N_{3.1} \ge 4]$ in two different ways.

Sample Problems from Actuarial Exam P

16. An insurance policy reimburses dental expense, X, up to a maximum benefit of 250. The probability density function for X is: $f(x) = c\,e^{-.004\,x}$, $x > 0$ where c is a constant. Calculate the median benefit for this policy.

17. An insurance company sells two types of auto insurance policies: Basic and Deluxe. The time until the next Basic Policy claim is an exponential random variable with mean two days. The time until the next Deluxe Policy claim is an independent exponential random variable with mean three days. What is the probability that the next claim will be a Deluxe Policy claim?

4.6 Chi-Square, Student's *t*, and *F*-Distributions

In Example 9 of Section 4.3 we showed how to use the mean $\overline{X}$ of a random sample to do *statistical inference* on μ, the underlying mean parameter of the normal distribution from which the sample was taken. Specifically we wondered, if the default value of mean balance measurement was $\mu = 1$, how likely it would be that $\overline{X}$ could be as extreme as the data indicated. The smaller was that likelihood, the less faith we had in the assumed value

$\mu = 1$. Statistical inference is the problem of using random samples of data to draw conclusions in this way about population parameters like the normal mean μ.

There are other important inferential situations in statistics that depend on the three distributions that we will cover in this section. We will only scratch the surface of their statistical application; the rest is best left for a full course in statistics. But you will complete this section with a sense of how the *chi-square distribution* gives information about σ^2 in $N(\mu, \sigma^2)$, how the *Student's t-distribution* enables inference on μ in $N(\mu, \sigma^2)$ without the necessity of knowing the true value of σ^2, and how the *F-distribution* applies to problems where two population variances $\sigma_1{}^2$ and $\sigma_2{}^2$ are to be compared. So our plan in this section will be to define each of these distributions in turn, examine their main properties briefly, and then illustrate their use in statistical inference problems.

Chi-Square Distribution

If you observe the sample variance $S^2 = \sum_{i=1}^{n} (X_i - \overline{X})^2 \Big/ (n-1)$ of a random sample $X_1, X_2, \ldots, X_n$ from a normal distribution many times, you will find the observed s^2 values distributing themselves in a consistent pattern. The command below simulates a desired number of sample variances in a straightforward way. The argument *numvars* refers to the number of sample variances we wish to simulate, *sampsize* is the common sample size n for each sample variance, and μ and σ are the normal distribution parameters. In the electronic version of the text you should try running it yourself several times. The right-skewed shape of the empirical distribution in Figure 28 should remind you of the gamma distribution. In fact, we argue in a moment that a rescaled version of S^2 has a probability distribution which is a special case of the gamma family, defined below.

```
Needs["KnoxProb7`Utilities`"];
```

```
SimSampleVariances[numvars_, sampsize_, μ_, σ_] :=
  Table[Variance[RandomReal[NormalDistribution[μ, σ],
    sampsize]], {i, 1, numvars}]
```

```
SeedRandom[984 521];
```

```
Histogram[SimSampleVariances[200, 20, 0, 1],
  8, "ProbabilityDensity",
  ChartStyle → Gray, BaseStyle → 8]
```

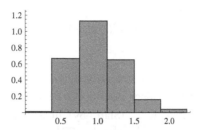

Figure 4.28 - Histogram of 200 sample variances of samples of size 20 from $N(0, 1)$

Definition 1. The *chi-square distribution* with *degrees of freedom parameter* r, abbreviated $\chi^2(r)$, is the $\Gamma(r/2, 2)$ distribution. Hence, the $\chi^2(r)$ density function is

$$f(x) = \frac{1}{2^{r/2}\,\Gamma(r/2)}\, x^{r/2-1}\, e^{-x/2}, \quad x > 0 \tag{1}$$

Mathematica has an object in the kernel for this distribution, whose syntax is

ChiSquareDistribution[r]

Then, for example, the following defines the density function in terms of both the variable x and the parameter r. Figure 29 graphs several densities for values of r from 5 to 8. Notice how, in line with the fact that you will observe in Activity 1, the probability weight shifts to the right as r increases (the mean is r) and also spreads out as r increases (the variance is $2\,r$). In the electronic version you may open up the closed cell that generated these graphs and try experimenting with other values of the r parameter.

```
f[x_, r_] := PDF[ChiSquareDistribution[r], x]
```

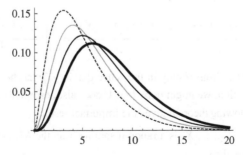

Figure 4.29 - $\chi^2(r)$ densities; $r = 5$ (dashed), $r = 6$ (gray), $r = 7$ (thin), $r = 8$ (thick)

Activity 1 Why is the mean of the $\chi^2(r)$ distribution equal to r? What is the variance of the distribution?

Exercise 2 of Section 4.3 instructed you to show that if $Z \sim N(0, 1)$, then the random variable $U = Z^2$ has a density which turns out to be $\chi^2(1)$. If you did not do that exercise yet, you should do it now. (Start by finding $P[U \le u]$, which is equal to $P[-\sqrt{u} \le Z \le \sqrt{u}] = P[Z \le \sqrt{u}] - P[Z \le -\sqrt{u}]$ as an integral, then use the Fundamental Theorem of Calculus to differentiate with respect to u. Note that $\Gamma(1/2) = \sqrt{\pi}$.) This is the key to understanding what the χ^2-distribution has to do with the sample variance S^2.

Recall that by the normal standardization theorem,

$$Z_i = \frac{X_i - \mu}{\sigma} \sim N(0, 1) \tag{2}$$

for each sample member X_i; hence, by the result mentioned above,

$$U_i = Z_i^2 = \left(\frac{X_i - \mu}{\sigma}\right)^2 \sim \chi^2(1) \tag{3}$$

for each $i = 1, 2, \ldots, n$. Moreover, in Exercise 2 of this section you are asked to use moment-generating functions to show that the sum of independent chi-square random variables is also chi-square distributed, with degrees of freedom equal to the sum of the degrees of freedom of the terms. So, the random variable

$$U = \sum_{i=1}^{n} U_i = \sum_{i=1}^{n}\left(\frac{X_i - \mu}{\sigma}\right)^2 \sim \chi^2(n) \tag{4}$$

Now look at the random variable

$$V = \frac{(n-1)S^2}{\sigma^2} = \frac{n-1}{\sigma^2}\frac{\sum_{i=1}^{n}(X_i - \overline{X})^2}{n-1} = \sum_{i=1}^{n}\left(\frac{X_i - \overline{X}}{\sigma}\right)^2 \tag{5}$$

Observe that V differs from U only in the sense that in V, μ has been estimated by $\overline{X}$. Since U has a χ^2-distribution, we might hope that V does also, especially in view of the simulation evidence. The following theorem gives the important result.

Theorem 1. If $X_1, X_2, \ldots, X_n$ is a random sample from the $N(\mu, \sigma^2)$ distribution, and S^2 is the sample variance, then

$$\frac{(n-1)S^2}{\sigma^2} \sim \chi^2(n-1) \tag{6}$$

and moreover S^2 is independent of $\overline{X}$. ∎

The full proof of Theorem 1 requires more information about multidimensional normal distributions than we have now. The discussion prior to the statement of the theorem provided good motivation for the fact that the distribution of $(n-1)\,S^2/\sigma^2$ should be of χ^2 type, but why is the degrees of freedom parameter $r = n-1$ (as opposed to n, for instance)?

To shed some light on this, consider the case $n = 2$. Then the sample variance is

$$
\begin{aligned}
S^2 = \tfrac{1}{1}\textstyle\sum_{i=1}^{2}(X_i - \overline{X})^2 \;&=\; (X_1 - \overline{X})^2 + (X_2 - \overline{X})^2 \\
&=\; \left(X_1 - \tfrac{X_1+X_2}{2}\right)^2 + \left(X_2 - \tfrac{X_1+X_2}{2}\right)^2 \\
&=\; \left(\tfrac{2X_1-X_1-X_2}{2}\right)^2 + \left(\tfrac{2X_2-X_1-X_2}{2}\right)^2 \\
&=\; \tfrac{(X_1-X_2)^2}{2}
\end{aligned}
\tag{7}
$$

after a bit of simplification. Also, $X_1 \sim N(\mu, \sigma^2)$ and $X_2 \sim N(\mu, \sigma^2)$, which means that $X_1 - X_2 \sim N(0, 2\sigma^2)$. Consequently,

$$
\frac{X_1 - X_2}{\sqrt{2\sigma^2}} \sim N(0, 1)
$$

Recalling that the square of a standard normal random variable is $\chi^2(1)$, it follows that for the case $n = 2$,

$$
\frac{(n-1)\,S^2}{\sigma^2} = \frac{(2-1)\,S^2}{\sigma^2} = \frac{(X_1 - X_2)^2}{2\sigma^2} \sim \chi^2(1)
$$

The estimation of μ by $\overline{X}$ and subsequent reduction of terms in S^2 from 2 to 1 in formula (7) resulted in the loss of one "degree of freedom" from the sample size $n = 2$. This argument can be made more elegant and extended to the case of larger n.

Activity 2 Adapt the command SimSampleVariances from earlier to simulate observations of $(n-1)\,S^2/\sigma^2$. Choose your own n and σ^2 and superimpose a graph of a scaled histogram of the simulated observations on a graph of the χ^2 density to observe the fit.

When we talk about the t-distribution in the next subsection, the second part of Theorem 1 becomes important: S^2 is independent of $\overline{X}$. At first glance this almost seems ludicrous, because both S^2 and $\overline{X}$ depend functionally on the sample values $X_1, X_2, \ldots, X_n$. To see that it is possible for S^2 and $\overline{X}$ to be independent, consider again the case $n = 2$, where S^2 is a function of $X_1 - X_2$ by formula (7), while $\overline{X}$ is a function of $X_1 + X_2$. We will argue using moment-generating functions that $Y_1 = X_1 - X_2$ and $Y_2 = X_1 + X_2$ are independent. (The following paragraph may be omitted on first reading.)

For two jointly distributed random variables Y_1 and Y_2 there is a natural two-variable analogue of the m.g.f., namely the *joint moment-generating function*, defined by

$$M(t_1, t_2) = \mathrm{E}\left[e^{t_1 Y_1 + t_2 Y_2}\right] \tag{8}$$

It has the same basic properties as the single variable m.g.f.; most importantly, it is unique to the distribution. In particular, if we find that a joint m.g.f. happens to match the joint m.g.f. of two independent random variables (which is the product of their individual m.g.f.'s), then the original two random variables must have been independent. This observation plays out very nicely with $Y_1 = X_1 - X_2$ and $Y_2 = X_1 + X_2$, where X_1 and X_2 are independent $N(\mu, \sigma^2)$ random variables. In this case by formula (8), the joint m.g.f. of Y_1 and Y_2 is

$$
\begin{aligned}
M(t_1, t_2) = \mathrm{E}\left[e^{t_1 Y_1 + t_2 Y_2}\right] &= \mathrm{E}\left[e^{t_1(X_1 - X_2) + t_2(X_1 + X_2)}\right] \\
&= \mathrm{E}\left[e^{(t_1 + t_2)X_1 + (t_2 - t_1)X_2}\right] \\
&= \mathrm{E}\left[e^{(t_1 + t_2)X_1}\right]\mathrm{E}\left[e^{(t_2 - t_1)X_2}\right] \\
&= e^{\mu(t_1 + t_2) + \frac{1}{2}\sigma^2(t_1 + t_2)^2}\, e^{\mu(t_2 - t_1) + \frac{1}{2}\sigma^2(t_2 - t_1)^2} \\
&= e^{0 \cdot t_1 + \frac{1}{2}(2\sigma^2)t_1^2}\, e^{2\mu \cdot t_2 + \frac{1}{2}(2\sigma^2)t_2^2}
\end{aligned}
$$

In the first line we use the defining formulas for Y_1 and Y_2, we gather like terms and then use independence of X_1 and X_2 to factor the expectation in line 3, we use the known formula for the normal m.g.f. in line 4, and the fifth line is a straightforward rearrangement of terms in the exponent of the fourth line, which you should verify. But the final result is that the last formula is the product of the $N(0, 2\sigma^2)$ m.g.f. and the $N(2\mu, 2\sigma^2)$ m.g.f. So, Y_1 and Y_2 must be independent with these marginal distributions. It follows that the sample mean $\overline{X}$ and the sample variance S^2 are independent when $n = 2$, and the result can be extended to higher n.

Example 1 The data below are percentage changes in population between 2000 and 2007 in a sample of 29 towns in west central Illinois.

```
popchanges = {-9.2, -7.4, -1.3, -6.9, -5.5, -6.4, -7.6,
    -7.5, -10.0, -8.0, -7.8, -3.5, -7.2, -7.1,
    -4.7, -9.4, -7.5, -2.7, -7.4, -9.2, -7.8, -7.1,
    -8.7, -9.9, -8.9, -8.3, -7.1, -1.9, -7.6};
```

Assume that these values form a random sample of population changes from all towns in the region, which we suppose are approximately normally distributed. Can we obtain a reasonable estimate of the variability of population changes, as measured by σ? Is there a way of measuring our confidence in the estimate?

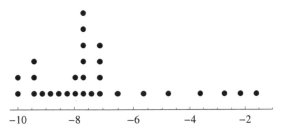

Figure 4.30 - Distribution of a sample of percentage changes in population

This rather small data set is displayed in the form of a dot plot in Figure 30. There are a few rather large observations indicating a right skewness in the empirical distribution, but this is possible in a small sample and there are at least no gross departures from normality that are visible here; so we will proceed under the assumption that percentage population changes are normally distributed. The sample variance and standard deviation are:

```
Ssquared = Variance[popchanges]
S = StandardDeviation[popchanges]
```

5.07384

2.25252

Hence, the observed $s = 2.25252$ is a point estimate of the population variability σ. But as you have probably experienced in such contexts as opinion polls, a more informative estimate would be equipped with a margin of error, so that we can say with high confidence that the true σ is in an interval. This gives us information about the precision of the point estimate s.

This is where Theorem 1 comes in. The transformed random variable $Y = (n-1)S^2/\sigma^2$ has the $\chi^2(28)$ distribution for this sample of size 29, so that we can find two endpoints a and b such that the interval $[a, b]$ encloses Y with a desired high probability. Then we will be able to say something about σ^2 itself, and in turn about σ. Let us demand 90% confidence that Y lands in $[a, b]$. Even under this specification, a and b can be chosen in infinitely many ways, but a natural way to choose them is to set $P[Y > b] = .05$ and $P[Y < a] = .05$, i.e., to split the 10% = .10 error probability into two equal pieces. Then we can find a and b by computing

```
a = Quantile[ChiSquareDistribution[28], .05]
b = Quantile[ChiSquareDistribution[28], .95]
```

16.9279

41.3371

Figure 31 illustrates the situation.

Figure 4.31 - 90% of the area under the $\chi^2(28)$ density

Now we have that

$$.90 = P[a \leq Y \leq b] \;=\; P\left[a \leq \frac{(n-1)S^2}{\sigma^2} \leq b\right]$$
$$=\; P\left[\frac{(n-1)S^2}{b} \leq \sigma^2 \leq \frac{(n-1)S^2}{a}\right] \tag{9}$$

by rearranging the inequalities in the middle expression. The interval $\left[(n-1)S^2/b, (n-1)S^2/a\right]$ is called a *confidence interval* of level 90% for the parameter σ^2, and a confidence interval for σ itself follows by just taking the square roots of the endpoints. For our data the interval for σ is:

$$\left\{ \frac{\sqrt{28} \ast S}{\sqrt{b}}, \; \frac{\sqrt{28}\,S}{\sqrt{a}} \right\}$$

{1.85386, 2.89699}

The interpretation is that it is 90% likely for the random interval $[\sqrt{n-1}\,S/\sqrt{b}, \sqrt{n-1}\,S/a]$ to contain the parameter σ. For our observed value of S, we say loosely that we are 90% confident that σ lies between 1.85386 and 2.89699. When you take a course in statistics you will hear much more about confidence intervals. But the main idea is always as in derivation (9). Choose endpoints to achieve a target probability that a transformed random variable is between the endpoints, then rearrange inequalities to isolate the desired parameter in order to get an interval estimate of the parameter. ■

Student's *t*-Distribution

In 1908, statistician W. S. Gossett, an employee of the Guinness brewery, studied the distribution of the sample mean. Recall that the expected value of $\overline{X}$ is the population mean μ, and the variance of $\overline{X}$ is σ^2/n, but Gossett considered standardizing $\overline{X}$ by an estimate $S/\sqrt{n}$ of its standard deviation, rather than the usually unknown $\sigma/\sqrt{n}$. He was led to discover a continuous probability distribution called the *t-distribution*, which we study now. But for proprietary reasons he was unable to publish his result openly. In the end he did publish under the pseudonym "Student," and for this reason the distribution is sometimes called the *Student's t-distribution*.

In Figure 32 is the result of a simulation experiment in which 200 observations of the random variable

$$T = \frac{\overline{X} - \mu}{S/\sqrt{n}} \tag{10}$$

are obtained and plotted, where in each of the 200 sampling experiments a random sample of size $n = 20$ is drawn from the normal distribution with mean 10 and standard deviation 2. The closed cell below this paragraph contains the code, but do not open it until you have tried Exercise 13, which asks you to write such a simulation program. This is an important exercise for your understanding of sampling for inference about a normal mean.

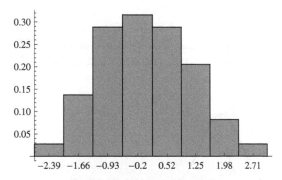

Figure 4.32 - 200 simulated *t*-statistics

Note in particular the approximate symmetry of the empirical distribution of T about 0, and the bell-shape similar to the shape of the normal density. The distribution defined below has a density with these very properties.

Definition 2. The *Student's t-distribution* is the probability distribution of the transformed random variable

$$T = \frac{Z}{\sqrt{U/r}} \tag{11}$$

where $Z \sim N(0, 1)$, $U \sim \chi^2(r)$, and Z and U are independent. We abbreviate the distribution as $t(r)$. The parameter r is called the *degrees of freedom* of the distribution.

Mathematica has an object called

StudentTDistribution[r]

in its kernel, to which we can apply functions like PDF, CDF, Quantile, and RandomReal in the usual way. Here for instance is a definition of the density function, and an evaluation of it for general x so that you can see the rather complicated form. (The standard mathematical function Beta[a,b] that you see in the output is defined as $(\Gamma[a]\,\Gamma[b])/\Gamma[a + b]$.)

```
f[x_, r_] := PDF[StudentTDistribution[r], x];
f[x, r]
```

$$\frac{\left(\frac{r}{r+x^2}\right)^{\frac{1+r}{2}}}{\sqrt{r}\ \text{Beta}\left[\frac{r}{2}, \frac{1}{2}\right]}$$

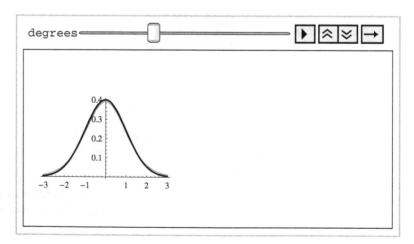

Figure 4.33 - $N(0, 1)$ density (bold) and *t*-density

It can be shown that the t-density converges to the standard normal density as $r \to \infty$. The cell above Figure 33 contains code for generating graphs of the t-density and the $N(0, 1)$ density for degrees of freedom $r = 2, ..., 25$. In the electronic file you can execute the command to observe the convergence of the t-density to the standard normal density as r becomes large. Observe that the t-density is a little heavier in the tails, but the probability weight shifts toward the middle as r increases. And the two distributions are very hard to tell apart for even such moderate r as 20.

Activity 3 Use *Mathematica* to find out how large r must be so that the probability that a $t(r)$ random variable lies between -1 and 1 is within .01 of the probability that a standard normal random variable lies between -1 and 1.

How is the formula for the $t(r)$ density obtained? Using the c.d.f. technique of Section 4.3, you can write an integral for the c.d.f. $P[T \le t] = P\left[Z / \sqrt{U/r} \le t \right]$, then differentiate with respect to t. You are asked to do this in Exercise 8. But the density formula is of secondary importance to what we turn to now.

To see the connection of the t-distribution to statistical inference on the mean, consider the sample mean $\overline{X}$ and sample variance S^2 of a random sample of size n from the $N(\mu, \sigma^2)$ distribution. We know from previous work that

$$\overline{X} \sim N(\mu, \sigma^2/n) \quad \text{and} \quad \frac{(n-1) S^2}{\sigma^2} \sim \chi^2(n-1) \tag{12}$$

and also $\overline{X}$ and S^2 are independent. So, we can standardize $\overline{X}$ and use the result as Z in the defining formula (11) for the t-distribution, and we can take $U = (n-1) S^2/\sigma^2$ and $r = n-1$ in that formula. The result is that the random variable in formula (13) has the $t(n-1)$ distribution:

$$T = \frac{\dfrac{\overline{X} - \mu}{\sigma/\sqrt{n}}}{\sqrt{\dfrac{(n-1) S^2}{\sigma^2 (n-1)}}} \sim t(n-1) \tag{13}$$

Simplifying the complex fraction, we obtain the following theorem.

Theorem 2. If $\overline{X}$ and S^2 are the sample mean and sample variance of a random sample of size n from the $N(\mu, \sigma^2)$ distribution, then the random variable

$$T = \frac{\overline{X} - \mu}{S/\sqrt{n}} \tag{14}$$

has the $t(n-1)$ distribution. ∎

The transformation given by formula (14) is similar to what we do when we standardize the sample mean, and some statisticians refer to this process as "studentizing" the sample mean. The random variable T is striking in the sense that it does not depend on σ^2, nor does the t-distribution have σ^2 as a parameter. So, if we want to infer whether an observed $\overline{X}$ is reasonably consistent with an hypothesized μ, we need not make any assumptions about the value of σ^2. We illustrate statistical reasoning about μ using the t-distribution in the next example.

Example 2 Suppose that an auto maker has a new model that is claimed to average 28 miles per gallon of gasoline on the highway. What do you think of the veracity of the claim, if 25 cars were driven under similar conditions, and the average mileage was 26.5 with a sample standard deviation of 2.7?

If the claim is true, then the random variable

$$T = \frac{\overline{X} - 28}{s/\sqrt{n}}$$

would have the $t(24)$ distribution, since the sample size is 25. If we observe a T of unusually large magnitude, then doubt is cast on the claim that $\mu = 28$. And we can quantify how much doubt by computing the probability that a $t(24)$ distributed random variable is as large or larger in magnitude than the actual value we observe. The smaller is that probability, the less we believe the claim. Our particular data give an observed value

$$t = \frac{26.5 - 28}{2.7/\sqrt{25}}$$

-2.77778

The probability that $T \le -2.7778$ is

```
CDF[StudentTDistribution[24],-2.7778]
```

0.00522692

So if truly $\mu = 28$, we would only observe such an extremely small T with probability about .005. It is therefore very likely that the true value of μ is less than 28. ∎

Activity 4 The sample average mileage of 26.5 in the last example was within one S of the hypothesized $\mu = 28$. In view of this, why was the evidence so strong against the hypothesis? Is there strong evidence against the hypotheses that $\mu = 27$?

F-Distribution

The last of the three special probability distributions for statistical inference is defined next.

Definition 3. Let $U \sim \chi^2(m)$ and $V \sim \chi^2(n)$ be two independent chi-square distributed random variables. The F *distribution* with *degree of freedom parameters* m and n (abbreviated $F(m, n)$) is the distribution of the transformed random variable

$$F = \frac{U/m}{V/n} \tag{15}$$

Mathematica's version of the F-distribution goes by the name

FRatioDistribution[m, n]

and is contained in the kernel. We define the density function below and give a typical graph in Figure 34 for degrees of freedom $m = n = 20$. Its shape is similar to that of the gamma density. Note in particular that the state space is the set of all $x \geq 0$.

```
f[x_, m_, n_] := PDF[FRatioDistribution[m, n], x];
Plot[f[x, 20, 20], {x, 0, 6}, PlotStyle → Black]
```

Figure 4.34 - $F(20, 20)$ density

Activity 5 Hold the m parameter fixed at 10, and use the Manipulate command to plot the F-density for integer values of the n parameter ranging from 5 to 25 and observe the effect of changing the n parameter on the shape of the density. Similarly, hold n at 10 and let m range from 5 through 25 to see the effect on the density of changes to m.

We use *Mathematica* to show the exact form of the $F(m, n)$ density below, which can again be obtained by the c.d.f. technique, but that is less interesting than the use of the F-distribution in doing inference on two normal variances, which we discuss in the next paragraph.

```
f[x, m, n]
```

$$\frac{m^{m/2} \, n^{n/2} \, x^{-1+\frac{m}{2}} \, (n + m\,x)^{\frac{1}{2}\,(-m-n)}}{\text{Beta}\left[\frac{m}{2}, \frac{n}{2}\right]}$$

There may be many reasons to be interested in comparing the variances σ_x^2 and σ_y^2 of two normal populations. If for example random variables X and Y are physical measurements on items manufactured using two different processes, then for quality purposes the process with smaller variability is the more desirable one. Or, perhaps the application of a treatment such as a blood sugar regulator has as its main goal the maintenance of more even sugar concentrations, so that a population of readings without the regulator should have greater variance than a population of readings with it.

Statistical comparisons of variances may be done in the following way. Let $X_1, X_2, \ldots, X_m$ and $Y_1, Y_2, \ldots, Y_n$ be independent random samples from $N(\mu_x, \sigma_x^2)$ and $N(\mu_y, \sigma_y^2)$ distributions, respectively. Let the sample variances be S_x^2 and S_y^2. From Theorem 1 we know that

$$U = \frac{(m-1)\,S_x^2}{\sigma_x^2} \sim \chi^2(m-1) \quad \text{and} \quad V = \frac{(n-1)\,S_y^2}{\sigma_y^2} \sim \chi^2(n-1)$$

and also, by the independence of the two samples, S_x^2 and S_y^2 are independent; hence U and V are independent. Using this U and V in the definition of the F-distribution, and dividing by their respective degrees of freedom yields the following result.

Theorem 3. If $X_1, X_2, \ldots, X_m$ is a random sample from the $N(\mu_x, \sigma_x^2)$ distribution with sample variance S_x^2, and $Y_1, Y_2, \ldots, Y_n$ is an independent random sample from the $N(\mu_y, \sigma_y^2)$ distribution with sample variance S_y^2, then the random variable

$$F = \frac{S_x^2 / \sigma_x^2}{S_y^2 / \sigma_y^2} \tag{16}$$

has the $F(m-1, n-1)$ distribution. ∎

Exercise 19 asks you to simulate a number of F-ratios as in (16) and compare the empirical distribution to the $F(m - 1, n - 1)$ density. You should see a good fit.

An important special case of the use of the F-random variable in (16) occurs when we want to make a judgment about whether it is feasible that two population variances σ_x^2 and σ_y^2 are equal to one another. Under the hypothesis that they are equal, they cancel one another out in (16), leaving $F = S_x^2 / S_y^2$. If F is either unreasonably large or unreasonably close to 0, i.e., if we observe an F too far out on either the left or right tail of the distribution in Figure 34, then we disbelieve the hypothesis of equality of the two population variances. The probability of observing such an F quantifies our level of disbelief. The next example illustrates the reasoning, and recalls the idea of confidence intervals from earlier in the section.

Example 3 I am the coordinator for a mathematics placement examination given every year to entering students at my college. Not only is the average level of mathematics preparation of students of concern in teaching, but also the variability of their preparation. Below are samples of placement test scores from two years. Assuming that the population of all scores is normally distributed for both years, let us see if there is significant statistical evidence that the variability of scores is different from one year to another based on this data.

```
year1 = {9, 14, 16, 8, 19, 7, 14, 17, 16, 7, 13, 16,
    11, 9, 30, 14, 9, 9, 9, 11, 13, 16, 6, 17, 18, 11,
    12, 18, 3, 12, 20, 8, 13, 14, 21, 11, 27, 26, 5};
year2 = {16, 21, 12, 19, 13, 14, 9, 29, 21, 5, 7,
    11, 4, 8, 17, 5, 10, 8, 13, 19, 14, 16, 15, 8,
    6, 8, 15, 9, 6, 23, 19, 8, 19, 7, 16, 27, 13,
    13, 28, 9, 7, 18, 12, 9, 10, 7, 8, 20, 19};
Length[year1]
Length[year2]
VarX = N[Variance[year1]]
VarY = N[Variance[year2]]
```

39

49

35.4103

39.5323

The first sample has size $m = 39$ and the second has size $n = 49$, and so the F-ratio in (16) has the $F(38, 48)$ distribution. The two sample variances of 35.4103 and 39.5323 seem to be close, but are they sufficiently close for us to believe that σ_x^2 for year 1 is the same as σ_y^2 for year 2?

There are two bases upon which we can give an answer, which are somewhat different statistical approaches, but which turn out to be equivalent. The first approach is to give an interval estimate of the ratio σ_x^2 / σ_y^2 in which we are highly confident that this ratio falls. Under the hypothesis of equal population variances, the ratio is 1, and thus 1 should fall into our confidence interval. If it doesn't, then we disbelieve that the population variances are the same, and the level of the confidence interval measures our probability of being correct. To find a confidence interval for σ_x^2 / σ_y^2 of level 90% for example, we can use *Mathematica* to find, for the $F(38, 48)$ distribution, two numbers a and b such that $P[F < a] = .05$ and $P[F > b] = .05$; then $P[a \le F \le b] = .90$.

```
a = Quantile[FRatioDistribution[38, 48], .05]
b = Quantile[FRatioDistribution[38, 48], .95]
```

0.594734

1.65219

You might say that $a \approx .59$ and $b \approx 1.65$ are our extremes of reasonableness. It is not often that an F-random variable with these degrees of freedom takes a value beyond them. Then we can write

$$.90 = P[a \le F \le b] = P\left[a \le \frac{S_x^2 / \sigma_x^2}{S_y^2 / \sigma_y^2} \le b\right] = P\left[\frac{S_x^2}{b\,S_y^2} \le \frac{\sigma_x^2}{\sigma_y^2} \le \frac{S_x^2}{a\,S_y^2}\right] \qquad (17)$$

after rearranging the inequalities. So the two endpoints of the confidence interval are formed by dividing the ratio of sample variances by b and a, respectively, which we do below.

```
{ VarX        VarX    }
{ ------- ,  ------- }
{ b * VarY   a * VarY }
```

{0.542148, 1.5061}

Since 1 is safely inside this 90% confidence interval, we have no statistical evidence against the hypothesis $\sigma_x^2 = \sigma_y^2$.

The second approach is to specify a tolerable error probability for our decision as to whether the population variances are equal, such as 10%. Specifically, we want a rule for deciding, such that if the two variances are equal, we can only err by deciding that they are unequal with probability .10. If $\sigma_x^2 = \sigma_y^2$, then as noted earlier, $F = S_x^2/S_y^2$ has the $F(38, 48)$ distribution. We already know two numbers a and b, namely $a = .594734$ and $b = 1.65219$, such that the chance that F can be as extreme as these or more so is .10 in total. So if our decision rule is to accept the hypothesis of equal variance as long as F stays between a and b, then the chance of rejecting the hypothesis incorrectly is .10. We can compute

```
F = VarX / VarY
```

0.895729

and since .895729 lies between a and b, we do accept the hypothesis $\sigma_x^2 = \sigma_y^2$. Notice that

$$a \leq \frac{S_x^2}{S_y^2} \leq b \iff \frac{S_x^2}{b\,S_y^2} \leq 1 \leq \frac{S_x^2}{a\,S_y^2}$$

Referring to (17), we see that our decision rule for accepting the hypothesis $\sigma_x^2 = \sigma_y^2$ using this second approach is exactly the same as the first approach using confidence intervals. ∎

Activity 6 In the last example check to see whether 1 is inside confidence intervals of level .80 and .70. If it is, why is this information more helpful than merely knowing that 1 is inside a 90% confidence interval?

Exercises 4.6

1. (*Mathematica*) If X has the $\chi^2(10)$ distribution, find:
   (a) $P[8 \leq X \leq 12]$;
   (b) $P[X \geq 15]$;
   (c) the 25th percentile of the distribution, i.e., the point q such that $P[X \leq q] = .25$;
   (d) two points a and b such that $P[a \leq X \leq b] = .95$.

2. Find the moment-generating function of the $\chi^2(r)$ distribution. Use it to show that if $X_1 \sim \chi^2(r_1)$, $X_2 \sim \chi^2(r_2)$, ..., $X_n \sim \chi^2(r_n)$ are independent, then the random variable $Y = X_1 + X_2 + \cdots + X_n$ has the $\chi^2(r_1 + r_2 + \cdots + r_n)$ distribution.

3. (*Mathematica*) Derive a general formula for a 95% confidence interval for the variance σ^2 of a normal distribution based on a random sample of size n. Then write a *Mathematica* command to simulate such a confidence interval as a function of the normal distribution parameters and the sample size n. Run it 20 times for the case $\mu = 0$, $\sigma^2 = 1$, $n = 50$. How many of your confidence intervals contain the true $\sigma^2 = 1$? How many would you have expected to contain it?

4. (*Mathematica*) In Example 2 of Section 4.1 we worked with a data set of demands for an item, repeated below. Assuming that the measurements form a random sample from a normal distribution, find a 90% confidence interval estimate of the population standard deviation σ.

```
demands = {1844, 1845, 2299, 2229, 2143,
    2133, 2476, 1539, 2152, 2344, 2017, 2016};
```

5. (*Mathematica*) In Example 2 of Section 4.2 we introduced a data set of estimated percentages of young children that had been immunized in each state. The data for the MMR vaccine is reproduced below. Suppose that in the past, studies of variability in vaccination coverage had indicated an estimated standard deviation of 2.1% for these percentages. Do you find statistically significant evidence that the standard deviation has changed? Explain.

```
MMR = {95.0, 89.7, 89.0, 92.5, 94.6, 91.2, 95.3, 94.8,
    95.2, 92.3, 91.4, 93.8, 86.1, 93.1, 90.4, 93.0, 93.1,
    90.8, 92.9, 90.2, 97.1, 93.3, 89.5, 94.9, 87.2, 89.,
    89.6, 94.0, 86.3, 96.6, 91.2, 90.6, 93.6, 96.9, 95.2,
    90.7, 89.9, 88.9, 93.8, 94.7, 92.5, 95.0, 94.5,
    90.4, 90.9, 93.6, 90.9, 90.5, 96.2, 91.4, 87.5};
```

6. (*Mathematica*) If $\overline{X}$ and S^2 are the sample mean and sample variance of a random sample of size 20 from the $N(\mu, \sigma^2)$ distribution, find $P[\overline{X} \geq \mu, S^2 \geq \sigma^2]$.

7. (*Mathematica*) For the population change data in Example 1, do you find statistically significant evidence that the mean percentage change among all towns in the region is less than -6%? Explain.

8. Derive the formula for the *t*-density using the approach suggested in the section.

9. (*Mathematica*) If $T \sim t(23)$, find: (a) $P[T > 2.1]$; (b) $P[T < -1.6]$; (c) a point t such that $P[-t \leq T \leq t] = .80$.

```
ppg = {111.03, 110.65, 110.07, 108.58, 106.22,
    104.47, 103.98, 102.53, 100.85, 100.69, 100.54,
    100.41, 100.18, 98.82, 98.22, 97.54, 97.46,
    97.32, 97.06, 97.02, 96.92, 96.72, 96.62,
    96.37, 95.81, 95.59, 95.36, 95.36, 93.8, 91.35};
```

10. (*Mathematica*) In Exercise 4 of Section 4.2 we saw a data set of points per game scored by the 30 NBA teams in 2007-2008. The data is repeated here for your convenience. If over the last ten years the average points scored per game has been 98.7, do you find statistically significant evidence that scoring has increased? Explain.

11. Derive the form of a 90% confidence interval for the mean μ of a normal distribution based on a sample of size n and the t-distribution.

12. Show that the t-distribution is symmetric about 0.

13. (*Mathematica*) Simulate a list of 200 T random variables as in formula (14), where samples of size 20 are taken from the $N(0, 1)$ distribution. Superimpose a scaled histogram of the data on a graph of the appropriate t-density to observe the fit.

14. In Exercise 6 of Section 4.2 were data on sulfur dioxide pollution potentials for a sample of 60 cities. They are reproduced below for your convenience. You should have found in that exercise that the variable $X = \log(SO_2)$ had an approximate normal distribution. Comment as a probabilist on the reasonableness of the claim that the mean of X is 3.

```
SO2 = {59, 39, 33, 24, 206, 72, 62, 4, 37, 20, 27,
    278, 146, 64, 15, 1, 16, 28, 124, 11, 1, 10, 5,
    10, 1, 33, 4, 32, 130, 193, 34, 1, 125, 26, 78,
    8, 1, 108, 161, 263, 44, 18, 89, 48, 18, 68, 20,
    86, 3, 20, 20, 25, 25, 11, 102, 1, 42, 8, 49, 39};
```

15. (*Mathematica*) Find the smallest value of the degree of freedom parameter r such that the $t(r)$ density is within .01 of the $N(0, 1)$ density at every point x.

16. (*Mathematica*) If a random variable F has the $F(12, 14)$ distribution, find: (a) $P[F > 1]$; (b) $P[1 \leq F \leq 3]$; (c) numbers a and b such that $P[F < a] = .05$ and $P[F > b] = .05$.

```
statsdata = {hwk, quiz, final} =
    {{47.1, 27.7, 73.1, 41.5, 85.6, 65.2, 58.5, 43.1,
      87.5, 44.6, 54.0, 41.0, 86.0, 68.5, 67.1, 84.2,
      75.2, 61.5, 54.2, 86.5, 46.3, 86.5, 79.6},
     {65.8, 72.5, 76.7, 50.8, 86.7, 67.5, 75.0, 55.0,
      98.3, 51.7, 75.0, 65.0, 88.3, 75.8, 81.7, 61.7,
      66.7, 46.7, 68.3, 91.7, 72.5, 78.3, 80.0},
     {111, 110, 141, 89, 162, 137, 120, 109,
      191, 93, 152, 104, 165, 147, 171, 145,
      140, 117, 105, 169, 104, 165, 142}};
```

17. (*Mathematica*) In Example 1 of Section 4.2 was a data set of scores in a statistics class on three items: homework, quiz, and final exam, as shown above. Comment first on whether it would make sense to test whether there is a significant difference in variability between the quiz and final scores. Then consider the homework and quiz scores. View them as random samples from the universe of possible scores that students could have earned. Using the F ratio that was discussed in the section, do you find significant statistical evidence that there is a difference in variability of these item scores? Comment on the appropriateness of doing this.

18. Discuss how you would go about testing statistically whether one normal variance σ_1^2 is two times another σ_2^2, or whether alternatively σ_1^2 is more than $2\sigma_2^2$.

19. (*Mathematica*) Write a *Mathematica* command to simulate a desired number of F-ratios as in formula (16). Then superimpose a histogram of 200 such F-ratios on the appropriate F-density, where the two random samples used to form S_x^2 and S_y^2 are both of size 20 and are taken from $N(0, 4)$ and $N(0, 9)$ distributions, respectively.

20. (*Mathematica*) A cable television network is being pressured by its sponsors to show evidence that its Nielsen ratings across its program offerings shows the same consistency as the ratings for the same time period last year. Thus, they are interested not in average Nielsen rating (which may be strongly affected by a hit program that rates highly, but carries only a small amount of the sponsors' advertising), but variability of the ratings over the range of shows. Find a 90% confidence interval for the ratio of population variances, assuming that the ratings data below are independent random samples from normal populations. Conclude from your interval whether the data are consistent with the hypothesis of equality of the two variances.

```
NielsenThisYear = {3.9, 3.3, 4.7, 3.5, 4.1, 6.4, 4.5,
    4.9, 2.4, 4.0, 2.4, 4.1, 6.0, 6.0, 2.7, 4.4,
    4.1, 2.9, 5.9, 5.8, 3.9, 8.2, 5.6, 4.0, 2.3};
NielsenLastYear = {3.9, 6.1, 5.1, 4.1, 5.7, 6.6,
    6.6, 5.9, 7.2, 4.7, 3.2, 5.3, 7.6, 4.8, 4.8, 3.0,
    4.9, 5.9, 5.4, 1.1, 3.0, 5.9, 4.9, 2.2, 4.2};
```

21. If a random variable F has the $F(m, n)$ distribution, what can you say about the random variable $1/F$, and why?

4.7 Transformations of Normal Random Variables

This section is meant to prepare you for more advanced study of statistics, especially in the multivariate realm. Our goal is to set down two major theorems about linear and quadratic transformations acting on a group of random variables having the multivariate normal distribution. It may be skipped on a first pass through probability, but my hope is that you will at least get a general sense of what the theorems say, because not only are they very interesting and not only do they fill in some of the unanswered questions in Section 4.6, but they illustrate very well the unity of mathematics. Linear algebra in particular is required here, and we will state but not prove the results that we need from that subject. But there are also tendrils of the theory that extend from calculus to differential equations, to real and complex analysis, together with applications to statistics and financial mathematics.

Linear Transformations

If you did Exercises 17–18 in Section 4.2 you encountered the multivariate generalization of the normal distribution, called the *multivariate normal distribution*. This was a relatively straightforward extension of the bivariate normal distribution, couched in vector and matrix terms. The state space of a multivariate normal random vector $\mathbf{X} = (X_1, X_2, \ldots, X_n)$ consists of vectors $\mathbf{x} = (x_1 \ x_2 \ldots x_n) \in \mathbb{R}^n$, and the joint density function can be written in matrix form as

$$f(x) = \frac{1}{(2\pi)^{n/2} \sqrt{\det(\Sigma)}} \, e^{-1/2 \, (\mathbf{x}-\mu)^t \, \Sigma^{-1}(\mathbf{x}-\mu)} \tag{1}$$

where we use "det" to denote the determinant of a matrix, the superscript -1 to denote the inverse of a matrix, and the superscript t to indicate transpose of a vector or matrix. The two parameters governing the distribution are:

$$\mu = \begin{pmatrix} \mu_1 \\ \mu_2 \\ \vdots \\ \mu_n \end{pmatrix} \text{ and}$$

$$\Sigma = \begin{pmatrix} \sigma_1^2 & \rho_{12}\,\sigma_1\,\sigma_2 & \rho_{13}\,\sigma_1\,\sigma_3 & \cdots & \rho_{1n}\,\sigma_1\,\sigma_n \\ \rho_{12}\,\sigma_1\,\sigma_2 & \sigma_2^2 & \rho_{23}\,\sigma_2\,\sigma_3 & \cdots & \rho_{2n}\,\sigma_2\,\sigma_n \\ \rho_{13}\,\sigma_1\,\sigma_3 & \rho_{23}\,\sigma_2\,\sigma_3 & \sigma_3^2 & \cdots & \rho_{3n}\,\sigma_3\,\sigma_n \\ \vdots & \vdots & \cdots & \ddots & \vdots \\ \rho_{1n}\,\sigma_1\,\sigma_n & \rho_{2n}\,\sigma_2\,\sigma_n & \rho_{3n}\,\sigma_3\,\sigma_n & \cdots & \sigma_n^2 \end{pmatrix} \tag{2}$$

respectively, the vector of means of the X_i's and the matrix of variances and paired covariances of the X_i's. We refer to these as the *mean vector* and *covariance matrix* of the random vector $\mathbf{X}$, terms that have general usage outside of the domain of normal random variables.

Activity 1 Check in the case $n = 1$ that the multivariate normal density reduces to the single variable normal density from Section 4.1.

We would like to show that if $\mathbf{X}$ has a multivariate normal distribution and A is a constant matrix of n columns, then $A\mathbf{X}$ also has a multivariate normal distribution, thereby generalizing earlier results from Section 4.3 (See Example 7 of that Section). The most straightforward way of doing this is to introduce the multivariate generalization of the moment-generating function.

Definition 1. The *moment-generating function* of the random vector $\mathbf{X} = (X_1, X_2, \ldots, X_n)$ is the following function of a vector-variable $\mathbf{t} = (t_1, t_2, \ldots, t_n)$, defined for all $\mathbf{t}$ such that the expectation is finite:

$$M(\mathbf{t}) = M(t_1, t_2, \ldots, t_n) = \mathrm{E}\left[e^{t_1 X_1 + t_2 X_2 + \cdots + t_n X_n}\right] = \mathrm{E}\left[e^{\mathbf{t} \cdot \mathbf{X}}\right] \tag{3}$$

Let us now compute the m.g.f. of the multivariate normal distribution with mean vector μ and covariance matrix Σ. The derivation will involve completing the square in a similar vein to the single variable case. The two facts from linear algebra that we need for this derivation are:

$$\text{If } A \text{ and } B \text{ are matrices, then } (A\,B)^t = B^t\,A^t \tag{4}$$

$$\text{If } A \text{ is a matrix, then } A \cdot A^{-1} = A^{-1} \cdot A = I, \text{ where } I \text{ is an identity matrix} \tag{5}$$

Note also that Σ is a symmetric matrix, hence $\Sigma^t = \Sigma$.

Now, we write the definition of the m.g.f. of X, multiply and divide by $e^{\mathbf{t} \cdot \mu}$, write the n-fold integral representing the expectation, and make the simple change of variables $z_i = x_i - \mu_i$ for each i to get:

$$
\begin{aligned}
M(\mathbf{t}) &= \mathrm{E}\left[e^{\mathbf{t}^t \cdot \mathbf{X}}\right] \\
&= e^{\mathbf{t}^t \cdot \mu}\,\mathrm{E}\left[e^{\mathbf{t}^t \cdot (\mathbf{X}-\mu)}\right] \\
&= e^{\mathbf{t}^t \cdot \mu} \int_{-\infty}^{\infty}\cdots\int_{-\infty}^{\infty}\int_{-\infty}^{\infty} e^{\mathbf{t}^t \cdot (\mathbf{x}-\mu)}\, \frac{1}{(2\pi)^{n/2}\sqrt{\det(\Sigma)}}\, e^{-1/2\,(\mathbf{x}-\mu)^t\,\Sigma^{-1}(\mathbf{x}-\mu)}\, d\,x_n\, d\,x_{n-1} \\
&= e^{\mathbf{t}^t \cdot \mu} \int_{-\infty}^{\infty}\cdots\int_{-\infty}^{\infty}\int_{-\infty}^{\infty} e^{\mathbf{t}^t \cdot \mathbf{z}}\, \frac{1}{(2\pi)^{n/2}\sqrt{\det(\Sigma)}}\, e^{-(1/2)\,\mathbf{z}^t\,\Sigma^{-1}\mathbf{z}}\, d\,z_n\, d\,z_{n-1}\cdots d\,z_1
\end{aligned}
$$

In the integrand we can bring the two exponentials together; the exponent of e can then be written

$$
\begin{aligned}
\mathbf{t}^t \cdot \mathbf{z} - (1/2)\,\mathbf{z}^t\,\Sigma^{-1}\mathbf{z} &= -\frac{1}{2}\left(\mathbf{z}^t\,\Sigma^{-1}\mathbf{z} - 2\,\mathbf{t}^t \cdot \mathbf{z}\right) \\
&= -\frac{1}{2}\left(\mathbf{z}^t\,\Sigma^{-1}\mathbf{z} - 2\,\mathbf{t}^t \cdot \mathbf{z} + \mathbf{t}^t\,\Sigma\mathbf{t}\right) + \frac{1}{2}\mathbf{t}^t\,\Sigma\mathbf{t} \\
&= -\frac{1}{2}\left(\mathbf{z} - \Sigma\,\mathbf{t}\right)^t\,\Sigma^{-1}\left(\mathbf{z} - \Sigma\,\mathbf{t}\right) + \frac{1}{2}\mathbf{t}^t\,\Sigma\mathbf{t}
\end{aligned}
$$

In line 2 we added and subtracted $\frac{1}{2}\mathbf{t}^t\,\Sigma\mathbf{t}$ from the expression, and in line 3 we noticed that the expression in parentheses in line 2 factored. (Check this by multiplying out $(\mathbf{z} - \Sigma\,\mathbf{t})^t\,\Sigma^{-1}(\mathbf{z} - \Sigma\,\mathbf{t})$ in line 3 and using properties (4) and (5); note that the product $\mathbf{t}^t \cdot \mathbf{z}$

is the same as the product $\mathbf{z}^t \cdot \mathbf{t}$ because both are dot products of the same two vectors.) Returning to the moment-generating function of $\mathbf{X}$, we now have

$$M(\mathbf{t}) = E\left[e^{\mathbf{t}^t \cdot \mathbf{X}}\right]$$

$$= e^{\mathbf{t}^t \cdot \mu} \int_{-\infty}^{\infty} \cdots \int_{-\infty}^{\infty} \frac{1}{(2\pi)^{n/2} \sqrt{\det(\Sigma)}} e^{-\frac{1}{2}(\mathbf{z}-\Sigma\mathbf{t})^t \Sigma^{-1}(\mathbf{z}-\Sigma\mathbf{t}) + \frac{1}{2}\mathbf{t}^t \Sigma\mathbf{t}} \, dz_n \cdots dz_1$$

$$= e^{\mathbf{t}^t \cdot \mu + \frac{1}{2}\mathbf{t}^t \Sigma\mathbf{t}} \int_{-\infty}^{\infty} \cdots \int_{-\infty}^{\infty} \frac{1}{(2\pi)^{n/2} \sqrt{\det(\Sigma)}} e^{-\frac{1}{2}(\mathbf{z}-\Sigma\mathbf{t})^t \Sigma^{-1}(\mathbf{z}-\Sigma\mathbf{t})} \, dz_n \cdots dz_1$$

$$= e^{\mathbf{t}^t \cdot \mu + \frac{1}{2}\mathbf{t}^t \Sigma\mathbf{t}} \cdot 1$$

The last line follows because the integral is taken over all states of a valid multivariate normal p.d.f. with mean vector $\Sigma\mathbf{t}$ and covariance matrix Σ. We have therefore proved:

Lemma 1. If $\mathbf{X}$ has the multivariate normal distribution with parameters μ and Σ, then the moment-generating function of $\mathbf{X}$ is

$$M(\mathbf{t}) = e^{\mathbf{t}^t \cdot \mu + \frac{1}{2}\mathbf{t}^t \Sigma\mathbf{t}} \qquad \blacksquare \qquad (6)$$

It now becomes very easy to prove our first main theorem.

Theorem 1. If $\mathbf{X}$ has the multivariate normal distribution with parameters μ and Σ and A is a matrix for which the product $A\mathbf{X}$ is defined, then $A\mathbf{X}$ has the multivariate normal distribution with parameters $A\mu$ and $A \cdot \Sigma \cdot A^t$.

Proof. The m.g.f. of $A\mathbf{X}$ is

$$M_{A\mathbf{X}}(\mathbf{t}) = E\left[e^{\mathbf{t}^t \cdot A\mathbf{X}}\right] = E\left[e^{(\mathbf{t}^t \cdot A)\mathbf{X}}\right]$$

$$= M_{\mathbf{X}}(A^t \cdot \mathbf{t})$$

$$= e^{(\mathbf{t}^t \cdot A)\mu + \frac{1}{2}(A^t \cdot \mathbf{t})^t \Sigma(A^t \cdot \mathbf{t})}$$

$$= e^{\mathbf{t}^t \cdot (A\mu) + \frac{1}{2}\mathbf{t}^t \cdot (A\Sigma A^t)\mathbf{t}}$$

The last expression is the moment-generating function of the multivariate normal distribution with parameters $A\mu$ and $A \cdot \Sigma \cdot A^t$, hence $A\mathbf{X}$ must have that distribution. $\blacksquare$

Example 1 Consider a random vector $\mathbf{X}$ with the bivariate normal distribution with mean vector $\mu = (0, 0)^t$ and covariance matrix

$$\Sigma = \begin{pmatrix} 1 & .5 \\ .5 & 1 \end{pmatrix}$$

which means that each of X_1 and X_2 have variance and standard deviation equal to 1. The correlation between them is therefore

$$\rho = \frac{\text{Cov}(X_1, X_2)}{1 \cdot 1} = .5.$$

Let a new random vector $\mathbf{Y}$ be formed by $\mathbf{Y} = A\,\mathbf{X}$ where $A = \begin{pmatrix} 1 & 1 \\ 1 & -1 \end{pmatrix}$. Then $Y_1 = X_1 + X_2$ and $Y_2 = X_1 - X_2$. Clearly $A \cdot \mu_x = \mathbf{0}$, hence Theorem 1 implies that $\mathbf{Y}$ is multivariate normal with mean vector $\mathbf{0}$ and covariance matrix $A \cdot \sum \cdot A^t$, which is

```
covY = {{1, 1}, {1, -1}}.
     {{1, .5}, {.5, 1}}.{{1, 1}, {1, -1}};
MatrixForm[covY]
```

$$\begin{pmatrix} 3. & 0. \\ 0. & 1. \end{pmatrix}$$

We have the surprising result that even though the original variables X_1 and X_2 were correlated, the new variables Y_1 and Y_2 are not. Their variances are 3 and 1, respectively, the diagonal elements of the matrix covY above.

Let us check this by simulation. Here we simulate 200 vectors $\mathbf{X}$, transform each by multiplying by A, and then plot them in the plane. We show the contour plot of the density of $\mathbf{Y}$ next to the scatter plot of simulated points, and we see good agreement. ∎

```
Needs["MultivariateStatistics`"]
```

```
Xsimlist = RandomReal[MultinormalDistribution[
      {0, 0}, {{1, .5}, {.5, 1}}], {200}];
A = {{1, 1}, {1, -1}};
Ysimlist = Table[A.Xsimlist[[i]], {i, 1, 200}];
g1 = ListPlot[Ysimlist,
      PlotStyle → {Black, PointSize[0.015]}];
f[y1_, y2_] := PDF[MultinormalDistribution[
      {0, 0}, {{3, 0}, {0, 1}}], {y1, y2}];
g2 = ContourPlot[f[y1, y2], {y1, -4, 4}, {y2, -2, 2},
      AspectRatio → .6, ContourStyle → Black,
      ContourShading → {GrayLevel[0], GrayLevel[.1],
         GrayLevel[.2], GrayLevel[.3], GrayLevel[.4],
         GrayLevel[.5], GrayLevel[.6], GrayLevel[.7],
         GrayLevel[.8], GrayLevel[.9], GrayLevel[1]}];
GraphicsRow[{g1, g2}]
```

(a) Simulated transformed normal vectors;   (b) Contours of theoretical distribution
Figure 4.35 - Transformed normal random vectors Y_1 and Y_2

Activity 2 In the last example, use the elementary rules for variance and covariance to check the form of covY obtained by *Mathematica*.

Example 2 One of the reasons for the importance of Theorem 1 is the fact that linear combinations of correlated normal random variables do come up in applications. For example, suppose that there are three risky assets whose rates of return are R_1, R_2, and R_3, which together have a trivariate normal distribution with mean vector and covariance matrix:

$$\mu = \begin{pmatrix} .04 \\ .07 \\ .09 \end{pmatrix}, \quad \sum = \begin{pmatrix} .02 & -.02 & .001 \\ -.02 & .04 & -.01 \\ .001 & -.01 & .08 \end{pmatrix}$$

Find the distribution, mean and variance of the rate of return on the portfolio that places half of the wealth into the first asset, and splits the remaining wealth evenly between the others. What is the probability that the rate of return on the portfolio of assets will be at least .07?

Let A be the 1×3 matrix $(1/2, 1/4, 1/4)$, so that if we denote by $\mathbf{R}$ the column vector of the three rate of return variables, then the product $A \cdot \mathbf{R}$ becomes the portfolio rate of return $R_p = \frac{1}{2} R_1 + \frac{1}{4} R_2 + \frac{1}{4} R_3$. By Theorem 1, R_p is normally distributed. The mean of R_p is $A \cdot \mu = \frac{1}{2}(.04) + \frac{1}{4}(.07) + \frac{1}{4}(.09) = .06$. The variance of R_p is $A \cdot \sum \cdot A^t$, computed in *Mathematica* below.

```
sigma = {{.02, -.02, .001},
    {-.02, .04, -.01}, {.001, -.01, .08}};
A = {{1 / 2, 1 / 4, 1 / 4}};
A.sigma.Transpose[A]
```

{{0.0065}}

Hence the rate of return on the portfolio, R_p has the $N(.06, .0065)$ distribution. The probability that $R_p \geq .07$ is about .45 as shown in the computation below. ■

```
N[1 - CDF[NormalDistribution[.06, Sqrt[.0065]], .07]]
```

0.450644

Quadratic Forms in Normal Random Vectors

One of the most important theorems in statistics is Theorem 1 from Section 4.6, which stated that if $X_1, X_2, \ldots, X_n$ is a random sample from the $N(\mu, \sigma^2)$ distribution, and S^2 is the variance of the sample, then the random variable $(n-1)S^2/\sigma^2$ has the $\chi^2(n-1)$ distribution, and is independent of the sample mean $\overline{X}$. From this we constucted the T statistic using which statistical inference on the population mean μ is possible. We motivated but did not prove this crucial result. It actually follows from a pair of more general results about *quadratic forms* $X^t Q X$ of multivariate normal random vectors X, where Q is a constant matrix. We will show these results now, and then illustrate how they specialize to our Theorem 1. These theorems also have important consequences to the statistical areas of regression and experimental design, but we will leave that for another course.

From linear algebra we will be needing a few ideas and theorems. An *idempotent matrix* is a square matrix Q such that $Q^2 = Q$. The *rank* of a matrix is the number of linearly independent rows that it has. (Recall that vectors $x_1, x_2, \ldots, x_n$ are *linearly independent* if, whenever the linear combination $a_1 x_1 + a_2 x_2 + \cdots + a_n x_n$ is the zero vector, then all of the coefficients a_i must equal zero.) Lastly, we have a diagonalization result: if Q is a symmetric, $n \times n$ idempotent matrix of rank r, then there is an orthogonal matrix N (i.e., a matrix satisfying the identities $N \cdot N^t = N^t \cdot N = I$, where I is the identity matrix of dimension n) such that

$$N^t \cdot Q \cdot N = A = \begin{pmatrix} I_r & 0 \\ 0 & 0 \end{pmatrix} \tag{7}$$

In formula (7) I_r denotes an $r \times r$ identity matrix, and the 0's indicate blocks of zeros that fill out matrix A to give it dimension $n \times n$. By multiplying on the left by N and the right by N^t, we obtain the equivalent relationship:

$$Q = N A N^t \text{ where } A = \begin{pmatrix} I_r & 0 \\ 0 & 0 \end{pmatrix} \tag{8}$$

Example 3 A very simple example of a quadratic form is the sum of squares of independent standard normal random variables $Z_1, Z_2, \ldots, Z_n$. To see this, consider the identity matrix I of dimension n and the product

$$\mathbf{Z}^t I \mathbf{Z} = (Z_1 \ Z_2 \ \cdots \ Z_n) \begin{pmatrix} 1 & 0 & \cdots & 0 \\ 0 & 1 & \cdots & 0 \\ \vdots & \vdots & \ddots & \vdots \\ 0 & 0 & \cdots & 1 \end{pmatrix} \begin{pmatrix} Z_1 \\ Z_2 \\ \vdots \\ Z_n \end{pmatrix}$$

$$= (Z_1 \ Z_2 \ \cdots \ Z_n) \begin{pmatrix} Z_1 \\ Z_2 \\ \vdots \\ Z_n \end{pmatrix}$$

$$= \sum_{i=1}^{n} Z_i^2$$

By earlier results, this quadratic form has the $\chi^2(n)$ distribution, since the Z_i^2 variables are independent and chi-square(1) distributed. Notice that $I^2 = I$, so that I is idempotent, and also I has rank n. A similar computation shows that if X_i, $i = 1, 2, \ldots, n$ are independent normal random variables with means μ_i and common variance σ^2, then the quadratic form in the random vector $\mathbf{X} - \mu$ whose components are $X_i - \mu_i$, defined by $(\mathbf{X} - \mu)^t I (\mathbf{X} - \mu)/\sigma^2$, equals $\sum_{i=1}^{n} \frac{(X_i - \mu_i)^2}{\sigma^2}$, and it also has the $\chi^2(n)$ distribution. As a third example, let $\mathbf{1}$ denote an $n \times n$ matrix whose every entry is 1, and consider the matrix $Q = \frac{1}{n}\mathbf{1}$. You can check that $Q^2 = \left(\frac{1}{n}\mathbf{1}\right)\cdot\left(\frac{1}{n}\mathbf{1}\right) = \frac{1}{n^2}\mathbf{1}\cdot\mathbf{1} = \frac{1}{n}\mathbf{1} = Q$, so that Q is a symmetric, idempotent matrix. The rows of Q are identical and so the rank of Q is 1. For simplicity let $Z_i = (X_i - \mu_i)/\sigma$, and consider the quadratic form

$$(\mathbf{X} - \mu)^t Q (\mathbf{X} - \mu)/\sigma^2 = \frac{1}{n}\cdot(Z_1 \ Z_2 \ \cdots \ Z_n) \begin{pmatrix} 1 & 1 & \cdots & 1 \\ 1 & 1 & \cdots & 1 \\ \vdots & \vdots & \ddots & \vdots \\ 1 & 1 & \cdots & 1 \end{pmatrix} \begin{pmatrix} Z_1 \\ Z_2 \\ \vdots \\ Z_n \end{pmatrix}$$

$$= \frac{1}{n}\cdot(Z_1 \ Z_2 \ \cdots \ Z_n) \begin{pmatrix} \sum_{i=1}^{n} Z_i \\ \sum_{i=1}^{n} Z_i \\ \vdots \\ \sum_{i=1}^{n} Z_i \end{pmatrix}$$

Each of the entries in the column vector on the right is the same as $n\cdot\overline{Z}$, and so completing the matrix multiplication gives that

$$(\mathbf{X} - \mu)^t Q (\mathbf{X} - \mu)/\sigma^2 = \frac{1}{n}\left(\sum_{i=1}^{n} Z_i\right)\cdot n\cdot\overline{Z} = n\cdot(\overline{Z})^2$$

$$= n\cdot\left(\sum_{i=1}^{n} \frac{X_i - \mu_i}{n\sigma}\right)^2 = \frac{n(\overline{X} - \overline{\mu})^2}{\sigma^2} \tag{9}$$

In the special case when all the means μ_i are equal to a common value μ, the right side is the square of the standard normally distributed random variable $\left(\overline{X} - \mu\right)\big/\left(\sigma\big/\sqrt{n}\right)$, which we know has the $\chi^2(1)$ distribution. So these examples are pointing at a theorem that says under some conditions that quadratic forms in normal random variables have the chi-square distribution, and the rank of the matrix in the form determines the degrees of freedom. That theorem is next. ■

Theorem 2. Assume that $\mathbf{X} = (X_1, X_2, \dots, X_n)$ has the multivariate normal distribution with mean parameter μ and diagonal covariance matrix $\Sigma = \sigma^2 \cdot I$.

(a) Suppose that Q is a symmetric, $n \times n$ idempotent matrix of rank r. Then the quadratic form

$$(\mathbf{X} - \mu)^t\, Q(\mathbf{X} - \mu)\big/\sigma^2 \tag{10}$$

has the $\chi^2(r)$ distribution.

(b) If Q_1 and Q_2 are two symmetric $n \times n$ matrices such that $Q_1 \cdot Q_2 = 0$, then the quadratic forms

$$(\mathbf{X} - \mu)^t\, Q_1(\mathbf{X} - \mu)\big/\sigma^2 \text{ and } (\mathbf{X} - \mu)^t\, Q_2(\mathbf{X} - \mu)\big/\sigma^2 \tag{11}$$

are independent.

Proof. (a) Because the determinant of Σ is $\left(\sigma^2\right)^n$ and the inverse of Σ is $\left(1/\sigma^2\right) \cdot I$, the m.g.f. of the real random variable $Y = (\mathbf{X} - \mu)^t\, Q(\mathbf{X} - \mu)\big/\sigma^2$ is the multiple integral

$$
\begin{aligned}
M(t) &= \mathrm{E}\!\left[e^{t \cdot Y}\right] \\
&= \int_{-\infty}^{\infty} \!\!\cdots \int_{-\infty}^{\infty} e^{t \cdot (\mathbf{x}-\mu)^t\, Q(\mathbf{x}-\mu)/\sigma^2} \frac{1}{(2\pi)^{n/2}\sqrt{\left(\sigma^2\right)^n}} \, e^{-1/2\,(\mathbf{x}-\mu)^t\, I(\mathbf{x}-\mu)/\sigma^2} \, d\,x_n \cdots d\,x_1 \\
&= \int_{-\infty}^{\infty} \!\!\cdots \int_{-\infty}^{\infty} \frac{1}{\left(2\pi\sigma^2\right)^{n/2}} \, e^{-1/2\,(\mathbf{x}-\mu)^t\,(I - 2t\,Q)\,(\mathbf{x}-\mu)/\sigma^2} \, d\,x_n \cdots d\,x_1 \\
&= \sqrt{\det(I - 2t\,Q)^{-1}} \int_{-\infty}^{\infty} \!\!\cdots \int_{-\infty}^{\infty} \frac{1}{\left(2\pi\sigma^2\right)^{n/2}\sqrt{\det(I - 2t\,Q)^{-1}}} \, e^{-(1/2)\,\mathbf{z}^t\,(I - 2t\,Q)\,\mathbf{z}/\sigma^2} \\
&\qquad\qquad\qquad\qquad\qquad\qquad\qquad\qquad\qquad d\,z_n \cdots d\,z_1
\end{aligned}
$$

In the last line we made the substitution $z_i = x_i - \mu_i$ for each i, and we multiplied and divided by $\sqrt{\det(I - 2 t Q)^{-1}}$. The integral is 1, because the integrand is the multivariate normal density with mean $\mathbf{0}$ and covariance matrix $\sigma^2 \cdot (I - 2 t Q)^{-1}$. Thus we have reduced the m.g.f. of Y to the expression $\sqrt{\det(I - 2 t Q)^{-1}}$. Moreover, it is also true that the determinant of the inverse of a matrix is the reciprocal of the determinant of the matrix, hence we can rewrite the expression in exponential form as $(\det(I - 2 t Q))^{-1/2}$. By the diagonalization formula (8), we can rewrite the m.g.f. as

$$
\begin{aligned}
M_Y(t) &= (\det(I - 2 t Q))^{-1/2} \\
&= (\det(I - 2 t N A N^t))^{-1/2} \\
&= (\det(N I N^t - 2 t N A N^t))^{-1/2} \\
&= (\det(N (I - 2 t A) N^t))^{-1/2} \\
&= (\det(N) \det(I - 2 t A) \det(N^t))^{-1/2} \\
&= (\det(I - 2 t A))^{-1/2}
\end{aligned}
$$

In lines 3 and 5 we used the fact that N and N^t are inverses, hence $I = N I N^t$ and also $\det(N)$ and $\det(N^t)$ are reciprocals. Notice that in block form the matrix in the last determinant is

$$
I - 2 t A = \begin{pmatrix} I_r & 0 \\ 0 & I_{n-r} \end{pmatrix} - 2 t \begin{pmatrix} I_r & 0 \\ 0 & 0 \end{pmatrix} = \begin{pmatrix} (1 - 2 t) I_r & 0 \\ 0 & I_{n-r} \end{pmatrix}
$$

The determinant of this diagonal matrix, that is the product of the diagonal elements, is $(1 - 2 t)^r$. Therefore, the m.g.f. of the quadratic form $Y = (\mathbf{X} - \mu)^t Q (\mathbf{X} - \mu)/\sigma^2$ simplifies to:

$$
M_Y(t) = (1 - 2 t)^{-r/2}
$$

We recognize this as the m.g.f. of the $\chi^2(r)$ distribution, which proves part (a).

Now suppose that the hypotheses of part (b) are in force. Let Y_1 and Y_2 be the two quadratic forms in question. We will show that the joint m.g.f. of Y_1 and Y_2 factors into the product of their individual m.g.f.'s, which is enough to show independence of the forms. Note that in the proof of part (a), we arrived at the formula $(\det(I - 2 t Q))^{-1/2}$ for the m.g.f. of a quadratic form in $\mathbf{X} - \mu$ without using the hypothesis of idempotency, which we are not making in part (b). Taking $t = 1$, this allows us to write the following formula for the joint m.g.f. of Y_1 and Y_2.

$$
\begin{aligned}
M(t_1, t_2) = E\left[e^{t_1 Y_1 + t_2 Y_2}\right] &= E\left[e^{t_1 (\mathbf{X}-\mu)^t Q_1 (\mathbf{X}-\mu)/\sigma^2 + t_2 (\mathbf{X}-\mu)^t Q_2 (\mathbf{X}-\mu)/\sigma^2}\right] \\
&= E\left[e^{(\mathbf{X}-\mu)^t (t_1 Q_1 + t_2 Q_2)(\mathbf{X}-\mu)/\sigma^2}\right] \\
&= (\det(I - 2 (t_1 Q_1 + t_2 Q_2)))^{-1/2}
\end{aligned}
$$

But notice that since $Q_1 Q_2 = 0$, we have

$$
\begin{aligned}
(I - 2 t_1 Q_1)(I - 2 t_2 Q_2) &= I - 2 t_1 Q_1 - 2 t_2 Q_2 + 4 t_1 t_2 Q_1 Q_2 \\
&= I - 2 (t_1 Q_1 + t_2 Q_2)
\end{aligned}
$$

Hence the joint moment-generating function of the forms Y_1 and Y_2 becomes

$$
\begin{aligned}
M(t_1, t_2) &= (\det(I - 2 t_1 Q_1)(I - 2 t_2 Q_2))^{-1/2} \\
&= (\det(I - 2 t_1 Q_1))^{-1/2} (\det(I - 2 t_2 Q_2))^{-1/2}
\end{aligned}
$$

which is the product of the m.g.f.'s of Y_1 and Y_2. This completes the proof. ∎

Example 4 It now remains to apply the previous theorem to our theorem about $\overline{X}$ and $(n-1) S^2 / \sigma^2$. We assume now that $X_1, X_2, \ldots, X_n$ is a random sample from the $N(\mu, \sigma^2)$ distribution, so the first hypothesis of Theorem 2 holds, and the mean vector μ has equal entries of μ. We already noted in Example 3 that if $Z = (\overline{X} - \mu) / (\sigma / \sqrt{n})$, then Z^2 can be written as a quadratic form $(\mathbf{X} - \mu)^t Q_1 (\mathbf{X} - \mu) / \sigma^2$ where $Q_1 = \frac{1}{n} \mathbf{1}$ is a symmetric, idempotent matrix of rank 1. Hence $Z^2 \sim \chi^2(1)$ by part (a) of Theorem 2. Now consider the matrix $Q_2 = \left(I - \frac{1}{n} \mathbf{1}\right)$ and the quadratic form $(\mathbf{X} - \mu)^t Q_2 (\mathbf{X} - \mu) / \sigma^2$. Written out in full,

$$
Q_2 = \begin{pmatrix}
1 - 1/n & -1/n & -1/n & \cdots & -1/n \\
-1/n & 1 - 1/n & -1/n & \cdots & -1/n \\
-1/n & -1/n & 1 - 1/n & \cdots & -1/n \\
\vdots & \vdots & \vdots & \ddots & \vdots \\
-1/n & -1/n & -1/n & \cdots & 1 - 1/n
\end{pmatrix}
\tag{12}
$$

Since the rows of Q_2 sum to zero, they cannot be linearly independent, hence the rank of Q_2 can be no more than $n - 1$. Actually, in Exercise 10 you will show that the rank is exactly $n - 1$. Also, Q_2 is idempotent, because:

$$
\begin{aligned}
Q_2 \cdot Q_2 &= \left(I - \frac{1}{n} \mathbf{1}\right)\left(I - \frac{1}{n} \mathbf{1}\right) \\
&= I - 2 \cdot \frac{1}{n} \cdot \mathbf{1} + \frac{1}{n^2} \cdot \mathbf{1} \cdot \mathbf{1} \\
&= I - 2 \cdot \frac{1}{n} \cdot \mathbf{1} + \frac{1}{n^2} \cdot n \cdot \mathbf{1} \\
&= I - \frac{1}{n} \mathbf{1} = Q_2
\end{aligned}
$$

By part (a) of the theorem again, the quadratic form $(\mathbf{X} - \mu)^t Q_2 (\mathbf{X} - \mu) / \sigma^2$ has the $\chi^2(n - 1)$ distribution. In Exercise 11 you will multiply out this quadratic form to find that

$$
(\mathbf{X} - \mu)^t Q_2 (\mathbf{X} - \mu) / \sigma^2 = (n - 1) S^2 / \sigma^2
\tag{13}
$$

So we have the distributional result that we wanted. The last item is the independence of $\overline{X}$ and S^2. Theorem 2(b) establishes the independence of Z^2 and $(n-1)S^2/\sigma^2$, because

$$Q_1 \cdot Q_2 = \frac{1}{n} \cdot 1 \cdot \left(I - \frac{1}{n} \mathbf{1}\right) = \frac{1}{n} \cdot 1 - \frac{1}{n^2} \cdot 1 \cdot 1 = \frac{1}{n} \cdot 1 - \frac{n}{n^2} \mathbf{1} = 0.$$

You will complete the argument that $\overline{X}$ must also be independent of S^2 in Exercise 12. ∎

Exercises 4.7

1. (*Mathematica*) Write out the expression for the three-variable normal density in which the means of X_1, X_2, and X_3 are, respectively, 2.5, 0, and 1.8; the variances are, respectively, 1, 1, and 4; and the correlations are as follows: between X_1 and X_2: 0; between X_1 and X_3: 0; and between X_2 and X_3: .2.

2. Simplify the expression for the multivariate normal density when the covariance matrix is diagonal.

3. If X and Y are independent $N(0, 1)$ random variables, find the joint distribution of $U = X + Y$ and $V = X - Y$.

4. (*Mathematica*) Let $\mathbf{X}$ be a random vector with the trivariate normal distribution with mean vector $\mu = (-1, 0, 1)'$ and covariance matrix

$$\sum = \begin{pmatrix} 4 & -.6 & -.4 \\ -.6 & 9 & 1.5 \\ -.4 & 1.5 & 1 \end{pmatrix}$$

Find the joint distribution of the random variables $Y_1 = X_1 - X_2$, $Y_2 = 2X_1 + X_3$, $Y_3 = X_1 + X_2 + X_3$.

5. A random vector $\mathbf{X} = (X_1, X_2, X_3)$ has joint moment-generating function $M(\mathbf{t}) = E\left[e^{\mathbf{t}' \cdot \mathbf{X}}\right] = e^{2t_1 + t_2 + \frac{3}{2}t_1^2 + \frac{1}{2}t_2^2 + 2t_3^2}$. What joint distribution does $\mathbf{X}$ have, and what are the marginal distributions of each of X_1, X_2, and X_3?

6. Let X_1 and X_2 be i.i.d. standard normal random variables. What distribution does the random variable $Y = \frac{1}{5}X_1^2 + \frac{4}{5}X_1 X_2 + \frac{4}{5}X_2^2$ have?

7. (*Mathematica*) Consider a quadratic form $Y_1 = \mathbf{X}' Q_1 \mathbf{X}/4$ in four independent $N(0, 4)$ random variables with matrix Q_1 as below. What distribution does Y_1 have? Use *Mathematica* to repeatedly simulate observations $\mathbf{X} = (X_1, X_2, X_3, X_4)$, transform them into observations of Y_1, and compare the scaled histogram of the Y_1 observations to the theoretical p.d.f. of Y_1.

$$Q_1 = \frac{1}{2} \begin{pmatrix} 1 & 1 & 0 & 0 \\ 1 & 1 & 0 & 0 \\ 0 & 0 & 1 & 1 \\ 0 & 0 & 1 & 1 \end{pmatrix}$$

8. (*Mathematica*) Under the same setup as Exercise 7, define the quadratic form $Y_2 = \mathbf{X}^t Q_2 \mathbf{X}/4$ where the matrix Q_2 is below. Are Y_1 and Y_2 independent? What is the probability distribution of Y_2? Jointly simulate 200 observations of Y_1 and Y_2 using the same underlying X_i simulated values for each, and construct a *Mathematica* list plot of the pairs to see whether the simulated pairs seem to obey independence. Finally, let 1_4 denote the 4×4 matrix of 1's. What can be said about the forms $Y_3 = \mathbf{X}^t Q_3 \mathbf{X}/4$ and $Y_4 = \mathbf{X}^t Q_4 \mathbf{X}/4$, where $Q_3 = Q_1 - \frac{1}{4} 1_4$ and $Q_4 = Q_2 - \frac{1}{4} 1_4$?

$$Q_2 = \frac{1}{2} \begin{pmatrix} 1 & 0 & 1 & 0 \\ 0 & 1 & 0 & 1 \\ 1 & 0 & 1 & 0 \\ 0 & 1 & 0 & 1 \end{pmatrix}$$

9. Suppose as in Theorem 2 that $\mathbf{X} = (X_1, X_2, \ldots, X_n)$ is multivariate normal with mean vector μ and covariance matrix $\Sigma = \sigma^2 \cdot I$. Let Q_1 and Q_2 be two $n \times n$ symmetric, idempotent matrices, of ranks r_1 and r_2, respectively, such that also $Q_1 Q_2 = 0$. Let Y_i, $i = 1, 2$ be the quadratic forms $(\mathbf{X} - \mu)^t Q_i (\mathbf{X} - \mu)/\sigma^2$. What is the moment generating function of $Y_1 + Y_2$? What is the corresponding probability distribution?

10. Show that the rank of the matrix Q_2 in formula (12) is $n - 1$.

11. Prove formula (13).

12. This finishes the argument in Example 4 that since the quadratic forms Z^2 and $(n-1) S^2/\sigma^2$ are independent of each other then so are $\overline{X}$ and S^2. (a) Show that if random variables X and Y are independent, so are the random variables $aX + b$ and $cY + d$. (b) Show that if random variables X^2 and Y are independent and the distribution of X is symmetric about 0, then X and Y are independent.

CHAPTER 5

ASYMPTOTIC THEORY

| 5.1 Strong and Weak Laws of Large Numbers

The basic facts, concepts, and computational methods of probability have mostly been covered in Chapters 1 to 4. This short chapter will attempt to set a context for our subject, by using the two most important theorems of probability, the Law of Large Numbers and the Central Limit Theorem, as rallying points around which we study historical development. The final chapter will continue the process of context development in a different way, by examining several problem areas in applied mathematics that arose mostly in the latter part of the 20th century, in which probability has taken a central place. Because our focus is on history and context rather than the development of new problem solving techniques, this chapter will be discursive and will have very few exercises. Yet it is an important one for you to read to complete your understanding.

The source material in this chapter comes from several places: some is just part of the oral tradition, and other information has come from some useful sources cited in the bibliography: Burton, Cooke, Hacking, Katz, and the fine probability book by Snell.

One might say that probability has always been with us, because humankind has always been engaged in games of chance involving dice, cards, and other media. According to Burton (p. 415), archaeological digs of prehistoric sites frequently produce examples of forerunners of today's dice called *astraguli*, which are made from the tarsal bones of hooved animals. There is evidence in Egypt reaching as far back as 3500 B.C. that people played a board game called "Hounds and Jackals" using these astraguli. Dice in today's form seem to have been with us since at least the early Roman empire. There and elsewhere gambling became such a problem that the authorities moved to limit or outlaw it, to little avail. Card playing was introduced into Europe by contact with Eastern civilizations: Chinese, Indian, and also Egyptian, and by the 1300's it had become very popular. In fact, according to Burton again (p. 417), after Johann Gutenberg printed his famous Bible in 1440, he immediately went to work printing cards.

Financial problems involving insurance, investment, and mortality also arose early in Western history, as the merchant class developed. Among the earliest and most common such problems was the practice adopted by communities of raising public money by borrowing lump sums from individuals, then agreeing to pay a lifetime annuity to the lender. In light of the importance of the problem, and its dependence on knowing the likelihoods that the lender dies in a given amount of time, it is strange that a theory of probability did not come about earlier than it did. These communities tended not to play the insurance game very well.

The mathematician Gerolamo Cardano (1501-1576) was one of the earliest to study problems of probability. Cardano was a brilliant and eccentric character: an astrologer, physician, and mathematician, who became Rector of the University of Padua. He did early work on the solution of the general cubic equation, and was involved in a dispute over what part was actually his work. But for us it is most interesting that he wrote a work called *Liber de ludo*, or *Book on Gambling*, in which among other things he introduced the idea of characterizing the probability of an event as a number p between 0 and 1, showed understanding in the context of dice of the theoretical concept of equiprobable outcomes, and anticipated that if the probability of an event is p and you perform the experiment a large number of times n, then the event will occur about $n\,p$ times. Thus, he predicted important results about binomial expectation and limiting probabilities that were more fully developed by others much later. But *Book on Gambling* was only published posthumously in 1663, by which time another very important event had already happened.

It is widely acknowledged that there is a specific time and place where probability as a well-defined mathematical area was born. Blaise Pascal (1623-1662) was a prodigy who at the age of 12 mastered Euclid's *Elements*. He was never in great health and died at the age of 39, but before that he gained acclaim not only as a mathematician, but as an experimental physicist, religious philosopher, and activist. In 1654 he began a correspondence with the great mathematician Pierre DeFermat (1601-1665) from which probability theory came. In those days in France, it was common for noblemen to dabble in academic matters as a part of their social duties and pleasures. One Antoine Gombaud, the Chevalier de Méré, was such a person. He was acquainted with Pascal and many other intellectuals of his day, and he enjoyed dicing. The problem he posed to Pascal was how to fairly divide the stakes in a game which is terminated prematurely. For instance, if a player has bet that a six will appear in eight rolls of a fair die, but the game is ended after three unsuccessful trials, how much should each player take away from the total stake? Pascal reported this to Fermat, and their interaction shows a clear understanding of repeated independent trials, and the difference between conditional and unconditional probability. Pascal reasoned that the probability of winning on throw i is $(1/6)(5/6)^{i-1}$ so that adding these terms together for $i = 4$, ..., 8 gives the total probability p of winning on trials 4 through 8, which is how the stake should be divided: a proportion p to the bettor under concern and $1 - p$ to the other. But Fermat responded that this analysis is only correct if you do not know the outcome of the first three rolls. If you do, then the chance that the bettor wins on the next roll is 1/6 (note this is a conditional probability), so that the bettor should take 1/6 of the stake in exchange for quitting at trial 4, in fact at any trial. Pascal was happy with Fermat's analysis. Later on as they studied independent trials problems, they both noticed the importance of the binomial coefficient $\binom{n}{k}$ (not written in this notation in those days) to the computation of probabilities involving independent trials. It should be added, though (Cooke (p. 201)), that combinatorics itself had its origin long before. At least as early as 300 B.C., Hindu scholars were asking such questions as how many philosophical systems can be formed by taking a certain number of doctrines from a larger list of doctrines, which is clearly our basic problem of sampling without order or replacement.

The second half of the seventeenth century was a very fertile one for probability, as many of the most famous mathematicians and scientists were making contributions. Christiaan Huygens (1629-1695) was a great Dutch scholar who wrote the first book on probability: *De ratiociniis in ludo aleae*, or *On Reasoning in a Dice Game* that compiled and extended the results of Pascal and Fermat. In particular, Huygens developed ways to compute multinomial probabilities, and crystallized the idea of expectation of a game. But it was James Bernoulli (1654-1705) who wrote the seminal work *Ars Conjectandi* or *The Art of Conjecturing*, released after his death in 1713, whose work would set new directions for the burgeoning subject. In *The Art of Conjecturing*, Bernoulli set down the combinatorial underpinnings of probability in much the form as they exist today. He is not the original source of most of these results, as previously mentioned, and probably benefited in this regard by the work of his mentor Gottfried Wilhelm Leibniz (1646-1716), who made contributions to combinatorics as well as co-inventing calculus. But the most important result to the development of probability was Bernoulli's proof of a special case of the theorem that we now describe.

Probability would never be able to have a firm foundation without an unambiguous and well-accepted definition of the probability of an event. And, this probability ought to have meaningful predictive value for future performances of the experiment to which the event pertains. One needs to know that in many repetitions of the experiment, the proportion of the time that the event occurs converges to the probability of the event. This is the Law of Large Numbers (a phrase coined by Simeon-Denis Poisson (1781-1840) for whom the Poisson distribution is named).

The Law of Large Numbers actually takes two forms, one called the *Weak Law* and the other called the *Strong Law*. Here are the modern statements of the two theorems.

Theorem 1. (Weak Law of Large Numbers) Assume that X_1, X_2, X_3, ... is a sequence of independent and identically distributed random variables. Let

$$\overline{X}_n = \frac{1}{n} \sum_{i=1}^{n} X_i$$

be the sample mean of the first n X's, and let μ denote the mean of the distribution of the X's. Then for any $\epsilon > 0$,

$$P\left[\left| \overline{X}_n - \mu \right| \geq \epsilon \right] \longrightarrow 0 \text{ as } n \longrightarrow \infty \quad \blacksquare \tag{1}$$

Theorem 2. (Strong Law of Large Numbers) Under the assumptions of the Weak Law of Large Numbers, for all outcomes ω except possibly some in a set of probability 0,

$$\overline{X}_n(\omega) \longrightarrow \mu \text{ as } n \longrightarrow \infty \quad \blacksquare \tag{2}$$

Both theorems talk about the convergence of the sequence of sample means to the population mean, but the convergence modes in (1) and (2) are quite different. To illustrate the difference let us use simulation.

Convergence mode (1), usually called *weak convergence* or *convergence in probability*, refers to an aspect of the probability distribution of $\overline{X}$. Fixing a small $\epsilon > 0$, as the sample size n gets larger, it is less and less likely for $\overline{X}$ to differ from μ by at least ϵ. We can show this by looking at the empirical distribution of $\overline{X}$ for larger and larger n, and observing that the relative frequencies above $\mu + \epsilon$ and below $\mu - \epsilon$ become smaller and smaller. Here is a command to simulate a desired number of sample means for a given sample size from a given distribution. You may remember that we used the same command earlier in Section 2.6. We apply the command to sample from the uniform(0,1) distribution, for which $\mu = .5$, taking $\epsilon = .02$ and sample sizes $n = 200$, 400, and 800.

```
Needs["KnoxProb7`Utilities`"]
```

```
SimulateSampleMeans[
   nummeans_, distribution_, sampsize_] :=
  Table[Mean[RandomReal[distribution, {sampsize}]],
   {nummeans}]
```

```
SeedRandom[18732]
list1 = SimulateSampleMeans[
    200, UniformDistribution[{0, 1}], 200];
list2 = SimulateSampleMeans[200,
    UniformDistribution[{0, 1}], 400];
list3 = SimulateSampleMeans[200,
    UniformDistribution[{0, 1}], 800];
```

```
Histogram[list1, {.4, .6, .02}, "Probability", Ticks →
   {{.44, .46, .48, .50, .52, .54, .56}, Automatic},
  PlotRange → {{.40, .60}, {0, .5}},
  AxesOrigin → {.40, 0},
  ChartStyle → Gray, BaseStyle → 6]
```

(a) Sample size = 200

```
Histogram[list2, {.4, .6, .02}, "Probability", Ticks →
   {{.44, .46, .48, .50, .52, .54, .56}, Automatic},
 PlotRange → {{.40, .60}, {0, .5}},
 AxesOrigin → {.40, 0},
 ChartStyle → Gray, BaseStyle → 6]
```

(b) Sample size = 400

```
Histogram[list3, {.4, .6, .02}, "Probability", Ticks →
   {{.44, .46, .48, .50, .52, .54, .56}, Automatic},
 PlotRange → {{.40, .60}, {0, .5}},
 AxesOrigin → {.40, 0},
 ChartStyle → Gray, BaseStyle → 6]
```

(c) Sample size = 300

Figure 5.1 - Weak convergence of sample means to μ

In Figure 1 the subintervals have been chosen so that the middle two categories range from .48 to .52, that is, from $\mu - \epsilon$ to $\mu + \epsilon$. The empirical proportion of sample means that fall outside of this interval is the total height of the outer rectangles. You see this proportion becoming closer to 0 as the sample size increases. This is what weak convergence means.

The mode of convergence characterized by (2), usually called *strong*, or *almost sure* convergence, is more straightforward. Idealize an experiment that consists of observing an infinite sequence X_1, X_2, X_3, ... of independent and identically distributed random variables. For fixed outcome ω the partial sample means $\overline{X}_1(\omega)$, $\overline{X}_2(\omega)$, $\overline{X}_3(\omega)$, ... form an ordinary sequence of numbers. For almost every outcome, this sequence of numbers converges in the calculus sense to μ. It turns out that in general strong convergence of a sequence of random variables implies weak convergence, but weak convergence does not imply strong convergence. Both modes of convergence occur for the sequence of partial means.

We can illustrate the Strong Law using the command SimMeanSequence from Exercise 10 of Section 2.6 and Example 2 of Section 3.3, which successively simulates observations from a discrete distribution, updating $\overline{X}$ each time m new observations have been generated. So we are actually just picking out a subsequence of $(\overline{X}_n(\omega))$, which hits every m^{th} sample mean, but that will let us plot the subsequence farther out than the whole sequence could have been plotted. The code for SimMeanSequence is repeated in the closed cell below.

```
SeedRandom[44 937];
SimMeanSequence[GammaDistribution[2, 3], 2000, 10]
```

Figure 5.2 - Strong convergence of sample means to μ

The connected list plot of sample means in Figure 2 is the result of calling this command on the $\Gamma(2, 3)$ distribution, whose mean is $\mu = 2 \cdot 3 = 6$, to simulate every tenth member of the sequence of sample means for a total of 2000 observations. We see the sequence of means converging to 6 as the Strong Law claims; however, the convergence is not necessarily rapid or smooth.

Activity 1 Read the code for SimulateSampleMeans and SimMeanSequence carefully to make sure you see how the algorithms relate to Theorems 1 and 2. Use these *Mathematica* commands on different probability distributions. Feel free to change the sample size and number of means. Do your experiments confirm the conclusions of the Laws of Large Numbers?

In Exercises 2 and 3 you are led through a fairly easy proof of the Weak Law of Large Numbers based on a lemma called *Chebyshev's Inequality*, which is interesting in itself. This states that for any random variable X with finite mean μ and variance σ^2,

$$P[\,|X - \mu| \geq k\sigma] \leq \frac{1}{k^2} \tag{3}$$

for any $k > 0$. In words, Chebyshev's inequality says that the chance that X is more than k standard deviations away from its mean is no more than $1/k^2$. The remarkable thing about the result is that it is a bound on the tail probability of the distribution that does not depend on what the distribution is.

The Russian mathematician who proved the inequality is worthy of special mention. Pafnutii L'vovich Chebyshev (1821–1894) formally introduced the concept of random variables and their expectation in their modern forms. Besides his proof of the weak law for means, he is responsible for stating the Central Limit Theorem using random variables, and he was the teacher of Andrei Andreevich Markov (1856–1922) who introduced the dependent trials analysis that was later to bear his name (see Section 6.1 on Markov chains).

James Bernoulli proved the weak law in *The Art of Conjecturing* in the following special case. Let the sequence $X_1, X_2, \ldots$ be a sequence of what we call today independent Bernoulli random variables, taking the value 1 with probability p and 0 with probability $1 - p$. Then $\mu = E[X] = p$. But the sample mean $\overline{X} = \sum_{i=1}^{n} X_i/n$ is just the proportion of successes in the sample. Therefore, the statement that $P[|\overline{X}_n - \mu| \geq \epsilon] \longrightarrow 0$ as $n \longrightarrow \infty$ says that as we repeat the experiment more and more often, the probability approaches 0 that the sample success proportion differs by a fixed positive amount from the true success probability p. The convergence of the sample proportion justifies the long-run frequency approach to probability, in which the probability of an event is the limit of the empirical proportion. It also is a precursor to the modern theory of estimation of parameters in statistics.

The strong law is much more difficult to show than the weak law. In fact, it was not until 1909 that Emile Borel (1871–1956) showed the strong law for the special case of a Bernoulli sequence described above. Finally in 1929, Andrei Nikolaevich Kolmogorov (1903–1987) proved the general theorem. Kolmogorov was one of the later members of the great Russian school of probabilists of the late 1800's and early 1900's, which included Chebyshev, Markov, and Alexsandr Mikhailovich Lyapunov (1857–1918) about whom we will hear again in the next section. That tradition has continued in the latter part of the twentieth century, as has the traditional role of the French in probability theory.

Exercises 5.1

1. (*Mathematica*) The Chevalier de Méré also posed the following problem to Pascal, which he thought implied a contradiction of basic arithmetic. Pascal expeditiously proved him wrong. How many rolls of two dice are necessary such that the probability of achieving at least one double six is greater than 1/2?

2. Prove Chebyshev's inequality, formula (3), for a continuous random variable X with p.d.f. f, mean μ, and standard deviation σ. (Hint: Write a formula for the expectation of $(X - \mu)^2 / (\sigma^2 k^2)$, and split the integral into the complementary regions of x values for which $|x - \mu| \geq k\sigma$ and $|x - \mu| < k\sigma$.)

3. Use Chebyshev's inequality to prove the weak law of large numbers.

4. (*Mathematica*) To what do you think the sample variance S^2 converges strongly? Check your hypothesis by writing a command similar to SimMeanSequence and running it for several underlying probability distributions.

5. (*Mathematica*) Is the Chebyshev bound in formula (3) a tight bound? Investigate by computing the exact value of $P[\,|X - \mu| \geq 2\sigma\,]$ and comparing to the Chebyshev bound when:

(a) $X \sim \Gamma(3, 4)$; (b) $X \sim$ uniform(0, 1); (c) $X \sim N(2, 4)$.

6. Consider a sample space Ω of an infinite sequence of coin flips in which we record $X_i = \pm 1/2^i$ on the ith trial according to whether the ith trial is a head or tail, respectively. Show that the sequence (X_n) converges strongly to 0. Show also that it converges weakly to 0.

7. Suppose that a random variable X has a finite mean μ and we have an a priori estimate of its standard deviation $\sigma \approx 1.6$. How large a random sample is sufficient to ensure that the sample mean $\overline{X}$ will estimate μ to within a tolerance of .1 with 90% probability?

5.2 Central Limit Theorem

In Section 4.3 we observed the very important fact that the sample mean $\overline{X}$ of a random sample from a normal population is itself a normally distributed random variable with mean μ and variance σ^2/n. From this we saw that we could assess the reasonableness of hypothesized values of μ; if $\overline{X}$ is unlikely to have taken its observed value for a given μ, then we disbelieve that μ. We saw these ideas again in Section 4.6 on the t-distribution. But in each case we made the apparently restrictive assumption that the population being sampled from is normally distributed. There is no theorem more useful in statistical data analysis than the *Central Limit Theorem*, because it implies that approximate statistical inference of this kind on the mean of a distribution can be done without assuming that the underlying population is normal, as long as the sample size is large enough. So, it is our purpose to explore this theorem in this section, and we will also take the opportunity to give some of the history of the normal distribution, the Central Limit Theorem itself, and the beginnings of the subject of statistics.

It was Abraham DeMoivre (1667-1754) who, in a 1733 paper, first called attention to an analytical result on limits which enables the approximation of binomial probabilities $\binom{n}{k} p^k (1 - p)^{n-k}$ for large n by an expression which turns out to be a normal integral. Although DeMoivre was born at the time when the great French probability tradition was developing, the fact that his family was Protestant made life in France uncomfortable, and even led to DeMoivre's imprisonment for two years. After he was freed in 1688 he spent the rest of his life in England where Newton's new Calculus was being worked out.

While DeMoivre had confined his work mostly to the case $p = 1/2$, Pierre de Laplace (1749-1827), motivated by an effort to statistically estimate p given the observed proportion of successes in a very large sample, generalized DeMoivre's work to arbitrary success probabilities in 1774. He essentially showed that (Katz, p. 550)

$$\lim_{n \to \infty} P[\, |\, \hat{p} - p\, | \, \le \epsilon] = \lim_{n \to \infty} \frac{2}{\sqrt{2\pi}} \int_0^{\epsilon/\sigma} e^{-u^2/2} \, du$$

where $\hat{p}$ is the sample proportion of successes and σ is the standard deviation of $\hat{p}$, namely $\sqrt{p(1-p)/n}$. Leonhard Euler had already shown that $\int_0^\infty e^{-u^2/2} \, du = \sqrt{\pi/2}$; so since σ is approaching 0, the limit above is 1. Notice that with this result Laplace has a novel proof of the Weak Law of Large Numbers by approximation, and is just inches from defining a normal density function. Laplace published in 1812 the first of several editions of his seminal work on probability, *Theorie Analytique des Probabilites*, which contained many results he had derived over a period of years. In this book he proved the addition and multiplication rules of probability, presented known limit theorems to date, and began to apply probability to statistical questions in the social sciences. But Laplace was actually more interested in celestial mechanics and the analysis of measurement errors in astronomical data than in social statistics. Because of the rise of calculus, celestial mechanics was a hot topic in European intellectual circles at that time, and the contemporary advances in probability combined with calculus formed the breeding ground for mathematical statistics.

Carl Friedrich Gauss (1777-1855), who rose from humble family beginnings in Brunswick, Germany to become one of the greatest and most versatile mathematicians of all time, took up Laplace's study of measurement errors in 1809. Gauss was interested in a function $f(x)$ which gave the probability (we would say probability <u>density</u> instead) of an error of magnitude x in measuring a quantity. Laplace had already given reasonable assumptions about such a function: it should be symmetric about 0 and the integral of $f(x)$ over the whole line should be 1; in particular, $f(x)$ should approach 0 as x approaches both $+\infty$ and $-\infty$ But it was Gauss who arrived at the formula for a normal density with mean zero, by considering the maximization of the joint probability density function of several errors, and making the further supposition that this maximum should occur at the sample mean error. He was able to derive a differential equation for f in this way, and to show a connection to the statistical problem of least squares curve fitting. Gauss' tremendous contributions have led to the common practice of referring to the normal distribution as the *Gaussian distribution*.

Let us return to the Central Limit Theorem, the simplest statement of which is next.

Theorem 1. (Central Limit Theorem) Let X_1, X_2, X_3, ... be a sequence of independent and identically distributed random variables with mean μ and variance σ^2. Let $\overline{X}_n$ be the mean of the first n X's. Then the cumulative distribution function F_n of the random variable

$$Z_n = \frac{\overline{X}_n - \mu}{\sigma / \sqrt{n}} \tag{1}$$

converges to the c.d.f. of the $N(0, 1)$ distribution as $n \longrightarrow \infty$. ∎

Figure 5.3 - Empirical and limiting distribution of $\overline{X}$, $n = 25$, population distribution $\Gamma(2, 1)$

Thus, the Central Limit Theorem allows us to say (vaguely) that the distribution of $\overline{X}_n$ is approximately $N(\mu, \sigma^2/n)$ for "large enough" n, despite the fact that the sample members X_1, X_2, ... , X_n may not themselves be normal. How large is "large enough" is a very good question. In Figure 3 for example is a scaled histogram of 500 simulated sample means from samples of size $n = 25$ from a $\Gamma(2, 1)$ distribution, with a normal density of mean $\mu = 2 \cdot 1 = 2$ and variance $\sigma^2/n = 2 \cdot 1^2/25 = 2/25$ superimposed. (We used the command SimulateSampleMeans from the last section; you should open up the closed cell above the figure to see the code.) The fit of the empirical distribution of sample means to the appropriate normal density is quite good, even for n as small as 25. You should try running this experiment with the gamma distribution again several times (intializing the random seed to a new value), to see whether this is a consistent result. Also, do the activity below.

Activity 1 Rerun the commands that produced Figure 3 using the uniform(0, 2) distribution instead. Experiment with sample sizes 25, 20, and 15. Does the histogram begin to depart drastically from the normal density as the sample size decreases?

Your experimental runs probably indicated a good fit of the empirical distribution of $\overline{X}$ to normality even for the smallest sample size of 15 for the uniform distribution. Experience has shown that $n = 25$ is fairly safe for most population distributions and, for some of the more symmetric population distributions, even fewer samples are necessary. This is because there is a general result (Loève (p. 300)) that gives a bound on the difference between the actual c.d.f. F_n of Z_n in (1) and the standard normal c.d.f. G:

$$\left| F_n(x) - G(x) \right| \leq \frac{c}{n^{1/2}\,\sigma^3} \cdot \mathrm{E}\!\left[\,|\,X\,|^3\right] \quad \text{for all } x, \tag{2}$$

where c is a constant. You can see that the maximum absolute difference decreases as n grows, as you would expect, but it increases as the third absolute moment $\mathrm{E}\!\left[\,|\,X^3\,|\right]$ of the distribution being sampled from increases. Actually, the ratio $\mathrm{E}\!\left[\,|\,X\,|^3\right]/\sigma^3$ can be used to measure the lack of symmetry of a distribution, called its *skewness*. Thus, we have the qualitative result that the more symmetric is the underlying distribution, the better the fit of the distribution of $\overline{X}$ to normality for fixed n, alternatively, the smaller n can be taken for an equally good fit.

Activity 2 If one distribution has a skewness that is half of the skewness of a second distribution, what is the relationship between the sample sizes required to have equal upper bounds on the absolute difference of c.d.f.'s in formula (2)?

The Central Limit Theorem can be proved by a moment-generating function argument. We will outline the proof that the moment-generating function of Z_n converges to the standard normal m.g.f. $M(t) = e^{t^2/2}$ as $n \longrightarrow \infty$, which must imply that the c.d.f. of Z_n converges to the standard normal c.d.f. at every point x. (The latter is actually a subtle theorem, whose proof we will omit.)

We can write the m.g.f. of Z_n as:

$$E[e^{tZ_n}] = E\left[e^{t\cdot\left(\overline{X}_n-\mu\right)/\left(\sigma/\sqrt{n}\right)}\right]$$

$$= E\left[e^{\frac{t\sqrt{n}}{\sigma}\cdot\frac{1}{n}\Sigma(X_i-\mu)}\right]$$

$$= E\left[e^{\frac{t}{\sigma\sqrt{n}}\cdot\Sigma(X_i-\mu)}\right] \tag{3}$$

$$= \prod_{i=1}^{n} E\left[e^{\frac{t}{\sigma\sqrt{n}}\cdot(X_i-\mu)}\right]$$

Let $M_1(t)$ be the m.g.f. of every $X_i - \mu$. Note that $M_1(0) = E[e^0] = 1$, $M_1'(0) = E[X_i - \mu] = 0$, and $M_1''(0) = E[(X_i - \mu)^2] = \text{Var}(X_i) = \sigma^2$. Expanding $M_1(t)$ in a second-order Taylor series about 0 we have for some $\xi \in (0, t)$,

$$M_1(t) = M_1(0) + M_1'(0)(t-0) + \frac{1}{2}M_1''(0)(t-0)^2 + \frac{1}{6}M_1'''(\xi)(t-0)^3$$

$$= 1 + \frac{1}{2}\sigma^2 t^2 + \frac{1}{6}M_1'''(\xi)t^3$$

Substituting this into (3), the m.g.f. of Z_n becomes

$$E[e^{tZ_n}] = \prod_{i=1}^{n} M_1\left(\frac{t}{\sigma\sqrt{n}}\right)$$

$$= \prod_{i=1}^{n}\left(1 + \frac{1}{2}\sigma^2\frac{t^2}{\sigma^2 n} + \frac{1}{6}M_1'''(\xi)\frac{t^3}{\sigma^3 n^{3/2}}\right)$$

$$= \left(1 + \frac{t^2/2}{n} + \frac{1}{6}M_1'''(\xi)\frac{t^3}{\sigma^3 n^{3/2}}\right)^n, \quad \xi \in \left(0, \frac{t}{\sigma\sqrt{n}}\right)$$

If the third derivative of M_1 is a well-behaved function near zero, then the third term in parentheses goes swiftly to zero as $n \longrightarrow \infty$. The limit of the exponential expression is therefore the same as that of $\left(1 + \frac{t^2/2}{n}\right)^n$ which from calculus is well known to be $e^{t^2/2}$. Thus the m.g.f. of Z_n converges to that of the standard normal distribution.

Though the basic result was known to DeMoivre and Laplace in the 1700's, it was not until the 1900's that the theorem as we know it was proved. And throughout much of the twentieth century, some of the best probabilistic minds have set to work improving the theorem by relaxing hypotheses. A.M. Lyapunov, working with bounds such as the one in formula (2), proved the first modern version of the Central Limit Theorem in 1901, making assumptions about the third moments of the random variables. J.W. Lindeberg proved a general version in 1922 of which ours is a corollary. Important contributions were also made by three men who can without hesitation be called the towering figures in probability in the 20th century: A.N. Kolmogorov who proved the strong law, the beloved Princeton mathematician and author Willy Feller (1906-1970), and the central figure in the modern French school of probability Paul Lévy (1886-1972). These and other probabilists have been interested in the more general question of what limiting distributions for sums of random variables are possible, if the restrictions of identical distributions, independence, and finiteness of the variance are removed.

The next example shows how our form of the Central Limit Theorem covers the special cases studied by DeMoivre and Laplace involving approximation of binomial probabilities.

Example 1 Suppose that we are conducting a poll to find out the proportion of the voting citizenry in a state that is in support of a proposition legalizing gambling. If we decide to sample 500 voters at random and 315 are in support, give an approximate 95% confidence interval for that proportion. How many voters should be sampled if we want to estimate the proportion to within 2% with 95% confidence?

To form a mathematical model for the problem, we suppose that there is an unknown proportion p of the voters who support the proposition. We sample $n = 500$ voters in sequence and with replacement, thereby forming 500 independent Bernoulli trials with success probability p. We can denote their responses by $X_1, X_2, \dots, X_{500}$ where each $X_i = 1$ or 0, respectively, according to whether the voter supports the proposition or not. Then the sample proportion $\hat{p}$ who are in favor is the same as the sample mean $\overline{X}$ of the random variables X_i. Therefore, the Central Limit Theorem applies to $\hat{p}$. By properties of the Bernoulli distribution, the common mean of the X_i's is $\mu = p$ and the variance is $\sigma^2 = p(1 - p)$. Thus, the appropriate mean and standard deviation with which to standardize $\hat{p}$ in the Central Limit Theorem are p and $\sqrt{p(1 - p)/n}$, respectively. We therefore have that

$$Z_n = \frac{\hat{p} - p}{\sqrt{p(1 - p)/n}} \approx N(0, 1) \tag{4}$$

Remember that to construct confidence intervals, the approach is to begin with a transformed random variable whose distribution is known, pick quantiles for that distribution that are suitable for the target level of confidence, then algebraically isolate the parameter of interest in the resulting inequalities. For a 95% confidence interval, each tail needs probability .025; so here is the suitable normal quantile.

```
z = Quantile[NormalDistribution[0, 1], .975]
```

1.95996

Then,

$$.95 \approx P[-z \le Z_n \le z] \;=\; P\!\left[-z \le \frac{\hat{p}-p}{\sqrt{p(1-p)/n}} \le z\right]$$

$$= P\!\left[\hat{p} - z\sqrt{p(1-p)/n} \;\le\; p \;\le\; \hat{p} + z\sqrt{p(1-p)/n}\,\right]$$

Hence to give an interval estimate of p we can start with the point estimate $\hat{p}$ and then add and subtract the margin of error term $z\sqrt{p(1-p)/n}$. The only trouble with this is that the margin of error depends on the unknown p. But since our probability is only approximate anyway, we don't lose much by approximating the unknown margin of error as well, by replacing p by the sample proportion $\hat{p}$.   We let *Mathematica* finish the computations for the given numerical data.

```
    315
p̂ = ———
    500

margin = z * √( p̂ (1 - p̂) / 500 )
{p̂ - margin, p̂ + margin}
```

```
 63
———
100

0.0423189

{0.587681, 0.672319}
```

So the sample proportion is 63% and we are 95% sure that the true proportion of all voters in support of the proposition is between about 59% and 67%.

The second question is important to the design of the poll. Our margin of error in the first part of the question was about 4%; what should we have done to cut it to 2%? Leaving n general for a moment, we want to satisfy

$$z\sqrt{p(1-p)/n} \;\le\; .02$$

We know that the quantile z is still 1.96 because the confidence level is the same as before, but prior to sampling we again do not know p, hence we cannot solve for n. There are two things that can be done: we can produce a pilot sample to estimate p by $\hat{p}$, or we can observe that the largest value that $p(1-p)$ can take on is $1/4$, when $p = 1/2$.   (Do the simple calculus to verify this.)   Then $z\sqrt{p(1-p)/n} \;\le\; z\sqrt{(1/4n)}$ and, as long as the right side is less than or equal to .02, then the left side will be as well. Thus,

$$\frac{z}{\sqrt{4n}} \le .02 \Longleftrightarrow \sqrt{n} \ge \frac{z}{2\,(.02)}$$

Squaring both sides of the last inequality gives the following threshhold value for the sample size n:

$$\left(\frac{z}{2*.02}\right)^2$$

2400.91

We would need to poll 2401 people in order to achieve the .02 margin with 95% confidence.

■

Exercises 5.2

1. Use the Central Limit Theorem for means, Theorem 1, to formulate a similar limit theorem for the partial sums $S_n = \sum_{i=1}^{n} X_i$.

2. The differential equation that Gauss arrived at for his density function of observed errors is

$$\frac{f'(x)}{x\,f(x)} = -k$$

where k is a positive constant. Solve this to obtain $f(x)$.

3. The DeMoivre-Laplace limit theorem says that

$$\sum_{k=a}^{b} \binom{n}{k} p^k (1-p)^{n-k} \xrightarrow[n\to\infty]{} \int_{(a-\mu)/\sigma}^{(b-\mu)/\sigma} \frac{1}{\sqrt{2\pi}} e^{-z^2/2}\, dz$$

where $\mu = np$ and $\sigma^2 = np(1-p)$. Explain how this result follows as a special case of the Central Limit Theorem.

4. The Central Limit Theorem has been extended to the case where $X_1, X_2, X_3, \ldots$ are independent but not necessarily identically distributed. Write what you think the theorem statement would be in this case.

5. (*Mathematica*) The Central Limit Theorem allows the distribution being sampled from to be discrete as well as continuous. Compare histograms of 300 sample means of samples of sizes 20, 25, and 30 to appropriate normal densities for each of these two distributions: (a) Poisson(3); (b) geometric(.5).

6. (*Mathematica*) Suppose that your probability of winning a game is .4. Use the Central Limit Theorem to approximate the probability that you win at least 150 out of 400 games.

7. (*Mathematica*) Do you find significant statistical evidence against the hypothesis that the mean of a population of SAT scores is 540, if the sample mean among 100 randomly selected scores is 550 and the sample standard deviation is 40?

Sample Problems from Actuarial Exam P

8. An insurance company issues 1250 vision care insurance policies. The number of claims filed by a policyholder under a vision care insurance policy during one year is a Poisson random variable with mean 2. Assume the numbers of claims filed by distinct policyholders are independent of one another. What is the approximate probability that there is a total of between 2450 and 2600 claims during a one-year period?

9. In an analysis of healthcare data, ages have been rounded to the nearest multiple of 5 years. The difference between the true age and the rounded age is assumed to be uniformly distributed on the interval from -2.5 years to 2.5 years. The healthcare data are based on a random sample of 48 people. What is the approximate probability that the mean of the rounded ages is within 0.25 years of the mean of the true ages?

CHAPTER 6

STOCHASTIC PROCESSES AND

APPLICATIONS

This chapter is meant to introduce you to a few topics in applied probability that have been the focus of much effort in the second half of the 20th century, and which will continue to flourish in the 21st century. To do this we will take advantage of many of the foundational results of probability from Chapters 1–5.

Probability theory has ridden the wave of the technological and telecommunications revolution. The analysis of modern systems for production, inventory, distribution, computing, and communication involves studying the effect of random inputs on complex operations, predicting their behavior, and improving the operations. Many of these applications involve modeling and predicting random phenomena through time.

A *stochastic process* is a family of random variables $(X_t)_{t \in T}$ thought of as observations of a random phenomenon through time. The random variable X_t is called the *state of the process* at time t. The set of times T is typically the non-negative integers 0, 1, 2, ... or the non-negative real numbers $\mathbb{R}^+ = [0, \infty)$. There are two points of view that can be taken on stochastic processes. We might be interested in the probability distribution of the state of the process at each particular time t, which emphasizes the role of the process as a collection of random variables. Or we might be interested in properties of the *path of the process*, that is, the function $t \mapsto X_t(\omega)$ for fixed outcome ω and varying time. Is the process continuous, does it reach a limit as $t \longrightarrow \infty$, what proportion of the time does it visit each state, etc.

6.1 Markov Chains

We will begin with the subject of *Markov chains*, which roughly are systems taking on various states as discrete time progresses, subject to random influences. Here are just a few systems which could be amenable to Markov chain modeling.

1. The state of deterioration of a piece of equipment
2. The popularity of a politician
3. The inventory level of an item in a store
4. The number of jobs waiting for processing by a computer server
5. The value of a retirement investment

You may also recall the random walk that we simulated in Section 1.3, which is a special kind of Markov chain.

Activity 1 Try to think of more examples of time dependent random processes in the world similar to 1−5 above.

A precise definition of a Markov chain is below. Note that the random states of the system are modeled as random variables, and we impose an extra condition on the conditional distributions that makes future states independent of the past, given the present.

Definition 1. A sequence X_0, X_1, X_2, ... of random variables is said to be a *Markov chain* if for every $n \geq 0$,

$$P[X_{n+1} = j \mid X_0 = i_0, X_1 = i_1, ... X_n = i_n] = P[X_{n+1} = j \mid X_n = i_n] \qquad (1)$$

In other words, at each time n, X_{n+1} is conditionally independent of X_0, X_1, ... , X_{n-1} given X_n.

This conditional independence condition (1) makes the prediction of future values of the chain tractable, and yet it is not unreasonably restrictive. Consider the chain of store inventory levels in the third example problem above. The next inventory level will depend on the current level, and on random sales activity during the current time period, not on the past sequence of levels. There are many systems in which past behavior has no bearing on the future trajectory, given the present state, and for those systems, Markov chain modeling is appropriate.

Many interesting results have been derived about Markov chains. But for our purposes we will be content to introduce you to three problems: (1) finding the probability law of X_n; (2) finding the limit as $n \longrightarrow \infty$ of this probability law, in order to predict the long term average behavior of the chain; and (3) for chains in which there are states that are able to absorb the chain forever, the likelihood that each such absorbing state is the final destination of the chain.

Figure 6.1 - The transition diagram of a Markov chain

Short Run Distributions

Now by formula (1), the probabilistic structure of a Markov chain can be represented by a graph called a *transition diagram*. Consider the chain X_0, X_1, X_2, ... with possible states 1, 2, 3, and 4 shown as the labeled bubbles in Figure 1. For example, if $X_n = 1$ at some time n, then the arrows in the transition diagram show that X_{n+1} can only take on the values 3 or 4, and it will do so with conditional probabilities 2/3 and 1/3, respectively. From state 2, the only possible moves that the chain can make in one step are to states 1 or 3, with equal likelihood. Because of (1), the probability law of the chain is completely determined by these single time step transition probabilities. The same transition probabilities can be assembled into a matrix called a *transition matrix*:

$$T = \text{now}: \begin{array}{c} \\ 1 \\ 2 \\ 3 \\ 4 \end{array} \begin{array}{cccc} \text{later}: & 1 & 2 & 3 & 4 \\ \left(\begin{array}{cccc} 0 & 0 & 2/3 & 1/3 \\ 1/2 & 0 & 1/2 & 0 \\ 0 & 1/4 & 0 & 3/4 \\ 0 & 0 & 1/2 & 1/2 \end{array} \right) \end{array} \tag{2}$$

where each row refers to a possible state "now" (at time n) and each column corresponds to a possible state "later" (at time $n + 1$). For example, the row 3-column 2 entry is

$$P[X_{n+1} = 2 \mid X_n = 3] = 1/4$$

Furthermore, the discrete probability distribution of the initial state X_0, if known, can be written as a row vector. For instance,

$$\mathbf{p}_0 = (1/2 \;\; 1/2 \;\; 0 \;\; 0)$$

is the initial distribution for which X_0 is equally likely to be in states 1 and 2. Special row vectors such as $\mathbf{p}_0 = (0 \; 1 \; 0 \; 0)$ apply when the inital state is certain; in this case it is state 2.

Activity 2 Try multiplying each row vector $\mathbf{p}_0$ above on the right by the transition matrix T. Make note of not only the final answers for the entries of $\mathbf{p}_0 T$, but also what they are composed of. Before reading on, try to guess at the meaning of the entries of $\mathbf{p}_0 T$.

The Law of Total Probability is the tool to prove the main theorem about the probability distribution of X_n. The nth power of the transition matrix gives the conditional distribution of X_n given X_0, and you can pre-multiply by the initial distribution $\mathbf{p}_0$ to get the unconditional distribution of X_n.

Theorem 1. (a) $T^n(i, j) = P[X_n = j \mid X_0 = i]$, where T^n denotes the nth power of the transition matrix.
(b) $\mathbf{p}_0 \cdot T^n(j) = P[X_n = j]$, where $\mathbf{p}_0$ denotes the distribution of X_0 in vector form.

Proof. (a) We work by induction on the power n of the transition matrix. When $n = 1$ we have for all i and j,

$$T^1(i, j) = P[X_1 = j \mid X_0 = i]$$

by the construction of the transition matrix. This anchors the induction. For the inductive step we assume for a particular $n - 1$ and all states i and j that $T^{n-1}(i, j) = P[X_{n-1} = j \mid X_0 = i]$. By the Law of Total Probability,

$$P[X_n = j \mid X_0 = i] = \sum_k P[X_{n-1} = k \mid X_0 = i] P[X_n = j \mid X_{n-1} = k, X_0 = i]$$

where the sum is taken over all states k. The first factor in the sum is $T^{n-1}(i, k)$ by the inductive hypothesis, and the second is $T(k, j)$ by the Markov property (1). Thus,

$$P[X_n = j \mid X_0 = i] = \sum_k T^{n-1}(i, k) T(k, j)$$

By the definition of matrix multiplication, the sum on the right is now the $i - j$ component of the product $T^{n-1} T = T^n$, which finishes the proof of part (a).

(b) By part (a) and the Law of Total Probability,

$$
\begin{aligned}
\mathbf{p}_0 T^n(j) &= \sum_k \mathbf{p}_0(k) T^n(k, j) \\
&= \sum_k P[X_0 = k] P[X_n = j \mid X_0 = k] \\
&= P[X_n = j]
\end{aligned}
$$

which completes the proof. ∎

Theorem 1 says that all we need to know about the probability distribution of the state X_n of the chain at time n is contained in the nth power of the transition matrix. Try the following activity.

Activity 3   For the Markov chain whose transition matrix is as in (2), find $P[X_2 = 3 \mid X_0 = 1]$. Compare your answer to the transition diagram in Figure 1 to see that it is reasonable. If $\mathbf{p}_0 = (1/4 \quad 1/4 \quad 1/4 \quad 1/4)$, find the probability mass function of X_2.

Example 1 A drilling machine can be in any of five different states of alignment, labeled 1 for the best, 2 for the next best, etc., down to the worst state 5. From one week to the next, it either stays in its current state with probability .95, or moves to the next lower state with probability .05. At state 5 it is certain to remain there the next week. Let us compute the probability distribution of the state of the machine at several times n, given that it started in state 1. What is the smallest n (that is, the first time) such that it is at least 2% likely that the machine has reached its worst state at time n?

By the assumptions of the problem, the chain of machine alignment states X_0, X_1, X_2, ... has the transition matrix

```
T = {{.95, .05, 0, 0, 0},
     {0, .95, .05, 0, 0}, {0, 0, .95, .05, 0},
     {0, 0, 0, .95, .05}, {0, 0, 0, 0, 1}};

MatrixForm[T]
```

$$\begin{pmatrix} 0.95 & 0.05 & 0 & 0 & 0 \\ 0 & 0.95 & 0.05 & 0 & 0 \\ 0 & 0 & 0.95 & 0.05 & 0 \\ 0 & 0 & 0 & 0.95 & 0.05 \\ 0 & 0 & 0 & 0 & 1 \end{pmatrix}$$

By Theorem 1, $P[X_n = j \mid X_0 = 1]$ is row 1 of the transition matrix T raised to the nth power. The conditional distribution of X_1 given $X_0 = 1$ is of course the first row $\{.95, .05, 0, 0, 0\}$ of the transition matrix itself. We compute these first rows for powers $n = 2$, 3, and 4 below, using *Mathematica*'s MatrixPower[mat, n] function. (You could also use the period for matrix multiplication.)

```
MatrixPower[T, 2][[1]]
MatrixPower[T, 3][[1]]
MatrixPower[T, 4][[1]]
```

$\{0.9025, 0.095, 0.0025, 0., 0.\}$

$\{0.857375, 0.135375, 0.007125, 0.000125, 0.\}$

$\left\{0.814506, 0.171475, 0.0135375, 0.000475, 6.25 \times 10^{-6}\right\}$

For $n = 4$, we notice that the likelihood is only 6.25×10^{-6} that the worst state 5 has been reached yet. The following command allows us to inspect the 1–5 element of some of the powers T^n.

```
Table[{n, (MatrixPower[T, n])[[1, 5]]}, {n, 20, 30}]
```

{{20, 0.0159015}, {21, 0.0188806},
 {22, 0.0221825}, {23, 0.0258145}, {24, 0.0297825},
 {25, 0.0340906}, {26, 0.0387414}, {27, 0.0437359},
 {28, 0.0490739}, {29, 0.0547534}, {30, 0.0607716}}

This output shows that it will be 22 weeks until it is at least .02 likely that the machine will be in state 5, given that it started in state 1. ∎

Long Run Distributions

Our second main goal is to find, if it exists, the vector of limiting probabilities

$$\pi(j) = \lim_{n \to \infty} P[X_n = j] \tag{3}$$

The jth entry of this limiting vector π has the interpretation of the long-run proportion of time the chain spends in state j. We will not prove it here, but a sufficient condition under which this π exists is the *regularity* of the chain.

Definition 2. A Markov chain is *regular* if there is a power T^n of the transition matrix that has all non-zero entries.

So, regularity of the chain means that there is a common time n such that all states can reach all other states by a path of length n with positive probability. The two chains with diagrams in Figure 2(a) and 2(b) are <u>not</u> regular. To simplify the diagrams we have suppressed the transition probabilities. The arrows display the transitions that have positive probability. Also in (b) we have suppressed a self-loop on state 4 indicating state 4 must return to itself. In (a), state 1 can only reach state 3 by a path of length 2, or 5, or 8, etc., while state 1 can only reach itself in a path of length 3, or 6, or 9, etc. There can be no common value of n so that both 1 reaches 3 and 1 reaches itself in n steps. In (b), state 4 does not reach any of the other states at all, so the chain cannot be regular.

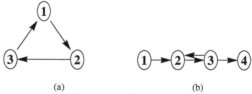

(a)   (b)

Figure 6.2 - Two non-regular Markov chains

The Markov chain whose transition diagram is in Figure 3 however is regular. You can check that there is a path of length 3 of positive probability from every state to every other state. (For example, paths 1,3,2,1; 1,3,3,2; and 1,3,3,3 are paths of length 3 from state 1 to all of the states.)

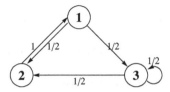

Figure 6.3 - A regular Markov chain

Activity 4 Use *Mathematica* to check that the Markov chain with transition matrix in (2) is regular.

Let us look at a few of the powers of the transition matrix of the chain corresponding to Figure 3.

```
T = {{0, .5, .5}, {1, 0, 0}, {0, .5, .5}};
MatrixForm[T]
```

$$\begin{pmatrix} 0 & 0.5 & 0.5 \\ 1 & 0 & 0 \\ 0 & 0.5 & 0.5 \end{pmatrix}$$

```
{MatrixPower[T, 2] // MatrixForm,
 MatrixPower[T, 3] // MatrixForm,
 MatrixPower[T, 4] // MatrixForm,
 MatrixPower[T, 5] // MatrixForm,
 MatrixPower[T, 6] // MatrixForm}
```

$$\left\{ \begin{pmatrix} 0.5 & 0.25 & 0.25 \\ 0. & 0.5 & 0.5 \\ 0.5 & 0.25 & 0.25 \end{pmatrix}, \begin{pmatrix} 0.25 & 0.375 & 0.375 \\ 0.5 & 0.25 & 0.25 \\ 0.25 & 0.375 & 0.375 \end{pmatrix}, \right.$$

$$\begin{pmatrix} 0.375 & 0.3125 & 0.3125 \\ 0.25 & 0.375 & 0.375 \\ 0.375 & 0.3125 & 0.3125 \end{pmatrix}, \begin{pmatrix} 0.3125 & 0.34375 & 0.34375 \\ 0.375 & 0.3125 & 0.3125 \\ 0.3125 & 0.34375 & 0.34375 \end{pmatrix},$$

$$\left. \begin{pmatrix} 0.34375 & 0.328125 & 0.328125 \\ 0.3125 & 0.34375 & 0.34375 \\ 0.34375 & 0.328125 & 0.328125 \end{pmatrix} \right\}$$

Rows 1 and 3 of T^n actually are identical for each n (look at the transition diagram to see why this is true), and row 2 is different from those rows. However all of the rows seem to be in closer and closer agreement as the power grows. The matrix T^n may be reaching a limit as $n \longrightarrow \infty$, which has the form of a matrix of identical rows. This would suggest that

$$\lim_{n \to \infty} T^n(i, j) = \lim_{n \to \infty} P[X_n = j \mid X_0 = i] = \pi(j) \tag{4}$$

independently of the initial state i. Let us find out what this limiting distribution must be, if it does exist.

If π is to be the limit in (4), then we must have

$$
\begin{aligned}
\pi(j) &= \lim_{n \to \infty} T^{n+1}(i, j) \\
&= \lim_{n \to \infty} \sum_k T^n(i, k) T(k, j) \\
&= \sum_k \lim_{n \to \infty} T^n(i, k) T(k, j) \\
&= \sum_k \pi(k) T(k, j)
\end{aligned} \tag{5}
$$

By computation (5), the limiting distribution π satisfies the system of matrix equations:

$$\pi = \pi T \quad \text{and} \quad \pi \mathbf{1} = 1 \tag{6}$$

Here $\mathbf{1}$ refers to a column vector of 1's of the same length as π so that the equation $\pi \mathbf{1} = 1$ means that the sum of the entries of π equals 1. We require this in order for π to be a valid discrete probability mass function.

We have motivated the following theorem.

Theorem 2. If a Markov chain X_0, X_1, X_2, ... is regular, then the limiting distribution $\pi(j)$ in (4) exists and satisfies the system of equations $\pi = \pi T$, $\pi \mathbf{1} = 1$. ∎

In the next example we see how Theorem 2 can be used in the long-run prediction of business revenue flow.

```
T = {{.52, .25, .11, .06, .03, .03},
     {.59, .22, .09, .03, .04, .03},
     {.50, .31, .12, .03, .02, .02}, {.55, .24, .09,
      .07, .03, .02}, {.60, .25, .08, .03, .02, .02},
     {.62, .28, .05, .03, .01, .01}};

MatrixForm[T]
```

$$
\begin{pmatrix}
0.52 & 0.25 & 0.11 & 0.06 & 0.03 & 0.03 \\
0.59 & 0.22 & 0.09 & 0.03 & 0.04 & 0.03 \\
0.5 & 0.31 & 0.12 & 0.03 & 0.02 & 0.02 \\
0.55 & 0.24 & 0.09 & 0.07 & 0.03 & 0.02 \\
0.6 & 0.25 & 0.08 & 0.03 & 0.02 & 0.02 \\
0.62 & 0.28 & 0.05 & 0.03 & 0.01 & 0.01
\end{pmatrix}
$$

Example 2   A plumbing contractor services jobs of six types according to their time demands, respectively 1 to 6 hours. The fee for the service is $40 per hour. Initially the contractor believes that the durations of successive jobs are independent of one another, and they occur with the following probabilities:

$$1 \text{ hr} : .56; 2 \text{ hrs} : .23; 3 \text{ hrs} : .10; 4 \text{ hrs} : .05; 5 \text{ hrs} : .03; 6 \text{ hrs} : .03$$

However, detailed data analysis suggests that the sequence of job times may instead be a Markov chain, with slightly different column entries (next job duration probabilities) for each row (current job duration) as in the matrix T above. By how much does the expected long-run revenue per job change if the job times do form a Markov chain?

First, in the default case where job times are independent and identically distributed, the average revenue per job is just $\mu = E[40\,X]$, where X is a discrete random variable modeling the duration of the job, with the distribution given above. By the Strong Law of Large Numbers, with probability 1 the actual average revenue per job $\overline{X}_n$ will converge to μ. We compute:

μ =
 40 (1 (.56) + 2 (.23) + 3 (.10) + 4 (.05) + 5 (.03) + 6 (.03))

74.

Now from Theorem 2 we can compute the long-run distribution of the Markov chain, then make a similar computation of the long-run expected revenue per job. By Exercise 6, the system of equations $\pi = \pi T$ is a dependent system; therefore, we can afford to discard one equation in it, and replace it with the condition that the sum of the entries of π is 1. We will discard the sixth equation. By setting $\pi = (x_1, x_2, x_3, x_4, x_5, x_6)$ and multiplying out $\pi\,T$ we get the following.

```
system =
   {x₁ == .52 x₁ + .59 x₂ + .5 x₃ + .55 x₄ + .6 x₅ + .62 x₆,
    x₂ == .25 x₁ + .22 x₂ + .31 x₃ + .24 x₄ + .25 x₅ + .28 x₆,
    x₃ == .11 x₁ + .09 x₂ + .12 x₃ + .09 x₄ + .08 x₅ + .05 x₆,
    x₄ == .06 x₁ + .03 x₂ + .03 x₃ + .07 x₄ + .03 x₅ + .03 x₆,
    x₅ == .03 x₁ + .04 x₂ + .02 x₃ + .03 x₄ + .02 x₅ + .01 x₆,
    x₁ + x₂ + x₃ + x₄ + x₅ + x₆ == 1};
Solve[system, {x₁, x₂, x₃, x₄, x₅, x₆}]
```

```
{{x₁ → 0.542039, x₂ → 0.249027, x₃ → 0.102504,
   x₄ → 0.0481887, x₅ → 0.0306065, x₆ → 0.0276343}}
```

Hence, the long-run average revenue per job under the Markov chain model is

> **40 (1 (.542039) + 2 (.249027) + 3 (.102504) +**
>
> **4 (.0481887) + 5 (.0306065) + 6 (.0276343))**

74.3679

So the more sophisticated Markov chain model did not make a great change in the revenue flow prediction; it predicts only about an extra 37 cents per job. This is perhaps to be expected, because the differences between the row probability distributions and the distribution assumed for X in the i.i.d. case are small. ∎

Absorbing Chains

The last problem type that we will study involves cases where the Markov chain has one or more states j such that $T(j, j) = 1$, so that once the chain enters that state it can never leave. Such a state is called *absorbing*. The specific question that we would like to answer is this: if a Markov chain has more than one absorbing state, how likely is it for the chain to be absorbed by each such state, given that the chain begins in a non-absorbing state? There are two types of states that we will consider here (not the only two types, by the way).

Definition 2. A state j in a Markov chain is called *transient* if, once the chain is there, there is a positive probability that it will never be visited again. A state j is called *absorbing* if, once the chain is there, with probability 1 the chain stays there forever.

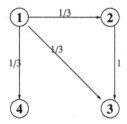

Figure 6.4 - A simple absorbing Markov chain

A simple example to illustrate the definition is in Figure 4. The absence of arrows emanating from states 3 and 4 indicates that the transition probabilities $T(3, 3)$ and $T(4, 4)$ equal 1, so states 3 and 4 are absorbing. State 2 is transient, because not only can it eventually be left forever, it must be left forever immediately in one step. Similarly state 1 is transient, because starting there the chain could either be absorbed by state 4 immediately, or by state 3 immediately, or the chain could pass through state 2 first, and then be absorbed by state 3.

Activity 5 Are any of the states 1, 2, or 3 in Figure 3 absorbing? Are any of them transient?

The simple structure of the chain in Figure 4 makes it especially simple to analyze. Starting from state 2, with certainty state 3 will absorb the chain. Starting from state 1, either state 4 will immediately absorb the chain, which happens with probability 1/3, or within two steps state 3 will absorb the chain, which will happen with total probability $1/3 + 1/3 = 2/3$. Since there were just two absorbing states, this probability is complementary to the state 4 absorption probability. Let V_j be the time (after time 0) of first visit to j; then if j is absorbing, the event $V_j < \infty$ occurs if and only if the chain is absorbed at j. If we denote

$$f_{ij} = P\left[V_j < \infty \mid X_0 = i\right] \tag{7}$$

for transient states i and absorbing states j, then for the chain in Figure 4 we have found that

$$f_{23} = 1; \; f_{13} = 2/3; \; f_{14} = 1/3$$

It is relatively easy to use recursive thinking and the Law of Total Probability to develop a system of linear equations using which we can compute the set of values f_{ij} for i transient and j absorbing. Activity 5 was attempting to get you to think about a third kind of state, the *recurrent state*, which is a non-absorbing state that is certain to be visited again and again by the chain. The states in a regular chain are recurrent. We consider chains that have no such states in our next theorem.

Theorem 3. Let TR be the set of transient states of the Markov chain (X_n) whose transition matrix is T. Assume that the state space of the chain consists only of transient states and absorbing states. Let $i \in TR$, and let j be an absorbing state of the chain. Then

$$f_{ij} = \sum_{k \in TR} T(i, k) f_{kj} + T(i, j) \tag{8}$$

Proof. We will condition and uncondition on X_1. Let ABS denote the set of absorbing states of the chain. If $X_0 = i$, then at time 1 the chain could move to another state in TR, or to a state in ABS. If that state to which it moves in ABS is state j then, of course, the chain is absorbed there, and if the state is any other absorbing state, it is impossible to be absorbed at j. Thus, we have:

$$
\begin{aligned}
f_{ij} &= P\left[V_j < \infty \mid X_0 = i\right] \\
&= \sum_{k \in E} P\left[V_j < \infty \mid X_1 = k, X_0 = i\right] P[X_1 = k \mid X_0 = i] \\
&= \sum_{k \in TR} P\left[V_j < \infty \mid X_1 = k, X_0 = i\right] T(i, k) + \sum_{k \in ABS} P\left[V_j < \infty \mid X_1 = k, X_0 = i\right] T(i, k) \\
&= \sum_{k \in TR} f_{kj} T(i, k) + 1 \cdot T(i, j).
\end{aligned}
$$

The Markov property justifies the fourth line. This proves formula (8). ∎

Formula (8) sets up a method for us to solve for the absorption probabilities f_{ij}. For each transient state i there is a linear equation in which all of the absorption probabilities f_{kj} are involved. We simply identify the coefficients and solve for the unknowns. It is particularly easy for the chain in Figure 4; there we have two transient states 1 and 2, and for absorbing state 3 we get the equations:

$$f_{13} = T(1, 1) f_{13} + T(1, 2) f_{23} + T(1, 3) = 0 + \frac{1}{3} f_{23} + \frac{1}{3}$$
$$f_{23} = T(2, 1) f_{13} + T(2, 2) f_{23} + T(2, 3) = 0 + 0 + 1 = 1$$

Substituting the value $f_{23} = 1$ into the top equation gives $f_{13} = 2/3$. (Try to write a similar system for absorbing state 4.)

We close with a less obvious example.

Example 3 Consider the six state random walk in Figure 5, in which states 0 and 5 are absorbing. Find the probability, starting at each of states 1 through 4 of being absorbed at state 5, and the probability of being absorbed at state 0.

Figure 6.5 - An absorbing random walk

Formula (8) can be applied at each state to give the system of four equations below in the unknowns f_{15}, f_{25}, f_{35}, f_{45}:

$$\left\{ \begin{array}{l} f_{15} = (1 - p) f_{05} + p f_{25} = p f_{25} \\ f_{25} = (1 - p) f_{15} + p f_{35} \\ f_{35} = (1 - p) f_{25} + p f_{45} \\ f_{45} = (1 - p) f_{35} + p f_{55} = (1 - p) f_{35} + p \end{array} \right.$$

(Compare the equations to the diagram and make sure that you understand the intuition.) In the first equation we have used the fact that $f_{05} = 0$ since state 0 is absorbing, and similarly in the last equation we note that $f_{55} = 1$ since state 5 is absorbing. We let *Mathematica* solve this system:

```
Solve[{f₁₅ == p * f₂₅,
  f₂₅ == (1 - p) f₁₅ + p * f₃₅, f₃₅ == (1 - p) f₂₅ + p * f₄₅,
  f₄₅ == (1 - p) f₃₅ + p}, {f₁₅, f₂₅, f₃₅, f₄₅}]
```

$$\left\{\left\{ f_{45} \rightarrow -\frac{-p + 2p^2 - 2p^3}{1 - 3p + 4p^2 - 2p^3 + p^4}, \quad f_{35} \rightarrow -\frac{-p^2 + p^3 - p^4}{1 - 3p + 4p^2 - 2p^3 + p^4}, \right.\right.$$

$$\left.\left. f_{15} \rightarrow \frac{p^4}{1 - 3p + 4p^2 - 2p^3 + p^4}, \quad f_{25} \rightarrow \frac{p^3}{1 - 3p + 4p^2 - 2p^3 + p^4} \right\}\right\}$$

By complementation, the absorption probabilities at zero are given by $f_{10} = 1 - f_{15}$, $f_{20} = 1 - f_{25}$, $f_{30} = 1 - f_{35}$, and $f_{40} = 1 - f_{45}$.

We can look at the dependence of the absorption probabilities on the probability p of taking a step to the right. For example, Figure 6 shows f_{35} as a function of p. The probability of being absorbed at state 5 increases, as one would expect, as the right step probability increases. At a value of p of .2, the chain has a strong tendency to drift left, and it is only about .05 likely for the chain to be absorbed at state 5. There is a rapid increase in the absorption probability between values of $p = .3$ and $p = .6$ followed by a leveling off. For right step probabilities of .8 and above it is nearly certain that the chain will be absorbed at state 5. ∎

```
f35[p_] := - -p² + p³ - p⁴ / (1 - 3 p + 4 p² - 2 p³ + p⁴);
Plot[f35[p], {p, 0, 1},
    AxesLabel → {p, None}, PlotStyle → Black]
```

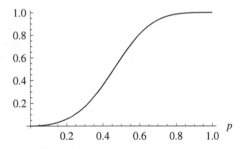

Figure 6.6 - Probability of absorption at state 5 starting at state 3

Activity 6 In Example 3, how large must p be so that the probability that the random walk is absorbed at state 5 given that it started at state 3 is at least .8? Approximate to at least two decimal places.

Exercises 6.1

1. (*Mathematica*) For the Markov chain whose transition diagram is as in Figure 1, compute the distribution of X_3 if X_0 is equally likely to equal each of the four states. Find the limiting distribution.

2. (*Mathematica*) A store keeps a small inventory of at most four units of a large appliance, which sells rather slowly. Each week, one unit will be sold with probability .40, or no units will be sold with probability .60. When the inventory empties out, the store is immediately able to order four replacements, so that from the state of no units, one goes next week to four units if none sell, or three if one unit sells. If the store starts with four units, find the distribution of the number of units in inventory at time 2. If the Markov chain of inventory levels is regular, find its limiting distribution.

3. (*Mathematica*) Build a command in *Mathematica* which takes an initial state X_0, a transition matrix T of a Markov chain, and a final time n, and returns a simulated list of states $X_0, X_1, \ldots, X_n$. Assume that states are numbered consecutively: 1, 2, 3, ...

4. Argue that an i.i.d. sequence of discrete random variables $X_0, X_1, X_2, \ldots$ is a special case of a Markov chain. What special structure would the transition matrix of this chain have?

5. (*Mathematica*) A communications system has six channels. When messages attempt to access the system when all channels are full, they are turned away. Otherwise, because messages neither arrive nor depart simultaneously, if there are free channels the next change either increases the number of channels in use by 1 (with probability .3) or decreases the number of channels in use by 1 (with probability .7). Of course, when all channels are free, the next change must be an increase and when all channels are busy the next change must be a decrease. Find the transition matrix for a Markov chain that models this situation. If the chain is regular, find its limiting distribution. If the chain is not regular, explain why it is not.

6. Show that for a regular Markov chain with transition matrix T, the system $\pi = \pi T$ has infinitely many solutions. (Hint: Consider a constant times π.)

7. (*Mathematica*) Consider the Markov chain in Figure 3. Find the limiting distribution, and find n large enough that each entry of T^n is within .01 of the corresponding entry of the limiting distribution.

8. Let (X_n) be a general two-state Markov chain such that $T(1, 1) = p$ and $T(2, 2) = q$. Find the distribution of X_3 if X_0 is equally likely to be each state. Find the limiting distribution in terms of p and q, and find, if possible, a combination of p and q values such that limiting distribution is $\{1/3, 2/3\}$.

9. (*Mathematica*) A short-haul rental moving van can be borrowed from and returned to three sites in a city. If it is borrowed from site 1, then it is returned to sites 1, 2, and 3 with probabilities 2/3, 1/6, and 1/6, respectively. If it is borrowed from site 2, then it is returned to sites 1, 2, and 3 with probabilities 1/2, 1/4, and 1/4, respectively. And if it is borrowed

from site 3, then it is returned to sites 1, 2, and 3 with equal probabilities. Given that the van starts at site 3, with what probability does it come to each site after four rentals? What is the long-run proportion of times that it is returned to each site?

10. A Markov chain has states 1, 2, 3, and 4, and the transition matrix below.

$$T = \begin{pmatrix} 1 & 0 & 0 & 0 \\ .3 & 0 & .7 & 0 \\ 0 & .6 & 0 & .4 \\ 0 & 0 & 0 & 1 \end{pmatrix}$$

Draw the transition diagram of the chain. Let

$$f(i) = P[\text{chain hits state 1 before state 4} \mid X_0 = i], i = 1, 2, 3, 4$$

Set up and solve a system of equations to find $f(i)$ for all states i. Explain where the equations in the system come from.

11. Let (X_n) be a Markov chain with the transition matrix below. Classify all states as transient, absorbing, or neither.

$$\begin{pmatrix} 1/3 & 2/3 & 0 & 0 & 0 & 0 \\ 1/2 & 1/2 & 0 & 0 & 0 & 0 \\ 0 & 1/2 & 0 & 1/6 & 1/3 & 0 \\ 0 & 0 & 0 & 1/3 & 1/3 & 0 \\ 0 & 0 & 0 & 0 & 1 & 0 \\ 0 & 0 & 0 & 0 & 0 & 1 \end{pmatrix}$$

12. Consider the general four state absorbing random walk below. Find all probabilities of absorption to state 4 in terms of p. For what p is the absorption probability to state 4 from state 2 increasing most rapidly?

Exercise 12

13. A *symmetric* random walk is one in which steps to the right and steps to the left have equal probability. If a 5-state symmetric random walk with an absorbing state on each end is such that states are numbered sequentially 0, 1, 2, 3, 4 show that the absorption probability at state 4 is a linear function of the state.

14. In a manufacturing process, items go through four stages of assembly before potentially being sent out for sale. At each stage, the item is inspected, and either goes on to the next stage with probability .9, or is disposed of because of defects. Model the situation as a

Markov chain. Find the proportion of items that complete all stages successfully and are sent out for sale rather than being disposed of.

15. For the drilling machine in Example 1, compute the p.m.f. of the time to absorption at state 5 given that the machine starts in state 1: $g(n) = P[T_5 = n \mid X_0 = 1]$, where T_5 is the first time that the chain hits state 5.

6.2 Poisson Processes

In Section 2.4 we informally introduced the Poisson process. Recall that this is essentially a counting process that gives the total number of arrivals of some entity that have come by each fixed time t. Looked at as a function of time t for fixed experimental outcome ω, the sample path of a Poisson process looks like the graph in Figure 7, which reprises Figure 13 of Chapter 2.

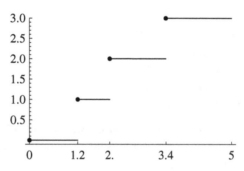

Figure 6.7 - Sample path of a Poisson process

Activity 1 Give some examples of physical phenomena for which a Poisson process model might be appropriate.

Here is a more precise definition than we gave earlier.

Definition 1. A *Poisson process* $N = N(t, \omega)$ with rate λ is for each fixed t a random variable and for each $\omega \in \Omega$ a function from $[0, \infty)$ to the non-negative integers, such that $N(0, \omega) = 0$ and

(a) the function $t \to N(t, \omega)$ increases by jumps of size 1 only;

(b) there is an independent, identically distributed sequence of $\exp(\lambda)$ random variables S_1, S_2, S_3, $\cdots$ such that the times of jumps are

$$T_n = S_1 + S_2 + \cdots + S_n, \quad n = 1, 2, \cdots$$

Following our usual convention for stochastic processes, we typically write N_t for the random variable $N(t, \omega)$ for fixed t. So, a Poisson arrival counting process begins at state 0, waits for an exponentially distributed length of time S_1, then at time $T_1 = S_1$ jumps to 1, where it waits another exponential length of time S_2 until at time $T_2 = S_1 + S_2$ it jumps to state 2, etc. Such processes are often used in the field of queueing theory to model arrivals of customers to a service facility. The memorylessness of the exponential distribution means that at any instant of time, regardless of how long we may have been waiting, we will still wait an exponential length of time for the next arrival to come.

Since the inter-jump times determine the process completely, one would suppose that it is now possible to derive the result stated earlier, namely that the probability distribution of N_t, the total number of arrivals by time t, is Poisson. In fact, the following interesting interplay between the exponential, gamma, and Poisson distributions comes about.

Theorem 1. (a) The distribution of the nth jump time T_n is $\Gamma(n, 1/\lambda)$.
(b) The distribution of $N_t =$ number of arrivals in $[0, t]$ is Poisson(λt).

Proof. (a) Since the exp(λ) distribution is also the $\Gamma(1, 1/\lambda)$ distribution, Theorem 2(b) of Section 4.5 implies that

$$T_n = S_1 + S_2 + \cdots + S_n \sim \Gamma(n, 1/\lambda)$$

(b) It is easy to see that the event $N_t \geq n$ is the same as the event $T_n \leq t$, since both of these are true if and only if the nth arrival has come by time t. Then,

$$
\begin{aligned}
P[N_t = n] &= P[N_t \geq n] - P[N_t \geq n + 1] \\
&= P[T_n \leq t] - P[T_{n+1} \leq t].
\end{aligned}
$$

Because T_{n+1} has the $\Gamma(n + 1, 1/\lambda)$ distribution, integration by parts yields

$$
\begin{aligned}
P[T_{n+1} \leq t] &= \int_0^t \frac{\lambda^{n+1}}{n!} x^n e^{-\lambda x} \, dx \\
&= -\frac{\lambda^{n+1}}{n!} \cdot \frac{1}{\lambda} \cdot e^{-\lambda x} \cdot x^n \Big|_0^t + \int_0^t \frac{\lambda^{n+1}}{n!} \frac{1}{\lambda} n x^{n-1} e^{-\lambda x} \, dx \\
&= -\frac{(\lambda t)^n}{n!} e^{-\lambda t} + \int_0^t \frac{\lambda^n}{(n-1)!} x^{n-1} e^{-\lambda x} \, dx \\
&= -\frac{(\lambda t)^n}{n!} e^{-\lambda t} + P[T_n \leq t]
\end{aligned}
$$

Substituting this into the previous equation yields

$$P[N_t = n] = \frac{(\lambda t)^n}{n!} e^{-\lambda t}$$

which is the nth term of the Poisson(λt) distribution, as desired. ∎

Example 1 Suppose that users of a computer lab arrive according to a Poisson process with rate $\lambda = 15$/hr. Then the expected number of users in a 2-hour period is $\lambda t = 15(2) = 30$ (recall that the Poisson parameter is equal to the mean of the distribution). The probability that there are 10 or fewer customers in the first hour is $P[N_1 \leq 10]$, i.e.:

$$\mathbf{N}\left[\sum_{n=0}^{10} \frac{\mathbf{15^n}}{\mathbf{n!}} \mathbf{E^{-15}}\right]$$

0.118464

The probability that the second arrival comes before time .1 (i.e., in the first 6 minutes) is $P[T_2 < .1]$, which is

$$\mathbf{N}\left[\int_0^{.1} \frac{\mathbf{15^2}}{\mathbf{1!}} \mathbf{x^{2-1}} \mathbf{E^{-15\,x}} \, \mathbf{dx}\right]$$

0.442175

■

Activity 2 For the computer lab example above, what is the standard deviation of the number of users in a 2-hour period? What is the probability that there are 5 or fewer customers in the first half hour? What is the probability that the third arrival comes somewhere between the first 10 and the first 20 minutes?

Example 2 Again in the context of the computer lab, let us find the smallest value of the arrival rate λ such that the probability of having at least 5 arrivals in the first hour is at least .95. Measuring time in hours, we want the smallest λ such that

$$P[N_1 \geq 5] \geq .95 \implies P[N_1 \leq 4] \leq .05$$

by complementation. Since N_1 has the Poisson($\lambda \cdot 1$) distribution, the probability that there are four or fewer arrivals is the sum of the following terms:

$$\mathbf{P4orfewer}[\lambda\_] := \sum_{k=0}^{4} \frac{\mathbf{E^{-\lambda}} * \lambda^{\mathbf{k}}}{\mathbf{k!}}$$

To get some idea of where the appropriate value of λ lies, we graph the function for values of λ between 5 and 15.

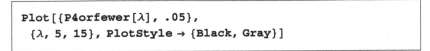

```
Plot[{P4orfewer[λ], .05},
  {λ, 5, 15}, PlotStyle → {Black, Gray}]
```

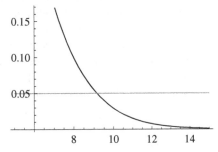

Figure 6.8 - Probability of 4 or fewer arrivals in first hour as a function of arrival rate

So the solution seems to be about $\lambda = 9$. We can find a root of the equation P4orfewer[λ] $- .05 = 0$ with an initial guess of 9 in order to get closer. We find that to the fifth decimal place $\lambda = 9.15352$. ∎

```
FindRoot[P4orfewer[λ] - .05, {λ, 9}]
```

$\{\lambda \rightarrow 9.15352\}$

The Poisson process satisfies a pair of other conditions that are both interesting and useful in computations. We introduce these now.

Definition 2. (a) A stochastic process $(X_t)_{t\geq 0}$ is called *stationary* if for all $t \geq 0$, $s > 0$, the distribution of $X_{t+s} - X_t$ does not depend on t;

(b) A stochastic process $(X_t)_{t\geq 0}$ is said to have *independent increments* if for all $t \geq 0$, $s > 0$ $X_{t+s} - X_t$ is independent of X_r for all $r \leq t$.

The stationarity property says that the distribution of the change in value of the process over a time interval does not depend on when that time interval begins, only on its duration. Applying this at $t = 0$, the distribution of $X_{t+s} - X_t$ is the same as that of $X_s - X_0$. For the Poisson process, since $N_0 = 0$, it follows that $N_{t+s} - N_t \sim \text{Poisson}(\lambda s)$ for all s, t. The independent increments property says that changes in value of the process on disjoint time intervals are independent.

Theorem 2. The Poisson process has stationary, independent increments. ∎

A formal proof of Theorem 2 is rather technical. But an informal argument is easy to outline. Consider the increment $N_{t+s} - N_t$. At time t we can reinitialize the Poisson process, counting arrivals that occur after that time. By the memoryless property of the exponential distribution, the probability distribution of the time we will wait until the next arrival after time t is an exponential(λ) random variable, as are all subsequent interarrival intervals. And, the initial waiting time and subsequent interarrival intervals are also independent of previous interarrival intervals before time t. Looked at as a process indexed by s, $N_{t+s} - N_t$ starts at 0 at $s = 0$, and increases by 1 at each new arrival time. It is therefore a Poisson process in its own right, independent of the random variables N_r, $r \leq t$. The independent increments property follows from this, and stationarity follows from the observation that

$$P[N_{t+s} - N_t = k] = \frac{e^{-\lambda s}(\lambda s)^k}{k!} = P[N_s = k]$$

Example 3 Suppose that a mall department store has two entrances, one on the east and the other on the west end of the store. Customers enter from the east according to a Poisson process with rate 2 per minute, and they enter from the west by a Poisson process with rate 1.5 per minute. Furthermore, the two Poisson processes are independent. Compute the joint distribution of the numbers N_e and N_w of customers who have entered from the two directions by time t, and compute the distribution of the total number of customers N who have entered by time t. What is the conditional distribution of the total number of customers by time $t + s$, given that there have been n customers by time t?

The problem description indicates that we should assume that $N_e \sim$ Poisson($2t$), $N_w \sim$ Poisson($1.5t$), and N_e and N_w are independent random variables. The joint p.m.f. of these two random variables is the product of their marginals:

$$f(x, y) = p(x) \cdot q(y) = \frac{e^{-2t}(2t)^x}{x!} \cdot \frac{e^{-1.5t}(1.5t)^y}{y!}$$

Now let $N = N_e + N_w$ be the total number of customers by time t. Then its distribution is, by the law of total probability,

$$
\begin{aligned}
g(n) = P[N = n] &= \sum_{i=0}^{n} P[N_e = i, N_w = n - i] \\
&= \sum_{i=0}^{n} \frac{e^{-2t}(2t)^i}{i!} \cdot \frac{e^{-1.5t}(1.5t)^{n-i}}{(n-i)!} \\
&= \frac{e^{-3.5t}}{n!} \sum_{i=0}^{n} \frac{n!}{i!(n-i)!} (2t)^i (1.5t)^{n-i} \\
&= \frac{e^{-3.5t}}{n!} (2t + 1.5t)^n \\
&= \frac{e^{-3.5t}(3.5t)^n}{n!}
\end{aligned}
$$

Note that the fourth line of the derivation follows from the binomial theorem. This shows that N, the total number of arrivals, has the Poisson($3.5t$) distribution. We have an indication, though not a full proof, that the sum of two independent Poisson processes is also a Poisson process, whose rate is the sum of the rates of the two components. See the next activity for more evidence. We will assume this for the second part of the problem.

Write the total number of customers by time t as N_t. We want to compute for all $m \geq n$,

$$P[N_{t+s} = m \mid N_t = n] = \frac{P[N_{t+s} = m, \ N_t = n]}{P[N_t = n]}$$

Now in order for both $N_{t+s} = m$ and $N_t = n$ to occur, there must be exactly n arrivals during $[0, t]$ and $m - n$ arrivals during $(t, t + s]$. Since these time intervals are disjoint, by the independent increments property of the Poisson process the probability of the intersection of the two events must factor. Therefore,

$$
\begin{aligned}
P[N_{t+s} = m \mid N_t = n] &= \frac{P[N_{t+s} - N_t = m-n, \ N_t = n]}{P[N_t = n]} \\
&= \frac{P[N_{t+s} - N_t = m-n] \ P[N_t = n]}{P[N_t = n]} \\
&= P[N_{t+s} - N_t = m - n] \\
&= \frac{e^{-3.5s} (3.5\,s)^{m-n}}{(m-n)!}, \quad m \geq n
\end{aligned}
$$

The last line of the computation follows from the stationarity property of the Poisson process. ∎

Activity 3 In Example 3, starting at a particular time t, find the distribution of the time until the next customer from either of the two directions (i.e., the minimum of the waiting time for the next customer from the east and the next customer from the west). Explain how your answer helps prove that the process that counts total customers is a Poisson process.

Example 4 If (N_t) is a Poisson process with rate λ, find $\mathrm{Cov}(N_t, N_{t+s})$.

We know that since the value of the process at time t has the Poisson distribution with parameter λt, $E[N_t] = \lambda t$ and $E[N_{t+s}] = \lambda(t+s)$. It remains to compute $E[N_t \cdot N_{t+s}]$. But also we note that $N_{t+s} - N_t$ is independent of N_t and has the Poisson(λs) distribution; hence,

$$
\begin{aligned}
E[N_t \cdot N_{t+s}] &= E[N_t \cdot (N_{t+s} - N_t + N_t)] \\
&= E[N_t] \cdot E[N_{t+s} - N_t] + E[N_t^2] \\
&= E[N_t] \cdot E[N_{t+s} - N_t] + \left(\mathrm{Var}(N_t) + E[N_t]^2\right) \\
&= \lambda t \cdot \lambda s + \lambda t + (\lambda t)^2
\end{aligned}
$$

Therefore,

$$\text{Cov}(N_t, \ N_{t+s}) = \lambda t \cdot \lambda s + \lambda t + (\lambda t)^2 - (\lambda t)(\lambda(t + s)) = \lambda t$$

after a little simplification. Oddly enough this does not depend on s, the length of the time interval. (Can the same be said for the correlation?) ∎

Here is one last interesting result about Poisson processes, and a nice application of it. We will not attempt to prove this theorem, although a very special case is the subject of Exercise 9.

Theorem 3. Conditioned on the event $\{N_t = n\}$, the random arrival times $T_1, T_2, \ldots, T_n$ have the joint distribution of the order statistics $U_{(1)}, U_{(2)}, \ldots, U_{(n)}$ of a random sample of size n from the uniform distribution on $[0, t]$. ∎

Roughly, Theorem 3 states that if we know how many arrivals occurred by time t, the times of arrival are scattered randomly and uniformly in $[0, t]$.

Example 5 Cars arrive by a Poisson process with rate λ to an essentially infinite parking garage, so that each is able to find a space. Each car stays an exponentially distributed amount of time V with parameter μ, independently of other cars and independent of its own arrival time, and then departs. Find the probability that the garage has exactly k cars in it at time t.

Each car has a particular arrival time T_i and an occupation time, say V_i, so that the ith car departs at time $T_i + V_i$. Condition and uncondition on the event $\{N_t = n\}$ to obtain:

$$\text{P[exactly } k \text{ cars at time } t] = \sum_{n=k}^{\infty} \text{P[exactly } k \text{ cars at time } t \mid N_t = n] \cdot \text{P}[N_t = n]$$

Note that in the sum, n must be at least k, else not enough cars arrived. There are exactly k cars still parked in the garage if and only if exactly k of the events $\{T_i + V_i > t\}$ occur. By Theorem 3, conditioned on the number of arrivals n, the probability that exactly k of these events occur is the same as the probability that exactly k of the events $\{U_i + V_i > t\}$ occur, where $U_1, U_2, \ldots, U_n$ are independent uniform random variables on $[0, t]$, independent also of the V_i's. We thereby form a binomial experiment with n trials and success probability:

$$p = \text{P}[U + V > t] \tag{1}$$

where $U \sim \text{unif}(0, t)$, $V \sim \exp(\mu)$, and U and V are independent. Therefore,

$$
\begin{aligned}
\text{P[exactly } k \text{ cars at time } t] &= \sum_{n=k}^{\infty} \text{P}[k \text{ successes} \mid N_t = n] \cdot \text{P}[N_t = n] \\
&= \sum_{n=k}^{\infty} \binom{n}{k} p^k (1 - p)^{n-k} \frac{e^{-\lambda t}(\lambda t)^n}{n!} \\
&= \frac{e^{-\lambda t}(\lambda t)^k}{k!} \cdot p^k \cdot \sum_{n=k}^{\infty} \frac{(1-p)^{n-k}(\lambda t)^{n-k}}{(n-k)!} \\
&= \frac{e^{-\lambda t}(\lambda t \, p)^k}{k!} \cdot \sum_{l=0}^{\infty} \frac{((1-p) \cdot \lambda t)^l}{l!}
\end{aligned}
$$

$$P[\text{exactly } k \text{ cars at time } t] = \frac{e^{-\lambda t}(\lambda t\, p)^k}{k!} \cdot e^{\lambda t(1-p)}$$

$$= \frac{e^{-\lambda t p}(\lambda t\, p)^k}{k!} \tag{2}$$

In the third line we remove factors having to do with k only; in the fourth line we change the variable of summation to $l = n - k$; and in the fifth line we use the well known Taylor series formula for the exponential function. The algebraic simplification in line 6 shows that the distribution of the number of cars remaining in the lot at time t is Poisson($\lambda t\, p$), where p is the probability in formula (1). ■

Activity 4 Find an explicit expression for p in Example 5.

Exercises 6.2

1. (*Mathematica*) If $N = N(t, \omega)$ is a Poisson process with rate 6, find:
(a) $P[N_{1.2} = 3]$; (b) $P[T_4 \le .25]$.

2. For a Poisson process of rate $\lambda = 1$, compute $P[T_3 - T_1 < 2, T_6 - T_4 > 2]$.

3. For a Poisson process of rate $\lambda = 2$, compute $P[N_{1.2} = 1, N_{2.7} = 3, N_{4.1} = 5]$.

4. (*Mathematica*) Assume that cars pull up to a drive-in window according to a Poisson process of rate λ per minute. What is the largest value of λ such that the probability that the second car arrival time is less than one minute is no more than .5?

5. (*Mathematica*) Implement a *Mathematica* command called SimulatePoissonProcess to simulate and graph a Poisson process.

6. (*Mathematica*) Use the SimulatePoissonProcess command of Exercise 5 to simulate around 40 paths of a Poisson process with rate 2 between times 0 and 4. Count how many of your paths had exactly 0, 1, and 2 arrivals by time 1. How well do your simulations fit the theoretical probabilities of 0, 1, and 2 arrivals?

7. Customer arrivals forming a Poisson process of rate 2/minute come to a single server. The server takes a uniformly distributed amount of time between 20 and 30 seconds for each customer to perform service. Compute the expected number of new arrivals during a service time. (Hint: condition and uncondition on the duration of the service time.)

8. Show that for a Poisson process, if you take the fact that $N_t \sim$ Poisson(λt) for all t as a primitive assumption, then it follows that the time of the first arrival must have the exponential distribution.

9. Show that given $N_t = 1$, T_1 is uniformly distributed on $[0, t]$.

10. Cars arrive to a T-intersection according to a Poisson process at an overall rate of 5/min. Each car turns left with probability 1/4 and right with probability 3/4, independently of other cars. If N_t^l is the number of cars that have turned left by time t, find the probability distribution of N_t^l.

11. Suppose that arrivals of customers to a discount store form a Poisson process with rate 5 per hour. Each customer buys $1, $2, $3, or $4 with probabilities 1/8, 1/8, 1/4, and 1/2, respectively. Successive customer purchases are independent. Let X_t be the total purchase amount up through time t. Find $E[X_t]$.

| 6.3 Queues

The word "queue" is a word used mostly in Great Britain and the former empire to refer to a waiting line. We have all been in them at stores, banks, airports and the like. We the customers come to a facility seeking some kind of service. Most often, our arrival times are not able to be predicted with certainty beforehand. There is some number of servers present, and we either join individual lines before them (as in grocery store checkout lanes) or we form a single line waiting for the next available server (as in most post office counters and amusement park rides). Some rule, most commonly first-in, first-out, is used to select the next customer to be served. Servers take some amount of time, usually random, to serve a customer. When a customer finishes, that customer may be directed to some other part of the queueing system (as in a driver's license facility where there are several stations to visit), or the customer may depart the system for good.

Predicting the behavior of human queueing systems is certainly an interesting problem, to which it is easy to relate. However, it should be noted that it is the study of very large, non-human queueing systems, especially those involved with computer and telecommunications networks, which gave probability great impetus in the latter 20th century. Because instead of human customers and servers we could just as easily consider computer jobs or messages in a queue waiting for processing by a server or communication channel.

There is a wide variety of problems in queueing theory, generated by variations in customer arrival and facility service patterns, breakdowns of servers, changes in the size of the waiting space for customers, rules other than first-in, first-out for selecting customers, etc. Our small section will just touch on the most basic of all queues: the so-called *M/M/1 queue.*

> **Activity 1** Think of three particular human queueing systems you have been involved in. What main features did they have; for example, how many servers were there, were customer arrivals and services random, was the queue a first-in, first-out queue, etc.? Then think of three examples of non-human queueing systems.

Here are the assumptions that identify a queueing system as M/M/1. Customers arrive singly at random times T_1, T_2, T_3, ... such that the times between arrivals

$$T_1, T_2 - T_1, T_3 - T_2, ... \tag{1}$$

are i.i.d. $\exp(\lambda)$ random variables. The parameter λ is called the *arrival rate* of the queue. Therefore, the process that counts the total number of customer arrivals by time t is a Poisson process with rate λ. There is a single server (the 1 in the notation M/M/1), and customers form one first-come, first-served line in front of the server. Essentially infinite waiting space exists to accommodate the customers in queue. The server requires an exponentially distributed length of time with rate μ for each service, independently of the arrival process and other service times. Customers who finish service depart the system. (By the way, the two "M" symbols in the notation stand for Markov, because the memoryless nature of the exponential distribution makes the arrival and departure processes into continuous-time versions of a Markov chain.)

One of the most important quantities to compute in queueing systems is the probability distribution of the number of customers in the queue or in service, at any time t. Such knowledge can help the operators of the system to know whether it will be necessary to take action to speed up service in order to avoid congestion. From the customer's point of view, another important quantity is the distribution of the customer's waiting time in the queue. We will consider each of these questions in turn for the M/M/1 queue, producing exact results for the long-run distributions. But exact computations become difficult even for apparently simple queueing situations, so very often analysts must resort to simulating the queue many times to obtain statistical estimates of such important quantities as these. Entire languages, such as GPSS, exist to make it easy to program such simulations, but as we will show, it is not so hard to write queueing simulation programs even in a very general purpose language like *Mathematica* which has not been specifically designed for queueing simulation.

Activity 2   How would you simulate the process of customer arrivals in an M/M/1 queue? Try writing a *Mathematica* command to do this, ignoring the service issue for the moment.

Long Run Distribution of System Size

To begin the problem of finding the long-run distribution of the number of customers in the queueing system, define M_t as the number of customers at time t, including the one being served, if any. Define also a sequence of real-valued functions of t:

$$f_k(t) = P[M_t = k], \quad k = 0, 1, 2, ... \tag{2}$$

If for each k, the limit as $t \longrightarrow \infty$ of $f_k(t)$ can be found, then these limits f_k form the limiting probability distribution of system size. Actually, the functions $f_k(t)$ can be computed themselves, which would give us the short run system size distribution, but this is much more difficult than finding the limiting distribution and we will not attempt this problem here.

The unique approach to the problem of finding the limiting distribution is to construct a system of differential equations satisfied by the functions $f_k(t)$. This illustrates beautifully the power of blending apparently different areas of mathematics to solve a real problem. In the interest of space we will leave a few details out and not be completely rigorous. The interested reader can try to fill in some gaps.

We first would like to argue that the probability that two or more arrival or service events happen in a very short time interval is negligibly small. Consider for example the probability that there are two or more arrivals in a short time interval $(t, t + h]$. Because arrivals to an M/M/1 queue form a Poisson process, we can complement, use the Poisson p.m.f., and rewrite the exponential terms in a Taylor series as follows:

$$
\begin{aligned}
P[2 \text{ or more arrivals in } (t, t + h]] &= 1 - P[0 \text{ or } 1 \text{ arrival in } (t, t + h]] \\
&= 1 - \frac{e^{-\lambda h}(\lambda h)^0}{0!} - \frac{e^{-\lambda h}(\lambda h)^1}{1!} \\
&= 1 - e^{-\lambda h} - \lambda h e^{-\lambda h} \\
&= 1 - \left(1 - \lambda h + \frac{1}{2}(\lambda h)^2 - \cdots\right) \\
&\quad - \lambda h\left(1 - \lambda h + \frac{1}{2}(\lambda h)^2 - \cdots\right) \\
&= \frac{1}{2}(\lambda h)^2 + \text{higher powers of } h
\end{aligned}
$$

So this probability is a small enough function of h such that if we divide by h and then send h to zero, the limit is 0. Recall from calculus that such a function is generically called $o(h)$. We can argue similarly that if 2 or more customers are present in the system at time t, then the probability that at least 2 customers will be served during $(t, t + h]$ is an $o(h)$ function. (Try it.) The activity below leads you through another case of two queueing events in a short time.

Activity 3 Suppose there is at least one customer in the queue and we look at a time interval $(t, t + h]$ again. Write one expression for the probability of exactly one arrival in $(t, t + h]$, and another for the probability of exactly one service in $(t, t + h]$. Use them to argue that the probability of having both an arrival and a service in $(t, t + h]$ is $o(h)$.

Assuming that you are now convinced that the probability of two or more queueing events in a time interval of length h is $o(h)$, let us proceed to derive a differential equation for $f_0(t) = P[M_t = 0]$ by setting up a difference quotient

$$
\frac{f_0(t + h) - f_0(t)}{h}
$$

Now $f_0(t + h) = P[M_{t+h} = 0]$, so let us consider how the system size could have come to be 0 at time $t + h$, conditioning on the system size at time t. There are two main cases, 1 and 2 below, and a collection of others described in case 3 that involve two or more queueing events in $(t, t + h]$, which therefore have probability $o(h)$:

1. $M_t = 0$, and there were no arrivals in $(t, t + h]$;
2. $M_t = 1$, and there was one service and no new arrivals in $(t, t + h]$;
3. Neither 1 nor 2 hold, and all existing and newly arriving customers are served in $(t, t + h]$.

The total probability $P[M_{t+h} = 0]$ is the sum of the probabilities of the three cases. But the probability of no arrivals in $(t, t + h]$ is $e^{-\lambda h} = 1 - \lambda h + o(h)$ and the probability of exactly one service is $\mu h e^{-\mu h} = \mu h(1 - \mu h + o(h)) = \mu h + o(h)$. By independence, the probability of one service and no new arrivals is therefore $(1 - \lambda h + o(h))(\mu h + o(h)) = \mu h + o(h)$. Thus, by the Law of Total Probability,

$$
\begin{aligned}
f_0(t + h) = P[M_{t+h} = 0] &= P[M_t = 0](1 - \lambda h) + P[M_t = 1](\mu h) + o(h) \\
&= (1 - \lambda h) f_0(t) + \mu h f_1(t) + o(h)
\end{aligned}
\tag{3}
$$

Subtracting $f_0(t)$ from both sides, and then dividing by h and sending h to zero yields

$$
\frac{f_0(t + h) - f_0(t)}{h} = -\lambda f_0(t) + \mu f_1(t) + \frac{o(h)}{h}
\tag{4}
$$
$$
\implies f_0'(t) = -\lambda f_0(t) + \mu f_1(t)
$$

Equation (4) is one of the equations in the infinite system of equations we seek. You are asked to show in Exercise 3 that the rest of the differential equations have the form:

$$
f_k'(t) = -(\lambda + \mu) f_k(t) + \lambda f_{k-1}(t) + \mu f_{k+1}(t), \quad k \geq 1
\tag{5}
$$

Intuitively, the first term on the right side of (5) represents the case of neither an arrival nor a service during $(t, t + h]$, the second term is for the case where there were $k - 1$ customers at time t and one new arrival appeared in $(t, t + h]$, and the third term corresponds to the case where there were $k + 1$ customers at time t, and one customer was served in $(t, t + h]$.

The system of differential equations (4) and (5) can be solved with a lot of effort (see [Gross and Harris, p. 129]), but we will just use the system to derive and solve an ordinary system of linear equations for the limiting probabilities $f_k = \lim_{t \to \infty} f_k(t)$ $= \lim_{t \to \infty} P[M_t = k]$.

Observe that if the $f_k(t)$ functions are approaching a limit, there is reason to believe that the derivatives $f_k'(t)$ approach zero. Presuming that this is so, we can send t to ∞ in (4) and (5) to obtain the system of linear equations:

$$
\begin{cases}
0 = -\lambda f_0 + \mu f_1 \\
0 = -(\lambda + \mu) f_k + \lambda f_{k-1} + \mu f_{k+1}, \quad k \geq 1
\end{cases}
\tag{6}
$$

We can adjoin to this system the condition that $f_0 + f_1 + f_2 + \cdots = 1$ in order to force the f_k's to form a valid probability distribution. Now the top equation in (6) implies

$$f_1 = \frac{\lambda}{\mu} f_0$$

When $k = 1$, the bottom equation in (6) implies

$$\mu f_2 = (\lambda + \mu) f_1 - \lambda f_0 = (\lambda + \mu) \frac{\lambda}{\mu} f_0 - \lambda f_0 = \frac{\lambda^2}{\mu} f_0$$

$$\implies f_2 = \frac{\lambda^2}{\mu^2} f_0$$

Activity 4 Check that $f_3 = \dfrac{\lambda^3}{\mu^3} f_0$.

In Exercise 4 we ask you to show inductively that for each $k \geq 1$,

$$f_k = \frac{\lambda^k}{\mu^k} f_0 \tag{7}$$

Clearly though, in order for the sum of the f_k's to be 1, or to be finite at all, the ratio $\rho = \lambda/\mu$ must be less than 1. This ratio ρ is called the *traffic intensity* of the M/M/1 queue, a good name because it compares the arrival rate to the service rate. If $\rho < 1$, then $\mu > \lambda$, and the server works fast enough to keep up with the arrivals. I will leave it to you to show that under the condition $\sum\limits_{k=0}^{\infty} f_k = 1$, the limiting M/M/1 queue system size probabilities are

$$f_k = \lim_{t \to \infty} P[M_t = k] = \rho^k (1 - \rho), \quad k = 0, 1, 2, \ldots \tag{8}$$

So we see that the limiting system size has the geometric distribution with parameter $1 - \rho$ (see formula (1) of Section 2.3). An interesting consequence is that the average number of customers in the system is

$$L = \frac{\rho}{1 - \rho} \tag{9}$$

From this expression, notice that as the traffic intensity ρ increases toward 1, the average system size increases to ∞, which is intuitively reasonable.

Example 1 At a shopping mall amusement area, customers wanting to buy tokens for the machines form a first-in, first-out line in front of a booth staffed by a single clerk. The customers arrive according to a Poisson process with rate 3/min., and the server can serve at an average time of 15 sec. per customer. Find the distribution and average value of the number of customers awaiting service. Suppose that replacing the booth clerk by a machine

costs $2000 for the machine itself plus an estimated $.50 per hour for power and maintenance, and with this self-service plan customers could serve themselves at an average time of
12 sec. apiece. The human server is paid $5/hr. Also, an estimate of the intangible cost of
inconvenience to customers is $.20 per waiting customer per minute. Is it worthwhile to
replace the clerk by the machine?

The given information lets us assume that the times between customer arrivals are
exponentially distributed with rate parameter $\lambda = 3$/min. If we are willing to assume that
service times are independent exponential random variables, which are furthermore independent of arrivals, then the M/M/1 assumptions are met. The service rate is the reciprocal of
the mean service time, which is $1/(1/4$ min.$) = 4$/min. Thus the queue traffic intensity is
$\rho = \lambda/\mu = 3/4$. By formula (8) the long-run probability of k customers in the system is

$$f_k = \frac{1}{4}\left(\frac{3}{4}\right)^k, \quad k = 0, 1, 2, \ldots$$

and by formula (9) the average number of customers in the system is

$$L = \frac{3/4}{1 - 3/4} = \frac{3/4}{1/4} = 3$$

To answer the second question, we must compute the average cost per minute C_h for
the human server system and the average cost per minute of the machine system C_m. Note
that the total cost for x minutes of operations under the human system is just $C_h x$, but
because of the fixed purchase cost of the machine, the total cost for x minutes is $C_m x + 2000$
under the machine scenario. For consistency let us express all monetary units in dollars and
all times in minutes. Then for instance the clerk's pay rate is ($5/hr.)/(60 min./hr.) =
$(1/12)/min., and the cost of operation of the machine is ($.50/hr.)/(60 min./hr) =
$(1/120)/min. Now in calculating C_h and C_m there are two aspects of cost to consider:
operating cost and the overall cost to waiting customers. The latter would be the average
number of customers waiting, times the cost rate of $.20/min. per customer. For the human
server system, the overall cost per minute would be

$$C_h = \$(1/12)/\text{min.} + (3 \text{ customers}) \times (\$.20/\text{customer})/\text{min.} = \$.683/\text{min.}$$

Since the machine average service time is 12 sec. $= 1/5$ min., its service rate is 5/min. and the
resulting queue traffic intensity is $\rho = \lambda/\mu = 3/5$. The average number of customers in the
system under the machine scenario is therefore

$$L = \frac{3/5}{1 - 3/5} = \frac{3/5}{2/5} = \frac{3}{2}$$

Thus the overall cost per minute for the machine system is

$$C_h = \$(1/120)/\text{min.} + (3/2 \text{ customers}) \times (\$.20/\text{customer})/\text{min.} = \$.308/\text{min.}$$

The cost rate for the machine is less than half that of the human server. And notice that the bigger factor in the improvement was not the running cost but the savings in customer waiting cost. Taking into account the purchase cost, the use of the machine becomes economical after it has been used for at least x minutes, where x satisfies

$$C_h\, x = C_m\, x + 2000 \Longrightarrow .683\, x = .308\, x + 2000 \Longrightarrow .375\, x = 2000$$

The solution value for x turns out to be about 5333 minutes, or about 89 hours. If the arcade is open for 10/hrs. per day, then before 9 working days have elapsed the machine will have begun to be less costly than the human server. ■

Waiting Time Distribution

The second major question that we will address is the distribution of the waiting time of a customer who arrives to an M/M/1 queue at some time t. There are two things that we might mean by this: the time W_q that a customer spends in queue prior to entering service, or the time W spent in the system in total, including the customer's time in service. We will work on W, saving the W_q problem for the exercises.

Assume that the queue has reached equilibrium, so that we are really looking at the limiting distribution as $t \longrightarrow \infty$ of the waiting time W of a customer who arrives at time t. This allows us to make use of the limiting system size probabilities in formula (8).

When our customer arrives, there is some number $n \geq 0$ of customers ahead. Because of the memoryless property of the exponential service time distribution, the distribution of the waiting time is the same as if the customer currently in service, if any, just began that service. So if n customers are ahead, the waiting time in system of the new customer will be the total $U_1 + U_2 + \cdots + U_{n+1}$ of $n + 1$ independent $\exp(\mu)$ distributed service times (including the new customer as well as the n customers ahead). Remember that the moment-generating function method implies that the distribution of the sum of $n + 1$ independent $\exp(\mu) = \Gamma(1, 1/\mu)$ random variables is $\Gamma(n + 1, 1/\mu)$. So, conditioned on the event that there is a particular number $n \geq 0$ of customers in the system upon the arrival of the new customer, $W \sim \Gamma(n + 1, 1/\mu)$. By the Law of Total Probability we can write the limiting c.d.f. of W as

$$
\begin{aligned}
F(s) = \mathrm{P}[W \leq s] \;&=\; \Sigma_{n=0}^{\infty}\, \mathrm{P}[W \leq s \mid n \text{ in system}]\, \mathrm{P}[n \text{ in system}] \\
&=\; \Sigma_{n=0}^{\infty}\left(\int_0^s \frac{\mu^{n+1}}{\Gamma(n+1)}\, x^n\, e^{-\mu x}\, d\,x \right) \rho^n (1 - \rho) \\
&=\; \mu(1 - \rho) \int_0^s e^{-\mu x} \left(\Sigma_{n=0}^{\infty} \frac{(\mu x \rho)^n}{n!} \right) d\,x \\
&=\; \mu(1 - \rho) \int_0^s e^{-\mu x}\, e^{\mu x \rho}\, d\,x \\
&=\; \mu(1 - \rho) \int_0^s e^{-\mu x(1 - \rho)}\, d\,x \\
&=\; 1 - e^{-\mu(1 - \rho)\, s}
\end{aligned}
\tag{10}
$$

Hence, W has density function

$$f(s) = F'(s) = \mu(1 - \rho) e^{-\mu(1-\rho)s} \tag{11}$$

which we recognize as the $\exp(\mu(1 - \rho))$ distribution. It follows that the long-run mean waiting time is $E[W] = 1/(\mu(1 - \rho))$, which is intuitively reasonable. As the traffic intensity ρ increases toward 1, the equilibrium queue length increases, thus increasing the average waiting time of the new customer. Also, as the service rate μ increases, the average waiting time decreases. Note also that $\mu(1 - \rho) = \mu(1 - \lambda/\mu) = \mu - \lambda$, so that the closer λ is to μ, the longer is the expected wait.

Example 2 Suppose that a warehouse processes orders for shipment one at a time. The orders come in at a rate of 4 per hour, forming a Poisson process. Under normal working conditions, a team of order fillers working on preparing the order can complete an order in an exponential amount of time with rate 4.2 per hour. Hence the team can just barely keep up with input. What is the average time for a newly arriving order to be completed? If more people can be assigned to the team, the service rate can be increased. To what must the service rate be increased so that orders require no more than one hour of waiting on average from start to finish?

We are given that $\lambda = 4.0$/hr., and the default service rate is $\mu = 4.2$/hr. Hence the traffic intensity is $\rho = 4.0/4.2 = .952$, and the expected waiting time in equilibrium is $1/(4.2 (1 - .952)) \approx 4.96$. This is an excessive amount of time for an order. To answer the second question, we need to choose μ such that

$$E[W] = \frac{1}{\mu - \lambda} = \frac{1}{\mu - 4} \le 1 \Longrightarrow \mu - 4 \ge 1 \Longrightarrow \mu \ge 5$$

So an increase in service rate from 4.2 to just 5 per hour is enough to cut the average waiting time from about 5 hours to 1 hour. ∎

Simulating Queues

Consider a single-server first-in, first-out queue as before except that we generalize: (1) the distribution of customer interarrival times; and (2) the distribution of customer service times. We have so far treated only the case where both of these distributions are exponential. It is much harder, and sometimes impossible, to get analytical results about waiting times and limiting distributions for general interarrival and service distributions; so it is important to be able to estimate these quantities by observing the results of simulation. Simulation can also be used in more complicated cases than single-server first-in, first-out queues, but we will not treat such problems here.

It is very easy in *Mathematica* to simulate a sequence of arrival times with a given interarrival distribution. Notice that the nth arrival time T_n is the sum of the first n interarrival times. The following utility function takes advantage of this observation by keeping a running sum of interarrival times, simulating the next one, and adjoining the updated running sum to the end of a list of arrival times. The arguments to the function are the interarrival distribution and the number of arrivals to simulate, and the output is the list of arrival times.

```
SimArrivals[arrdist_, numarrs_] :=
  Module[{arrtimelist, currtime, nextinterarr},
    arrtimelist = {};
    currtime = 0;
    Do[nextinterarr = RandomReal[arrdist];
       currtime = currtime + nextinterarr;
       AppendTo[arrtimelist, currtime], {numarrs}];
    arrtimelist]
```

```
SeedRandom[787 632]
```

```
SimArrivals[ExponentialDistribution[3], 4]
```

{0.519352, 0.770897, 0.933421, 1.08503}

It is even easier to simulate a list of service times for those customers. We just simulate a table of the desired number of observations from the desired distribution, as in the next command.

```
SimServiceTimes[servicedist_, numservices_] :=
  RandomReal[servicedist, {numservices}]
```

```
SimServiceTimes[ExponentialDistribution[4], 4]
```

{0.429965, 0.130504, 0.250825, 0.53062}

All of the behavior of the queue is determined by these lists of arrival and service times, but there is still the considerable problem of extracting information from the lists. As an example let us consider the relatively easy problem of finding from a simulated queueing process a list of customer waiting times.

Think of the queueing behavior for the simulated output above. Customer 1 arrives at time .519352. It takes .429965 time units to serve this customer, who therefore departs at time .519352 + .429965 = .949317. Customer 2 arrives at time .770897, at which time customer 1 is still in the queue, so he must wait. Since his service time was .130504, he departs at time .949317 + .130504 = 1.079821. Customer 3 arrives at time .933421 and must line up behind customer 2, because at that point customer 1 is still not finished with service. Customer 3 requires .250825 time units to serve, so this customer departs at time 1.079821 + .250825 = 1.330646. Meanwhile, customer 4 has come in at time 1.08503, so he has to wait until time 1.330646 to begin service. The departure time of customer 4 is then 1.330646 + .53062 = 1.861266.

Activity 5 If five successive customer arrival times are: 2.1, 3.4, 4.2, 6.8, 7.4 and their service times are: 1.0, 2.3, 1.7, 3.4, 0.6, find the departure times of each customer, and find the amount of time each customer is in the system. What is the longest queue length?

Notice that any customer's waiting time in the system is just his departure time minus his arrival time, so that producing a list of waiting times will be simple if we can produce a list of departure times. This too is not hard, because a customer's departure time is his service time plus either his arrival time, or the departure time of the previous customer, whichever is later. We can generate this list of departure times iteratively by this observation, beginning with the first customer, whose departure time is just his arrival time plus his service time. The next function does this. It takes the arrival time and service time lists as inputs, and returns the list of departure times using the strategy we have discussed.

```
DepartureTimes[arrtimelist_, servtimelist_] :=
 Module[
   {numcustomers, deptimelist, customer, nextdeptime},
   numcustomers = Length[servtimelist];
   deptimelist =
     {arrtimelist[[1]] + servtimelist[[1]]};
   Do[nextdeptime = servtimelist[[customer]] +
       Max[deptimelist[[customer - 1]],
         arrtimelist[[customer]]];
     AppendTo[deptimelist, nextdeptime],
     {customer, 2, numcustomers}];
   deptimelist]
```

Here we apply the function to the arrival and service lists in the example above.

```
arrlist = {0.519352, 0.770897, 0.933421, 1.08503};
servlist = {0.429965, 0.130504, 0.250825, 0.53062};
DepartureTimes[arrlist, servlist]
```

{0.949317, 1.07982, 1.33065, 1.86127}

Now waiting times are departure times minus arrival times; so we can build a function which calls on the functions above, plots a histogram of waiting times, and computes the mean and variance.

```
Needs["KnoxProb7`Utilities`"];
```

```
SimulateWaitingTimes[
  arrdist_, servdist_, numcustomers_] :=
 Module[{arrtimelist, servtimelist,
   deptimelist, waitimelist},
  arrtimelist = SimArrivals[arrdist, numcustomers];
  servtimelist =
   SimServiceTimes[servdist, numcustomers];
  deptimelist = DepartureTimes[
    arrtimelist, servtimelist];
  waitimelist = deptimelist - arrtimelist;
  Print["Mean waiting time: ", N[Mean[waitimelist]],
   " Variance of waiting times: ",
   N[Variance[waitimelist]]];
  Histogram[waitimelist, 6, "Probability",
   ChartStyle -> Gray, BaseStyle -> 8, PlotRange -> All]]
```

Here is an example run in a setting that we already know about: an M/M/1 queue with arrival rate $\lambda = 3$ and service rate $\mu = 4$. The distribution suggested by the histogram in Figure 9(a) does look exponential, and the mean waiting time of 1.0601 did come out near the theoretical value of $1/(\mu - \lambda) = 1$ in this run. You should try running the command several more times to get an idea of the variability. Also try both increasing and decreasing the number of customers. What do you expect to happen as the number of customers gets larger and larger?

```
SeedRandom[9 689 321]
```

```
g1 = SimulateWaitingTimes[ExponentialDistribution[3],
    ExponentialDistribution[4], 1000];
g2 = SimulateWaitingTimes[UniformDistribution[{0, 2}],
    UniformDistribution[{0, 1.5}], 1000];
GraphicsRow[{g1, g2}]
```

Mean waiting time: 1.0601 Variance of waiting times: 1.07599

Mean waiting time: 1.42062
  Variance of waiting times: 0.888548

(a) M/M/1 queue, $\lambda = 3$ and $\mu = 4$ (b) U/U/1 queue, $\rho = .75$

Figure 6.9 - Histograms of 1000 waiting times

Our command allows us to use any interarrival and service distribution we choose. To see another example, we repeat the simulation for a uniform[0, 2] interarrival distribution and a uniform[0, 1.5] service distribution. The result is shown in Figure 9(b). Note that the service rate is the reciprocal of the mean time 1/.75, and the arrival rate is the reciprocal of the interarrival mean 1/1, so the traffic intensity is $\rho = .75$. The distribution of waiting times is rather right-skewed, reminiscent of a gamma distribution, and the mean time in system is around 1.42; in particular, it is more than the mean time in the M/M/1 simulation even though the traffic intensity is the same.

Exercises 6.3

1. Customers waiting in a post office line experience a frustration cost of 10 cents a minute per customer while they wait. What target traffic intensity should the postal supervisors aim for so that the average frustration cost per unit time of the whole queue is 20 cents a minute?

2. Customers arrive to a department store service desk according to a Poisson process with rate 8 per hour. If it takes 5 minutes on average to serve a customer, what is the limiting probability of having at least 2 people in the system? What is the limiting probability that an arriving customer can go directly into service?

3. Derive the differential equation (5) for the short-run M/M/1 system size probabilities.

4. Verify by mathematical induction expression (7) for the limiting M/M/1 system size probabilities.

5. What, in general, should be the relationship between the expected waiting time in the system $E[W]$ and the expected waiting time in the queue $E[W_q]$?

6. Derive the long-run probability distribution of the time W_q that a customer spends in queue prior to entering service. Be careful: this random variable has a mixture of a discrete and continuous distribution, because there is non-zero probability that W_q equals 0 exactly, but for states greater than zero there is a sub-density, which when integrated from 0 to ∞ gives the complementary probability to that of the event $\{W_q = 0\}$.

7. People come to a street vendor selling hot pretzels at a rate of 1 every 2 minutes. The vendor serves them at an average time of 1 minute and 30 seconds. Find the long-run probability that a customer will take at least 2 minutes to get a pretzel.

8. The operator of a queueing system must solve an optimization problem in which he must balance the competing costs of speeding up service, and making his customers wait. There is a fixed arrival rate of λ, operating costs are proportional to the service rate, and customer inconvenience cost is proportional to the expected waiting time of a customer. Find the optimal service rate in terms of λ and the proportionality constants.

9. (*Mathematica*) Use the commands of the section to study the long-run waiting time distribution and average waiting time when the interarrival and service time distributions are approximately normal. What do you have to be careful of when you choose the parameters of your normal distributions?

10. (*Mathematica*) Build a command that takes a list of arrival and service times and a fixed time t, and returns the number of people in the system at time t. Combine your command with the arrival and service time simulators of this section to produce a command that simulates and plots the system size as a function of t.

11. One case in which it is possible to derive the short-run distribution of system size is the M/M/1/1 queue. The extra 1 in the notation stands for 1 waiting space for the customer in service; hence any new customer arriving when someone is already being served is turned away. Therefore, the system size M_t can only take on the values 0 or 1. Derive and solve a differential equation for $f_1(t) = P[M_t = 1] = 1 - f_0(t) = 1 - P[M_t = 0]$. (You will need to make the assumption that the system is empty at time 0.)

12. A self-service queueing system, such as a parking garage or a self-service drink area in a restaurant, can be thought of as an M/M/∞ queue, i.e., a queue with an unlimited number of servers. All customers therefore go into service immediately on arrival. If n customers are now in such a queue, what is the distribution of the time until the next customer is served? (Assume customers serve themselves at the same rate μ.) Obtain a system of differential equations analogous to (5) (case $k \geq 1$ only) for the short-run system size probabilities. You need not attempt to solve either the differential equations or the corresponding linear equations for the limiting probabilities.

6.4 Brownian Motion

To this point the stochastic processes that we have studied have had discrete state spaces. The Brownian motion process that we introduce in this section is appropriate when the state space of the process is continuous, the time set is continuous, and changes in value of the process over short time intervals are small and random. In this sense, the Brownian motion is a version of a random walk in continuous time. Applications are found in many areas, including genetics, population growth, diffusion of heat and, more recently, finance.

We begin by exploring an example data set. *Mathematica* is able to pull financial data from the Wolfram Research web site using the command

FinancialData[companycode, {startdate, enddate, period}]

(There are some variations in the arguments that are explained in the on line help system, but this will suffice for our needs.) Given the short code for a company, for instance "ATT," or "IBM," or "GE," and a list including the starting and ending date, and the period over which you want price data to be returned (such as "Day," "Week," or "Month"), this command returns a list of pairs of the form {closing date, closing price}. The dates in the input can be written out as text strings such as "Jan. 1, 2005." The format of the date in the output from the command is a list of elements of the form {year, month, day}. For example, the following is the February 2006 weekly price data for IBM stock. The output shows that observations of price are made on the 1st, 6th, 13th, 21st, and 27th of the month in this case.

```
IBMpricefulldata = FinancialData["IBM",
  {"Feb. 1, 2006", "Feb. 28, 2006", "Week"}]
```

```
{{{2006, 2, 1}, 76.48},
  {{2006, 2, 6}, 77.97}, {{2006, 2, 13}, 77.38},
  {{2006, 2, 21}, 76.79}, {{2006, 2, 27}, 76.93}}
```

The ith price in the list would be referred to as the second member of the ith sublist, as below. In this way we can strip off the date lists to create a list of only prices.

```
IBMpricedata =
  Table[IBMpricefulldata[[i, 2]], {i, 1, 5}]
IBMpricedata[[2]]
```

```
{76.48, 77.97, 77.38, 76.79, 76.93}
```

```
77.97
```

Now consider the series of weekly closing stock prices P_n for General Electric, and the ratios of successive prices $R_n = P_{n+1}/P_n$ graphed in Figure 10(a) and 10(b), respectively, for a period of three years beginning in January 2005. The two types of data show somewhat different characteristics, with the sequence of prices appearing to grow with time, perhaps superlinearly, and the ratios moving up and down in an apparently random way, centered near 1.

```
GEpricefulldata = FinancialData["GE",
    {"Jan. 1, 2005", "Dec. 31, 2007", "Week"}];
GEprices = Table[GEpricefulldata[[j, 2]],
    {j, 1, Length[GEpricefulldata]}];
GErates = Table[GEprices[[i + 1]] / GEprices[[i]],
    {i, 1, Length[GEprices] - 1}];
GraphicsRow[{ListPlot[GEprices,
    Joined → True, PlotStyle → Black],
    ListPlot[GErates, Joined → True, PlotStyle → Black]}]
```

(a) Weekly closing prices of GE, 2005−2007   (b) Weekly ratios of return on GE stock

Figure 6.10

Here we look for correlations between the ratios and their nearest neighbors, by forming a list GErate1 of all but the last ratio, and a lagged list GErate2 of all but the first ratio. So the GErate1 list would have price ratios $(R_1, R_2, ..., R_{n-1})$ and the GErate2 list would have price ratios $(R_2, R_2, ..., R_n)$ This enables us to study the pairs (R_i, R_{i+1}), looking particularly for statistical dependence. The computation below shows that the correlation between ratios and their nearest neighbors is very small.

```
GErate1 = Table[GErates[[i]],
    {i, 1, Length[GErates] - 1}]; GErate2 =
  Table[GErates[[i]], {i, 2, Length[GErates]}];
Correlation[GErate1, GErate2]
```

0.00790478

Activity 1 Use ListPlot to look at the graphical relationship between the GErate1 and GErate2 variables. Do you see any dependence?

In order to make a point below, notice from Figure 11 that the logs of the price ratios also seem to be approximately normally distributed.

```
Needs["KnoxProb7`Utilities`"]
Histogram[Log[GErates], {.01}, "Probability",
   ChartStyle → Gray, PlotRange → All, BaseStyle → 8]
```

Figure 6.11 - Histogram of logged GE price ratios

As above, denote by P_n the price at week n, and $R_n = P_{n+1}/P_n$ the price ratio between weeks n and $n + 1$. Then we can telescope P_n multiplicatively as

$$P_n = P_1 \cdot \frac{P_2}{P_1} \cdot \frac{P_3}{P_2} \cdots \frac{P_n}{P_{n-1}} = P_1 \cdot R_1 \cdot R_2 \cdots R_{n-1} \tag{1}$$

If we plot the logs of the prices as in Figure 12, we get a stochastic process that shows signs of linear increase on the average, with variability. By formula (1),

$$\log(P_n) = \log(P_1) \cdot \log(R_1 \cdot R_2 \cdots R_{n-1}) = \log(P_1) \cdot \sum_{k=1}^{n-1} \log(R_k) \tag{2}$$

So the logged price is a constant times a total of random variables that empirically seem to be independent and normally distributed. The increment of the logged price process is $\log(P_n) - \log(P_{n-1}) = \log(P_1) \log(R_{n-1})$. Roughly stated, between successive weeks the log price seems to change by a random amount that is normally distributed, independent of other increments of price.

```
ListPlot[Log[GEprices],
   Joined -> True, PlotStyle → Black]
```

Figure 6.12 - TIme series graph of logged GE prices

The preceding discussion should help to motivate the definition of a continuous-time stochastic process whose paths resemble Figure 12, and which has similar probabilistic characteristics. In order to state the definition, we need to recall two properties, *stationarity* and *independent increments*, that were introduced in Section 6.2 on Poisson processes. For ease of reference the definitions are repeated below.

Definition 1. (a) A stochastic process $(X_t)_{t \geq 0}$ is called *stationary* if for all $t \geq 0$, $s > 0$, the distribution of $X_{t+s} - X_t$ does not depend on t;

(b) A stochastic process $(X_t)_{t \geq 0}$ is said to have *independent increments* if for all $t \geq 0$, $s > 0$, $X_{t+s} - X_t$ is independent of X_r for all $r \leq t$.

Stationarity means that the probability law of the increment of the process over $[t, t + s]$ does not depend on the beginning time t, only on the length of the interval. When a process has independent increments, the change in value $X_{t+s} - X_t$ is independent of a similar change in value $X_{r+u} - X_r$ as long as the two time intervals $[t, t + s]$ and $[r, r + u]$ are disjoint.

A Brownian motion is a stochastic process with these two properties, and one more important additional property.

Definition 2. A stochastic process $(X_t)_{t \geq 0}$ is called a *Brownian motion* with drift rate μ and variance rate σ^2 if

(a) For all outcomes ω except possibly some in a set of probability zero, the path function $t \mapsto X_t(\omega)$ is continuous at every t;

(b) (X_t) has stationary, independent increments

(c) For all $t > 0$, $X_t - X_0$ is normally distributed with mean μt and variance $\sigma^2 t$.

The process is called a *standard Brownian motion* if $\mu = 0$ and $\sigma^2 = 1$.

We usually assume that the initial value X_0 is a known constant x, often 0. Then a standard Brownian motion with initial value 0 is such that $X_t \sim N(0, t)$. The paths of a Brownian motion will vary around the line $x + \mu t$, and the fact that the variance of X_t is proportional to t implies that the value of the process becomes less and less predictable as time goes on. In advanced texts on Brownian motion some of the peculiar properties of the path are discussed, particularly that for almost every outcome they are continuous, but nowhere differentiable.

Activity 2 What does the definition of Brownian motion imply about the probability distribution of $X_{t+s} - X_t$?

Example 1 It is a relatively easy matter to simulate paths of a Brownian motion, and in doing so, gain a better appreciation for its defining properties. Focus on simulating the path between times 0 and t by discretizing the time interval $[0, t]$ into a large number n of disjoint, equally sized subintervals, each of length t/n. The subintervals are formed by partition points of the form

$$t_i = i \cdot \frac{t}{n}, \quad i = 0, 1, \ldots, n$$

We will simulate values of the process at each of these time points. If you answered the question in the last Activity, it should be easy to understand that the increment $X_{t_{i+i}} - X_{t_i}$ has the same distribution as $X_{t/n} - X_0$, namely, $N\left(\mu \cdot \frac{t}{n}, \sigma^2 \cdot \frac{t}{n}\right)$. Also, by the independent increments property, the increment $X_{t_{i+i}} - X_{t_i}$ is independent of each of the other increments. The values of the Brownian motion simply accumulate these increments:

$$X_{t_1} = X_0 + \left(X_{t_1} - X_0\right), \ X_{t_2} = X_{t_1} + \left(X_{t_2} - X_{t_1}\right), \ X_{t_3} = X_{t_2} + \left(X_{t_3} - X_{t_2}\right) \ldots$$

This means that, beginning at a starting state x, we can simulate a sequence of new states by successively adding a random value sampled from the appropriate normal distribution.

Below is the algorithm. The input parameters are the starting state x, the final time t, the number of time intervals n, and the drift rate and variance rate. The function first computes the subinterval length, and then computes a list called *times* of all the partition points. It successively adjoins to a list called *states* the initial state, the intial state plus the first random normal increment, etc., by simply appending to the growing list the most recent state (i.e., Last[states]) plus a new normal increment.

Figure 13 shows the result of one simulation for the case where $x = 0$, $t = 1$, $n = 100$ time intervals, and $\mu = 0$, $\sigma^2 = 1$, that is, standard Brownian motion with initial state zero. In the electronic book, you should set the seed randomly and repeat the simulation to see how the paths behave. Sometimes variability can give the appearance of a trend, as in this particular picture. But for the standard Brownian motion, over many simulatons you will see false downward trends about as often as false upward trends, since the distribution of X_t is symmetric about zero. ∎

```
SimulateBrownianMotion[x_, t_, n_, μ_, σ2_] :=
  Module[{intlength, times, states},
    intlength = t / n;
    times = Table[k * intlength, {k, 0, n}];
    states = {x};
    Do[AppendTo[states,
      Last[states] + RandomReal[NormalDistribution[
        μ * intlength, √(σ2 * intlength) ]]], {n}];
    Transpose[{times, states}]];
```

```
SeedRandom[567 763];
path = SimulateBrownianMotion[0, 1, 100, 0, 1];
ListPlot[path, Joined → True, PlotStyle → Black]
```

Figure 6.13 - Simulated values of a standard Brownian motion with starting state 0

Example 2 Let $(X_t)_{t \geq 0}$ be a standard Brownian motion starting at $x = 0$. Find (a) $P[X_{1.2} > .5]$; (b) $P[X_{5.4} - X_{2.3} \leq -1.6, X_1 > 0]$; (c) $P[X_{2.6} > 2 \mid X_{1.1} = 1]$.

For part (a), we simply use Definition 2(c) to observe that $X_{1.2} \sim N(0, 1.2)$, since $\mu = 0$ and $\sigma^2 = 1$ here. Consequently, $P[X_{1.2} > .5]$ is:

```
1 - CDF[NormalDistribution[0, √1.2 ], .5]
```

```
0.324038
```

In part (b), the increment of the process $X_{5.4} - X_{2.3}$ in the time interval [2.3, 5.4] is independent of previous values of the process, so that

$$P[X_{5.4} - X_{2.3} \leq -1.6, \, X_1 > 0] = P[X_{5.4} - X_{2.3} \leq -1.6] \, P[X_1 > 0].$$

Since X is a standard Brownian motion, $P[X_1 > 0] = .5$, and also by the stationarity property, $X_{5.4} - X_{2.3}$ has the same distribution as $X_{3.1}$, namely, the $N(0, 3.1)$ distribution. Consequently, $P[X_{5.4} - X_{2.3} \leq -1.6, \, X_1 > 0]$ is:

$$\boxed{\texttt{CDF}\left[\texttt{NormalDistribution}\left[\texttt{0, } \sqrt{\texttt{3.1}}\,\right], \, \texttt{-1.6}\right] \texttt{ * .5}}$$

0.090872

Now in the conditional probability in part (c), subtract the given value $X_{1.1} = 1$ from both sides of the inequality and use stationarity and independent increments to get

$$
\begin{aligned}
P[X_{2.6} > 2 \mid X_{1.1} = 1] \quad &- \quad P[X_{2.6} - X_{1.1} > 2 - 1 \mid X_{1.1} = 1] \\
&= \quad P[X_{2.6} - X_{1.1} > 1 \mid X_{1.1} = 1] \\
&= \quad P[X_{2.6} - X_{1.1} > 1] \\
&= \quad P[X_{1.5} > 1]
\end{aligned}
$$

Since $X_{1.5}$ has the $N(0, 1.5)$ distribution, this probability is:

$$\boxed{\texttt{1 - CDF}\left[\texttt{NormalDistribution}\left[\texttt{0, } \sqrt{\texttt{1.5}}\,\right], \, \texttt{1}\right]}$$

0.207108

∎

Example 3 Suppose that the number of people in a large community who are infected by a certain variety of the flu can be approximately modeled by a Brownian motion process $(X_t)_{t \geq 0}$. If the initial state is 50, if the parameters are $\mu = 0$, $\sigma^2 = 4$, and if the flu disappears forever when the number who are infected reaches zero, find the probability that at time $t = 20$ there are at least 25 individuals still infected.

The problem statement indicates that we have a different version of the Brownian motion here than usual: the flu infection process stops at 0 once the population is clear. Define the Brownian motion (X_t) to be the unconstrained process with constant drift rate 0 and variance rate 4. It is convenient to introduce the time of first visit by the X process to state 0: $T_0 = \min\{s > 0 : X_s = 0\}$. If we define a new stochastic process whose t^{th} member is:

$$X_t' = \begin{cases} X_t & \text{if } t < T_0 \\ 0 & \text{otherwise} \end{cases} \tag{3}$$

then the X' process captures the absorption property at 0. We are interested in finding $P[X_{20}' \geq 25]$. Actually, we will compute more generally $P[X_t' \geq 25]$ for each t.

First, by complementation and the multiplication rule we have

$$
\begin{aligned}
P[X_t' \geq 25] &= P[X_t \geq 25, T_0 > t] \\
&= P[X_t \geq 25] - P[X_t \geq 25, T_0 \leq t] \\
&= P[X_t \geq 25] - P[X_t \geq 25 \mid T_0 \leq t] P[T_0 \leq t]
\end{aligned}
\tag{4}
$$

At time T_0, the value of X is zero. Since the X process has no drift, given the event $T_0 \leq t$, X is as likely to go up 25 units in the next $t - T_0$ time units as it is to go down 25 units. Thus, the conditional probability that $X_t \geq 25$ in the second term in formula (4) can be replaced by the conditional probability that $X_t \leq -25$. But then the event $\{X_t \leq -25\}$ is contained in the event $\{T_0 \leq t\}$, so that we can write

$$
\begin{aligned}
P[X_t' \geq 25] &= P[X_t \geq 25] - P[X_t \leq -25 \mid T_0 \leq t] P[T_0 \leq t] \\
&= P[X_t \geq 25] - P[X_t \leq -25, T_0 \leq t] \\
&= P[X_t \geq 25] - P[X_t \leq -25]
\end{aligned}
\tag{5}
$$

Applying formula (5) at time $t = 20$, and observing that $X_{20} \sim N(50, 4 \cdot 20)$ yields

$$
\texttt{N}\!\left[1 - \texttt{CDF}\!\left[\texttt{NormalDistribution}\!\left[50,\ \sqrt{4*20}\,\right],\ 25\right] - \right.
$$
$$
\left. \texttt{CDF}\!\left[\texttt{NormalDistribution}\!\left[50,\ \sqrt{4*20}\,\right],\ -25\right]\right]
$$

0.997406

Notice that with the given problem parameters, X_{20} has standard deviation a little less than 9, so it is very likely to take a value within 18 units of the mean of 50. Hence it is highly likely for X_{20} to exceed 25. In fact, we worked hard to cope with the possibility of absorption prior to time 20, which is what introduced the extra term $P[X_t \leq -25]$ in formula (5), but that possibility did not change the answer in any significant way, as shown in the computation below.

$$
\texttt{N}\!\left[\texttt{CDF}\!\left[\texttt{NormalDistribution}\!\left[50,\ \sqrt{4*20}\,\right],\ -25\right]\right]
$$

2.5308×10^{-17}

As time wears on however, variability increases the chance of absorption, as you will discover in the next activity. ∎

Activity 3   Use the general expression for $P[X_t' \geq 25]$ in Example 3 to graph it as a function of t for values of $t > 200$. On the same graph, show $P[X_t \leq -25]$.

Example 4 Assume that the logged GE prices as in Figure 12 do form a Brownian motion. Estimate the parameters μ and σ^2, and use the estimates to compute the probability that the price after 100 weeks is at least 35, given that the initial price is 32.

Let $(P_t)_{t\geq 0}$ denote the process of prices, and let $(X_t)_{t\geq 0}$ denote the process of logged prices, i.e., for each $t \geq 0$, $X_t = \log(P_t)$. We are assuming that $(X_t)_{t\geq 0}$ forms a Brownian motion with parameters μ and σ^2, hence the increments $X_2 - X_1$, $X_3 - X_2$, $X_4 - X_3$, ..., for which the length of the time interval is exactly 1 week, must have the $N(\mu \cdot 1, \sigma^2 \cdot 1)$ distribution, and they are independent of each other. We can therefore estimate the parameters by the sample mean and sample variance of these increments.

```
Xtprocess = Log[GEprices];
Increments =
  Table[Xtprocess[[i]] - Xtprocess[[i - 1]],
    {i, 2, Length[Xtprocess]}];
xbar = Mean[Increments]
var = Variance[Increments]
```

0.000738278

0.000315158

Now if the initial price is 32, then the log of the initial price is $\log(32)$, and at time $t = 100$ the log of the GE price has approximately the $N(\log(32) + \text{xbar} * 100, \text{var} * 100)$ distribution. Thus, the probability that the price is at least 35 is

$$P[P_t \geq 35] = P[X_t = \log(P_t) \geq \log(35)]$$

This probability is estimated as:

```
1 - CDF[NormalDistribution[
    Log[32] + xbar * 100, √‾‾‾‾‾‾‾‾‾‾‾ var * 100 ], Log[35]]
```

0.464576

■

Geometric Brownian Motion

Certain transformations of Brownian motions can be even more useful than Brownian motions themselves in probabilistic modeling, in particular in finance. To bring the discussion of this section around full circle, we now introduce one of these.

Definition 3. A stochastic process $(Y_t)_{t\geq 0}$ is called a *geometric Brownian motion* with parameters μ and σ^2 if its log forms a Brownian motion $(X_t)_{t\geq 0}$ with drift rate μ and variance rate σ^2. In other words, $(Y_t)_{t\geq 0}$ is a geometric Brownian motion if $Y_t = e^{X_t}$ for all $t \geq 0$, where $(X_t)_{t\geq 0}$ is a Brownian motion.

The data presented at the start of the section on the GE price process suggested that the logged prices might be well modeled as Brownian motion, hence the price process itself can be modeled as a geometric Brownian motion. We will say a bit more about this and give another example in a moment.

Example 5 We know the mean and variance of a Brownian motion as a function of time, namely $\mu_t = E[X_t] = x + \mu t$, $\sigma_t^2 = \mathrm{Var}(X_t) = \sigma^2 t$. What are the mean and variance functions for geometric Brownian motion?

Recall that the moment-generating function of a $N(\mu, \sigma^2)$ random variable is

$$M(s) = e^{\mu s + \sigma^2 s^2/2}, \ s \in \mathbb{R}$$

Suppose that $Y_t = e^{X_t}$ is our geometric Brownian motion, where the drift and variance parameters are μ and σ^2. Then $X_t \sim N(x + \mu t, \sigma^2 t)$, where x is the state at time 0. The mean function for the Y process can be considered as an instance of the m.g.f. of X_t, with real parameter $s = 1$:

$$
\begin{aligned}
E[Y_t] &= E\left[e^{X_t}\right] \\
&= E\left[e^{1 \cdot X_t}\right] \\
&= M_{X_t}(1) \\
&= e^{(x+\mu t)\cdot 1 + (\sigma^2 t) 1^2/2} \\
&= e^x \cdot e^{\left(\mu + \frac{1}{2}\sigma^2\right)t}
\end{aligned}
$$

Similarly, to find the variance function we can compute $E\left[Y_t^2\right] = E\left[\left(e^{X_t}\right)^2\right] = E\left[e^{2 X_t}\right]$. Following along the lines of the previous calculation,

$$
\begin{aligned}
E\left[Y_t^2\right] &= E\left[e^{2 \cdot X_t}\right] \\
&= M_{X_t}(2) \\
&= e^{(x+\mu t)\cdot 2 + (\sigma^2 t) 2^2/2} \\
&= e^{2x} \cdot e^{(2\mu + 2\sigma^2)t}
\end{aligned}
$$

Hence, $\mathrm{Var}(Y_t) = e^{2x} \cdot e^{(2\mu + 2\sigma^2)t} - \left(e^x \cdot e^{\left(\mu + \frac{1}{2}\sigma^2\right)t}\right)^2 = e^{2x}\left(e^{(2\mu + 2\sigma^2)t} - e^{(2\mu + \sigma^2)t}\right).$ ∎

It is easy to use the cumulative distribution function technique to find the p.d.f. of Y_t. It should be pointed out that even though Theorem 1 below gives a formula for the density, the m.g.f. approach that we used in Example 5 to compute the moments is simpler than the definition.

Theorem 1. If $(Y_t)_{t \geq 0}$ is a geometric Brownian motion, then the p.d.f. of Y_t is

$$g(y) = \frac{1}{y} \frac{1}{\sqrt{2\pi\sigma^2 t}} e^{-(\log(y)-(x+\mu t))^2/2\sigma^2 t} \tag{6}$$

Proof. First express $Y_t = e^{X_t}$ where X is a Brownian motion. The c.d.f. of Y_t is

$$G_{Y_t}(y) = P[Y_t \leq y] = P[e^{X_t} \leq y] = P[X_t \leq \log(y)] = F_{X_t}(\log(y))$$

where $F_{X_t}(x)$ is the c.d.f. of X_t. The p.d.f. of Y_t is the derivative of this c.d.f., which is $g(y) = f_{X_t}(\log(y)) \cdot (1/y)$, where f_{X_t} is the density of X_t. This density is $N(x + \mu t, \sigma^2 t)$, hence the density of Y_t is

$$g(y) = \frac{1}{y} \frac{1}{\sqrt{2\pi\sigma^2 t}} e^{-(\log(y)-(x+\mu t))^2/2\sigma^2 t} \quad \blacksquare$$

Example 6 An investor has entered into a deal in which it becomes profitable for him to sell 1000 shares of Microsoft stock if the price exceeds $42 at time 10 (weeks); otherwise, he will not sell. He is willing to assume that the stock price is in geometric Brownian motion. The current price of the stock is $40 and the investor has historical data on weekly rates of return as below (in the *Mathematica* list MicrosoftRofR, viewable in the electronic file) for a three year period in order to estimate parameters. Here, the *rate of return* on an investment over a time period refers to the ratio of the growth in value over that time period to the intial value. What is the probability that it will be profitable for him to sell? What is the estimated expected value of his sale proceeds (taking account of the chance that he will not sell at all)?

```
Microsoftfulldata = FinancialData["MSFT",
    {"Jan. 1, 2005", "Dec. 31, 2007", "Week"}];
Microsoftprices = Table[Microsoftfulldata[[j, 2]],
    {j, 1, Length[Microsoftfulldata]}];
MicrosoftRofR = Table[
    (Microsoftprices[[i + 1]] - Microsoftprices[[i]]) /
    Microsoftprices[[i]],
    {i, 1, Length[Microsoftprices] - 1}];
GraphicsRow[{ListPlot[Microsoftprices,
    Joined → True, PlotStyle → Black],
    ListPlot[MicrosoftRofR, Joined → True,
    PlotStyle → Black]}]
```

(a) Weekly closing prices of Microsoft, 2005-2007   (b) Weekly rates of return on Microsoft

Figure 6.14

Figure 14 shows time series graphs of both the weekly prices themselves and the weekly rates of return. The price graph shows signs of being an exponentiated Brownian motion, and the rate of return graph suggests noise centered near zero. (Do the Activity following this example to check that it is reasonable that the rates of return could be normally distributed with no significant dependence on one another.) First let us take a look back at the geometric Brownian motion model, and relate the model to rates of return on the stock.

If the price process is related to the underlying Brownian motion by $Y_t = e^{X_t}$, then the rate of return R_i for week i is

$$R_i = \frac{Y_{i+1} - Y_i}{Y_i} = \frac{e^{X_{i+1}} - e^{X_i}}{e^{X_i}} = e^{X_{i+1}-X_i} - 1 \Longrightarrow \log(1 + R_i) = X_{i+1} - X_i \tag{7}$$

Thus, by the defining properties of Brownian motion, the collection of random variables $\log(1 + R_i)$ form a random sample from the $N(\mu \cdot 1, \sigma^2 \cdot 1)$ distribution (the time interval is 1 week). Hence we can use the sample mean and variance of these data to estimate the drift rate μ and the variance rate σ^2, respectively:

```
logratesplus1 = Log[1 + MicrosoftRofR];
μ = Mean[logratesplus1]
sigsq = Variance[logratesplus1]
```

0.00210681

0.000812501

We can now answer the questions posed in the problem. The stock will be profitable to sell if the price Y_{10} at time 10 is greater than 42, and remember that the current price is 40. Because of the latter assumption, the initial value x of the Brownian motion satisfies $40 = e^x \Longrightarrow x = \log(40)$. Using the density derived in Theorem 1, we obtain:

```
(* P[Y₁₀ > 42 = *)
                 1              1
1 - NIntegrate[ ─ * ───────────────── *
                 Y    √2 π * sigsq * 10
   E^-(Log[y]-(Log[40]+μ*10))²/(2*sigsq*10) , {y, 0, 42}]
```

0.379213

The second question asks for the expected value of the sale proceeds. The investor earns $1000 \cdot Y_{10}$ if he sells, that is if $Y_{10} > 42$, and 0 otherwise. Thus, the usual integral that defines the expectation must be restricted to the interval of integration $(42, \infty)$, and we find:

```
(* Expected proceeds = *)
                        1          1
1000 * NIntegrate[y * ─ * ───────────────── *
                        y   √2 π * sigsq * 10
   E^-(Log[y]-(Log[40]+μ*10))²/(2*sigsq*10) , {y, 42, ∞}]
```

16 979.2

■

Activity 4 Use graphical and numerical techniques to check the assumption of independence of the weekly rates of return for Microsoft. Check also the assumption of normality of the distribution of the logs of the rates of return +1.

Exercises 6.4

1. Suppose that $(X_t)_{t\geq 0}$ is a Brownian motion starting at $x = 0$, with drift rate $\mu = 1$ and variance rate $\sigma^2 = 1$. Find: (a) $P[0 \leq X_{3.1} \leq 2]$; (b) $P[X_{3.1} \leq 3 \mid X_{1.4} = 2]$; (c) $P[X_{1.4} \leq 2, X_{3.1} \leq 3]$.

2. (*Mathematica*) In Example 1 we showed how to simulate paths of a Brownian motion X_t. Modify the SimulateBrownianMotion command to study the simulated distribution of X_t for fixed t. Specifically, for (a) a standard Brownian motion starting at $x = 0$, and (b) the Brownian motion in Exercise 1, simulate 500 replications of X_2 and produce a histogram. Comment on the relationship of the data histogram to the true distribution of X_2 in each case.

3. (*Mathematica*) A two-dimensional stochastic process $\mathbf{X} = (X_t, Y_t)_{t\geq 0}$ is called a *two-dimensional (independent) Brownian motion* if each of the component processes $(X_t)_{t\geq 0}$ and $(Y_t)_{t\geq 0}$ is a Brownian motion and all X observations are independent of all Y observations.

(a) Write a simulation program that simulates paths of a two-dimensional Brownian motion. Your program should return a list in which each element is a triple $\{t, x, y\}$ indicating the time, the state of the (X_t) process, and the state of the (Y_t) process. You may use the utility function ShowLineGraph below to plot the path in three-dimensional space, formatted as a list of triples.

(b) For the case where each component process is standard Brownian motion starting at 0, write another program that simulates 200 paths of the two-dimensional process up to the time when the process exits the open disk $S = \{(x, y) : x^2 + y^2 < 10\}$. Let the program compute the proportion of replications in which the departure from the circle occurred in each of the first, second, third, and fourth quadrants of the plane. What do you expect those proportions to be?

```
ShowLineGraph[datalist_] := Graphics3D[Join[
    Table[Line[{datalist[[i]], datalist[[i + 1]]}],
    {i, 1, Length[datalist] - 1}],
    Table[Point[{datalist[[i]]}],
    {i, 1, Length[datalist]}]],
    Axes → True, AxesLabel → {"t", "x", "y"}];
```

4.   Compute the *covariance function* of a Brownian motion $(X_t)_{t\geq 0}$, that is, $C(t, s) = \mathrm{Cov}(X_t, X_{t+s})$. (For simplicity, suppose that the starting state is $x = 0$. Does this matter?) Does this function depend on t?

5. Brownian motion can be viewed as a limit of discrete-time, discrete-state random walks as the time between jumps and the jump size approach zero together. Consider a sequence of random walks on the real line that start at 0, and such that the nth random walk has right move probability p, moves every $\Delta t = 1/n$ time units, and jumps either to the left of or the right of its current state by a distance $\Delta x = c \cdot \sqrt{\Delta t}$. Argue that the position of the random walk at time t is approximately normally distributed for large n, find the parameters of that distribution, and argue informally that each random walk has stationary, independent increments.

6. (*Mathematica*) A town whose population is currently 30,000 knows that when the popuation reaches 35,000 they will need to upgrade their water distribution network. If the population is in Brownian motion with drift rate 1,000/yr and variance rate 900/year, in how many years is the expected population 35,000? What is the smallest time t such that the probability that the population exceeds 35,000 at that time is at least .9?

7. In Example 3 we introduced the hitting time T_0 of state 0 by a Brownian motion with no drift. Now consider a Brownian motion with initial state $x = 0$, drift rate $\mu = 0$, and variance rate σ^2. Find the c.d.f. of T_a, the first time that the Brownian motion reaches state $a > 0$. (Hint: relate the event $X_t \geq a$ to the event $T_a \leq t$.)

8. (*Mathematica*) The solution to the epidemic problem in Example 3 depended critically on the assumption that $\mu = 0$. Using the same data as in that Example, except that now the drift rate is $\mu = -8$, write and run a simulation program to estimate the mean time until the epidemic dies out.

9. A stochastic process $(Y_t)_{t \geq 0}$ is called *(standard) Brownian motion reflected at 0* if for each t, $Y_t = |X_t|$, where $(X_t)_{t \geq 0}$ is a standard Brownian motion starting at zero. Compute the c.d.f., p.d.f., and mean of Y_t for a standard Brownian motion reflected at 0.

10. Let $(Y_t)_{t \geq 0}$ be a geometric Brownian motion with initial state 1 and parameters $\mu = 0$, $\sigma^2 = 1$. Use Definition 3, rather than Theorem 1, to compute $P[Y_6 / Y_3 \geq 2]$.

11. (*Mathematica*) As in Example 6, obtain data on the prices of IBM stock for the most recent three-year time period. Assuming that the prices are in geometric Brownian motion, use the data to estimate parameters. Find the probability that the current price gains in value by at least 10% in the next year.

12. (*Mathematica*) Repeat Exercise 11, except look at AT&T, and find the probability the the current price decreases by at least 1% in the next year.

13. (*Mathematica*) An investor owns two stocks, A and B, in a portfolio with 1000 shares apiece. The stocks are currently priced at $10 per share and $15 per share, respectively. Assume that the stocks are independent of each other, and are undergoing geometric Brownian motion with the following parameters, where the rates are in time units of years: $\mu_A = .05$, $\sigma_A^2 = .008$; $\mu_B = .04$, $\sigma_B^2 = .007$. Find the probability that the portfolio is worth at least $28,000 in two years.

6.5 Financial Mathematics

The comments that we made at the start of Section 6.3 about queueing theory and its effect on the rapid development of probability could also be made about the area of financial mathematics. Increasingly, economists have studied aspects of financial problems that are subject to random influences. Since time is usually a factor, the study of random processes like the Poisson process, Markov chains and others has become a key element in the education of finance specialists. Several Nobel prizes in economics have been awarded to people who have made key contributions to financial mathematics. Here are just a few example problems in which probabilistic analysis is important.

1. An investor must decide how to allocate wealth between a risky asset such as a stock, and a non-risky asset such as a bond.

2. An *option* on a stock is a contract that can be taken out which lets the option owner buy a share of the stock at a fixed price at a later time. What is the fair value of this option?

3. A firm must decide how to balance its equity (money raised by issuance of stock to buyers, who then own a part of the company) and debt (money raised by selling bonds which are essentially loans made by the bond-holders to the company). The main considerations are the tax-deductibility of the interest on the bonds, and the chance of default penalties if the company is unable to make payments on the bonds. The goal is to maximize the total value of the money raised.

4. How should an individual consume money from a fund which appreciates with interest rates that change randomly as time progresses, if the person wants a high probability of consumption of at least a specified amount for a specified number of years?

Activity 1 The insurance industry has always been filled with financial problems in which randomness is a factor. Write descriptions of at least two problems of this kind.

In the interest of brevity we will just look at problems 1 and 2 above, on finding an optimal combination of two assets in a portfolio, and pricing a stock option.

Optimal Portfolios

All other things being equal, you would surely prefer an investment whose expected return is 6% to one whose expected return is 4%. But all other things are rarely equal. If you knew that the standard deviation of the return on the first asset was 4% and that of the second was 1% it might change your decision. Chebyshev's inequality says that most of the time, random variables take values within 2 standard deviations of their means, so that the first asset has a range of returns in [−2%, 14%] most of the time, and the second asset will give a return in [2%, 6%] most of the time. The more risk averse you are, the more you would tend to prefer the surer, but smaller, return given by the second asset to the more variable return on the first.

Let us pose a simple mathematical model for investment. There are just two times under consideration: "now" and "later." The investor will make a decision now about what proportion of wealth to devote to each of two potential holdings, and then a return will be collected later. One of these two holdings is a bond with a certain rate of return r_1, meaning that the total dollars earned in this single period is certain to be r_1 times the dollars invested in the bond. The other potential investment is a risky stock whose rate of return R_2 is a random variable with known mean μ_2 and standard deviation σ_2. The investor starts with wealth W and is to choose the proportion q of that wealth to invest in the stock, leaving the proportion $1 - q$ in the bond. The dollar return on this portfolio of assets is a combination of a deterministic part and a random part:

$$W(1 - q) r_1 + W q R_2$$

Hence the rate of return on the portfolio of two assets, i.e., the dollar return divided by the number of dollars W invested, is

$$R = (1 - q) r_1 + q R_2 \tag{1}$$

Because R_2 is random, the investor cannot tell in advance what values this portfolio return R will take. So, it is not a well-specified problem to optimize the rate of return given by (1). The investor might decide to optimize expected return, but this is likely to be a trivial problem, because the risky stock will probably have a higher expected return $\mu_2 = E[R_2]$ than the return r_1 on the bond; so the investor should simply put everything into the stock if this criterion is used. Moreover, this optimization criterion takes no account of the riskiness of the stock, as measured by $\sigma_2{}^2$, the variance of the rate of return.

There are various reasonable ways to construct a function to be optimized which penalizes variability according to the degree of the investor's aversion to risk. One easy way is based on the following idea. Suppose we are able to query the investor and determine how many units of extra expected rate of return he would demand in exchange for an extra unit of risk on the rate of return. Suppose that number of units of extra return is a (for aversion). Then the investor would be indifferent between a portfolio with mean rate of return μ and variance σ^2 and another with mean $\mu + a$ and variance $\sigma^2 + 1$, or another with mean $\mu + 2a$ and variance $\sigma^2 + 2$, etc. Such risk averse behavior follows if the investor attempts to maximize

$$\mu_p - a \sigma_p{}^2 \tag{2}$$

where μ_p is the mean portfolio rate of return and $\sigma_p{}^2$ is the variance of the rate of return. This is because with objective (2), the value to this investor of a portfolio with mean $\mu + k a$ and variance $\sigma^2 + k$ is

$$(\mu + k a) - a(\sigma^2 + k) = \mu - a \sigma^2$$

which is the value of the portfolio of mean μ and variance σ^2. So we will take formula (2) as our objective function.

From formula (1), the mean portfolio rate of return as a function of q is

$$\mu_p = E[(1 - q) r_1 + q R_2] = (1 - q) r_1 + q \mu_2 \tag{3}$$

and the variance of R is

$$\sigma_p{}^2 = \text{Var}((1 - q) r_1 + q R_2) = q^2 \, \text{Var}(R_2) = q^2 \sigma_2{}^2 \tag{4}$$

Combining (2), (3), and (4), an investor with risk aversion constant a will maximize as a function of q:

$$f(q) = r_1 + (\mu_2 - r_1) q - a \sigma_2{}^2 q^2 \tag{5}$$

By simple calculus, the optimal value q^* for this objective is

$$q^* = \frac{\mu_2 - r_1}{2\,a\,\sigma_2^{\,2}}$$

(6)

Activity 2 Verify the formula for q^* in (6).

Notice from (6) the intuitively satisfying implications that as either the risk aversion a or the stock variance $\sigma_2^{\,2}$ increase, the fraction q^* of wealth invested in the stock decreases, and as the gap $\mu_2 - r_1$ between the expected rate of return on the stock and the certain rate of return on the bond increases, q increases.

Example 1 Suppose that the riskless rate of return is $r_1 = .04$, and the investor has the opportunity of investing in a stock whose expected return is $\mu_2 = .07$, and whose standard deviation is .03. Let us look at the behavior of the optimal proportion q^* as a function of the risk aversion a.

First, here is a *Mathematica* function that gives q^* in terms of all problem parameters.

```
qstar[a_, r1_, μ2_, σ2_] := ———————
                            μ2 - r1
                            ————————
                            2 * a * σ2²
```

The plot of this function in Figure 15 shows something rather surprising.   The optimal proportion q^* does not sink below 1 = 100% until the risk aversion becomes about 17 or greater. (Try zooming in on the graph to find more precisely where q^* intersects the horizontal line through 1.) There are two ways to make sense of this: investors whose risk aversions are less than 17 hold all of their wealth in the stock and none in the bond, or if it is possible to borrow cash (which is called holding a *short position* in the bond), then an investor with low risk aversion borrows an amount $|1 - q^*|\,W$ of cash and places his original wealth plus this borrowed amount in the stock. His overall wealth is still W, because he owns $W + (q^* - 1)\,W$ of stock and $-(q^* - 1)\,W$ of cash. He hopes that the stock grows enough to let him pay off his cash debt, and still profit by the growth of his holding in stock.

```
Plot[{qstar[a, .04, .07, .03], 1}, {a, 2, 20},
  PlotStyle → {Black, Gray}, AxesOrigin → {0, 0},
  AxesLabel → {"a", "q*"}, PlotRange → {0, 5}]
```

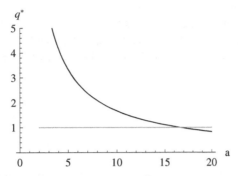

Figure 6.15- Optimal proportion of wealth in risky asset as a function of risk aversion

Let us simulate the experiment of making this investment 200 times, and keep track of the investor's profit under the optimal choice q^*, and then under a less than optimal choice. This will give us numerical evidence that this borrowing is a good thing for the investor. We will plot a histogram, and compute a sample mean and standard deviation of profit in each case. The distribution of the stock rate of return will be assumed normal. The rate of return information will be as described in the problem, and we will take an initial $100 investment and an investor with risk aversion parameter $a = 10$. First we define a simulator of total returns given initial wealth W, portfolio coefficient q, the rates of return r_1 and μ_2, the standard deviation of the stock return σ_2, and the number of simulations desired.

```
Needs["KnoxProb7`Utilities`"]
```

```
TotalReturn[W_, q_, r1_, μ2_, σ2_, numreps_] :=
  Table[W * ((1 - q) r1 + q * RandomReal[
        NormalDistribution[μ2, σ2]]), {numreps}]
```

```
SeedRandom[6 786 389]
```

Here is a typical result using the optimal strategy. The first output is the optimal q^*, and the sample mean and standard deviation of total returns corresponding to it. Then we let *Mathematica* compute the sample mean and standard deviation for the portfolio which puts 100% of wealth in the stock, without borrowing. Beneath these are histograms of total returns for each simulation.

```
q = qstar[10, .04, .07, .03]
returns1 = TotalReturn[100, q, .04, .07, .03, 200];
{Mean[returns1], StandardDeviation[returns1]}
```

1.66667

{8.47672, 4.83981}

```
q = 1
returns2 = TotalReturn[100, q, .04, .07, .03, 200];
{Mean[returns2], StandardDeviation[returns2]}
```

1

{6.73963, 3.03797}

(a) optimal strategy $q = 1.6667$, $a = 10$, $W = 100$ (b) $q = 1$(no borrowing), $a = 10$, $W = 100$

Figure 6.16 - Distribution of profits under two strategies

Since the first run began with more money in the stock, it is to be expected that the mean total return of about 8.5 exceeds the mean total return for the second scenario of about 6.7. It is also to be expected that the standard deviation of the return is a little higher in the first case than the second. But the distribution of returns in Figure 16 is noticeably more desirable in the optimal case than in the $q = 1$ case, since the center of the distribution is higher, and its highest category is higher than the one for the non-optimal case, while the lowest categories are comparable. ■

Activity 3 Compare the distribution, mean, and standard deviation in the optimal case to some suboptimal cases for an investor with high risk aversion of 30.

Extensions of this basic problem are possible. Exercise 4 investigates what happens when there are two stocks of different means and variances. An important result in the multiple stock case is the *Separation Theorem*, which says that risk aversion affects only the balance between the bond and the group of all stocks; all investors hold the stocks themselves in the same relative proportions. One can also consider multiple time periods over which the portfolio is allowed to move. Mathematical economists have also studied different objective criteria than the mean-variance criterion we have used. One of the main alternatives is to maximize a utility function $U(R)$ of the rate of return on the portfolio, where U is designed to capture both an investor's preference for high return, and aversion to risk.

Option Pricing

Our second problem is the option pricing problem. An asset called a *call option of European type* can be purchased, which is a contract that gives its owner the right to buy a share (actually, it is more commonly 100 shares) of another risky asset at a fixed price E per share at a particular time T in the future. The price E is called the *exercise price* of the option. The question is, what is the present value of such a call option contract? (The "European" designation means that it is not possible to exercise the option until time T, while options of *American type* permit the option to be exercised at any time up to T.)

The value of the European call option depends on how the risky asset, whose price at time t we denote by $P(t)$, moves in time. If the final price $P(T)$ exceeds E, the owner of the option can exercise it, buy the share at price E, then immediately resell it at the higher price $P(T)$ for a profit of $P(T) - E$. But if $P(T) < E$ then it does not pay to exercise the option to buy, so the option becomes valueless.

Suppose that the value of money discounts by a percentage of r per time period. This means that a dollar now is worth $1 + r$ times the worth of a dollar next time period; equivalently a dollar in the next time period is worth only $(1 + r)^{-1}$ times a dollar in present value terms. Similarly, a dollar at time T is only worth $(1 + r)^{-T}$ times a dollar now. Therefore, the present value of the call option is the expectation

$$E\left[\frac{1}{(1 + r)^T} \max(P(T) - E, 0)\right] \tag{7}$$

Activity 4 A *European put option* is a contract enabling the owner to <u>sell</u> a share of a risky asset for a price E at time T. Write an expression similar to (7) for the value of a put option.

To proceed, we need to make some assumptions about the risky asset price process $P(t)$. Nobel winners Fischer Black and Myron Scholes in their seminal 1973 paper chose a continuous time model in which $\log(P(t))$ has a normal distribution with known parameters, implying that $P(t)$ has a distribution called the *log-normal distribution*. (See also Section 6.4 on the Geometric Brownian motion process.) The Black-Scholes solution, which has dominated financial theory since its inception, is expressed in terms of the standard normal c.d.f. and some parameters which govern the stock price dynamics. But there is quite a lot of machinery involved in setting up the model process, and finding the expectation in (7) for this continuous time problem. So we will use the simpler Cox, Ross, and Rubinstein model (see Lamberton and LaPeyre, p. 12) in discrete time. It turns out that, in the limit as the number of time periods approaches infinity while the length of a period goes to 0, the Cox, Ross, and Rubinstein call option value converges to the Black-Scholes call option value.

In the Cox, Ross, and Rubinstein framework, we assume that from one time to the next the asset price changes by:

$$P(t + 1) = \begin{cases} (1 + a)\, P(t) & \text{with probability } p \\ (1 + b)\, P(t) & \text{with probability } 1 - p \end{cases} \tag{8}$$

where $a < b$. So there are two possible growth rates for the price in one time period: a or b, and these occur with probabilities p and $(1 - p)$. By the way, a need not be positive; in fact, it is probably useful to think of the a transition as a "down" movement of the price and the b transition as an "up" price movement. One other assumption is that the ratios $Y(t) = P(t)/P(t - 1)$ are independent random variables. By (8),

$$Y(t) = \begin{cases} (1 + b) & \text{with probability } p \\ (1 + a) & \text{with probability } 1 - p \end{cases} \quad \text{and} \quad P(t) = P(0)\, Y_1\, Y_2 \cdots Y_t \tag{9}$$

A tree diagram summarizing the possible paths of the asset for two time periods is in Figure 17.

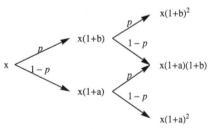

Figure 6.17- C-R-R asset price model

If the initial price of the risky asset is x, then in the time periods 1, 2, ... , T the price will undergo some number j of "ups" and consequently $T - j$ "downs" to produce a final price of $x(1 + b)^j (1 + a)^{T-j}$. By our assumptions, the number of "ups" has the binomial distribution with parameters T and p. Therefore, the expectation in (7), i.e., the present value of the call option, is a function of the initial price x, the exercise price E, and the exercise time T by

$$V(x, T, E) =$$

$$\frac{1}{(1 + r)^T} \sum_{j=0}^{T} \binom{T}{j} p^j (1 - p)^{T-j} \max\left(x(1 + b)^j (1 + a)^{T-j} - E, 0\right)$$

It appears that formula (10) solves the option valuation problem, but there is a little more to the story. Economic restrictions against the ability of investors to arbitrage, i.e., to earn unlimited risk-free returns at rates above the standard risk-free rate r, imply that we may assume not only that $a < r < b$, but also that

$$E\left[\frac{P(t)}{P(t - 1)}\right] = E[Y_t] = 1 + r$$

The argument that justifies this is actually rather subtle, and can be found in most finance books, including Lamberton and LaPeyre mentioned above. It hinges on the fact that if the option were valued in any other way, a portfolio of the stock and the option could be constructed which would hedge away all risk, and yet would return more than the riskless rate. We will omit the details. By construction of Y_t, we can use the anti-arbitrage condition $E[Y_t] = 1 + r$ to conclude that

$$(1 + b) p + (1 + a) (1 - p) = 1 + r \Longrightarrow p = \frac{r - a}{b - a} \tag{11}$$

after some easy algebra. So with the "up" and "down" percentages b and a, and the riskless rate r as given parameters, equations (10) and (11) form a complete description of the fair option value.

Also, the same reasoning as before applies to the problem of finding the present value at time t of the same option at the time t between 0 and T. All of the T's in formula (10) would be replaced by $T - t$, which is the remaining number of time steps.

Example 2 Let us define the option price function V as a *Mathematica* function that we call OptionValue, and examine some of its characteristics. In addition to current time t, current price x, T, and E, we will also put b, a, and r as arguments, and let the function compute p. We use variable name EP instead of E for the exercise price, since E is a reserved name.

```
OptValue [t_, x_, T_, EP_, b_, a_, r_] :=
Module[{ p, V }, p = r - a ;
                     b - a

    V =    1      *
        (1 + r)^T-t
       NSum[Binomial[T - t, j] * p^j * (1 - p)^T-t-j * Max[0,
            x * (1 + b)^j * (1 + a)^T-t-j - EP], {j, 0, T - t}];
    V]
```

Here is one example run. Think carefully about the question in the activity that follows.

```
OptValue[0, 22, 3, 18, .07, .03, .05]
```

6.45092

Activity 5 In the output above the difference between the stock price and the option exercise price is $P(0) - E = \$22 - \$18 = \$4$. Why is $6.45 a reasonable answer for the value of the option?

Figure 18(a) shows the graph of option price against stock price, for fixed time $t = 0$, time horizon $T = 8$, and exercise price $E = 18$. We take $b = .09$, $a = .03$, and $r = .05$. We observe that the option is worthless until the stock price is larger than 11 or so, which is to be expected because if the stock price is so much lower than E, and can only rise at best by a little less than 10% for each of 8 time periods, it has little if any chance of exceeding the exercise price by time T. As the initial stock price rises, the price has a much better chance of exceeding 18 at time $T = 8$, so the amount the buyer expects to receive increases, and the price of the option therefore increases.

```
g1 = Plot[OptValue[0, x, 8, 18, .09, .03, .05],
    {x, 5, 22}, AxesOrigin → {5, -.5},
    PlotRange → All, AxesLabel → {"stock", "option"},
    PlotStyle → Black];
g2 = Plot[OptValue[0, 22, 8, EP, .09, .03, .05],
    {EP, 15, 35}, AxesLabel → {"E", "option"},
    PlotStyle → Black];
GraphicsRow[{g1, g2}]
```

(a) $T = 8$, $E = 18$, $b = .09$, $a = .03$, $r = .05$ (b) $T = 8$, $b = .09$, $a = .03$, $r = .05$, stock price $22.

Figure 6.18 - (a) Option price at time 0 as a function of stock price; (b) Option price at time 0 as a function of exercise price

We see the same phenomenon in a different way in Figure 18(b), in which the option price is plotted against the exercise price E, for fixed initial stock price $22 (the other parameters have the same values as before). As the exercise price increases, it becomes more likely that the stock price will not exceed it, lowering the price of the option. Thus, the price of the option goes down as the exercise price increases. To complete the study, in Exercise 6 you are asked to study the effect of lengthening the time horizon on the option price. ∎

We can use the OptValue function in a simple way to simulate a stock price and its option price together. The function SimStockAndOption simulates a series of "ups" and "downs" of the stock price, and updates and returns lists of stock prices and corresponding option prices at each time up to T.

```
SimStockAndOption[initprice_, T_, EP_, b_, a_, r_] :=
Module[{p, stockprice,
    stocklist, optionprice, optionlist},
        r - a
    p = ─────;
        b - a
  stocklist = {{0, initprice}};
  optionlist =
    {{0, OptValue[0, initprice, T, EP, b, a, r]}};
  stockprice = initprice;
  Do[
    If[RandomReal[] ≤ p, stockprice = (1 + b) stockprice,
      stockprice = (1 + a) stockprice];
    AppendTo[stocklist, {t, stockprice}];
    optionprice =
      OptValue[t, stockprice, T, EP, b, a, r];
    AppendTo[optionlist, {t, optionprice}],
    {t, 1, T}];
  Print[ListPlot[{stocklist, optionlist},
    Joined → True, PlotStyle → {Black, Gray}]];
  {stocklist, optionlist}]
```

Here is an example simulation, with a starting stock price of $15, exercise time $T = 10$, exercise price $E = \$18$, multipliers $b = .08$, $a = -.02$, and riskless rate $r = .04$. Feel free to run the command several times with this choice of parameters, and then to change the parameters to find patterns. One thing that you will always see is that the difference between the stock price and the option price is the exercise price at the final time, if the option is valuable enough to be exercised. (Why?)

```
SimStockAndOption[15, 10, 18, .08, -.02, .04]
```

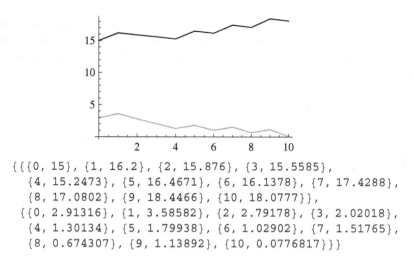

{{{0, 15}, {1, 16.2}, {2, 15.876}, {3, 15.5585},
    {4, 15.2473}, {5, 16.4671}, {6, 16.1378}, {7, 17.4288},
    {8, 17.0802}, {9, 18.4466}, {10, 18.0777}},
    {{0, 2.91316}, {1, 3.58582}, {2, 2.79178}, {3, 2.02018},
    {4, 1.30134}, {5, 1.79938}, {6, 1.02902}, {7, 1.51765},
    {8, 0.674307}, {9, 1.13892}, {10, 0.0776817}}}

Figure 6.19 - Simulated stock price (upper curve) and option price (lower curve) together with list of {time, stockprice} pairs and list of {time, optionprice} pairs

This chapter has only scratched the surface of the many interesting extensions and applications of probability. Other important areas, such as reliability theory, stochastic optimization, information theory, the analysis of algorithms, and inventory theory, await you. Or, you might decide you want to study Markov chains, Brownian motion and other random processes, queues, or finance more deeply. Probably the most important next step is the subject of statistics, which we have previewed a number of times. It is my hope that your experience with this book has prepared you well for all of these studies, as well as enhancing your problem-solving abilities, and giving you a new view of a world fraught with uncertainties.

Exercises 6.5

1. If your investment strategy is to pick just one stock from among the following three to invest in, and your risk aversion is $a = 5$, which stock would you pick?

Stock 1: $\mu_1 = .05$, $\sigma_1^2 = .008$; Stock 2: $\mu_2 = .06$, $\sigma_1^2 = .01$;
Stock 3: $\mu_1 = .09$, $\sigma_1^2 = .02$

2. Find the optimal portfolio of one stock and one bond for an investor with risk aversion $a = 12$, if the non-risky rate of return is 4%, and the stock rate of return is random with mean 5% and standard deviation 4%.

3. (*Mathematica*) Simulate the total returns for 200 replications of the investment problem in Exercise 2, for the optimal portfolio, for a portfolio consisting of 100% stock, and for a portfolio consisting of 50% stock and 50% bonds. Assume an initial wealth of $1000, and assume that the stock rate of return is normally distributed. Produce and compare histograms and summary statistics of the simulated returns.

4. Suppose that two independent stocks with mean rates of return μ_1 and μ_2 and standard deviations σ_1 and σ_2 are available, as well as the non-risky bond with rate of return r. Find expressions for the optimal proportions of wealth that an investor with risk aversion a is to hold in each asset. Does the relative balance between the two risky assets depend on the risk aversion?

5. Suppose that the non-risky rate of return is 5%, and we are interested in a two period portfolio problem in which we collect our profit at the end of the second time period. If the stock follows a binomial model as described in the second subsection, with $b = .10, a = 0$, and $p = .4$, and the investor's risk aversion is 15, find the optimal portfolio.

6. (*Mathematica*) In Example 2, produce a graph of the option price at time 0 as a function of the exercise time T, with the initial stock price $= \$14$, and other parameters as in the example: $E = 18, b = .09, a = .03, r = .05$. Then see how the graph changes as you vary the initial stock price.

7. A stock follows the binomial model with initial price $30, $b = .08$, $a = 0$, and the riskless rate is $r = .04$. Find (by hand) the probability p, the distribution of the stock price two time units later, and the value of an option whose exercise time is 2 and whose exercise price is $32.

8. Express the value of the European call option (10) in terms of the binomial cumulative distribution function.

9. An *American call option* is one that can be exercised at any time prior to the termination time T of the contract. Argue that a lower bound for the value of an American call option $V(S, t, E)$ is $\max(S - E, 0)$. Sketch the graph of this lower bound as a function of S. (Incidentally, it can be shown that the value of an American call is the same as that of a European call. So this lower bound holds for the European call option as well.)

10. Consider a single time period option problem with a general binomial model expressed by (8) holding for the stock upon which the option is based. Again let r be the non-risky rate. Show that the investment action of holding one option at time 0 can be *replicated* by holding a certain number N of shares of the stock, and borrowing an amount B at the non-risky rate, i.e., whether the stock goes up or down in the single time period, this replicating portfolio performs exactly as the option would perform. Find expressions for the N and B that do this replication.

11. In the context of options pricing it was claimed in the section that opportunities for limitless risk-free profits could ensue if the anti-arbitrage condition $a < r < b$ was not in force, where r is the interest rate on a risk-free asset and a and b are the parameters in formula (8) governing the possible states of the risky asset. Explain how this can happen.

12. The perception of a firm's value by its stockholders and potential stockholders is an influential factor in the price of a share of its stock on the market. There have been several models put forth by finance specialists for the value of a firm, but a relatively simple one involves the present value of the expected total profit. Consider a firm whose profits last year were \$40 million. They are expected to grow each year by a random factor $1 + X_i$, where X_i is normally distributed with mean .04 and variance .0004. Assume that these random variables are independent of each other. Suppose that the value of money discounts by a percentage of $r = .02$ per year. Compute the present value of the expected total profit through year 10.

13. (*Mathematica*) A man is planning to retire in two years, at which time he will have saved an amount W in his retirement account. Suppose that he is allowed to lock in the interest rate R on his retirement account at the time of retirement, but R is unknown to him now. He estimates that R is normally distributed with mean .035 and variance .000095. He would like to be able to consume a fixed income of at least \$50,000 for at least 20 years. How large must W be in order for him to achieve his financial goal with probability at least .9?

Appendices

Appendix A

Introduction to *Mathematica*

This text takes pedagogical advantage of many of the features of *Mathematica*, the computer algebra system from Wolfram Research, version 6.0 or higher. If you are a relative newcomer to *Mathematica*, this short introductory section focuses on the most important basic aspects of the software, including programming tools. Appendix B documents commands specific to probability that are defined in the book itself, in the accompanying package KnoxProb7`Utilities`, or in the *Mathematica* kernel or standard packages. This appendix is intended to be a brief review for those with at least some experience, not a comprehensive course, but it should suffice to prepare you for the way in which *Mathematica* is used in problem solving in this book.

Basic *Mathematica* Syntax and Predefined Objects

Mathematica has two parts: the *kernel*, which is the computational workhorse, and the *front-end*, or user interface. In the front-end, a *Mathematica* document (called a *notebook*), is divided into sections called *cells*, delineated by a bracket on the right side of the window. Click before, between, or after cells and start typing to form new cells. Under the Format menu you can change the cell type by first selecting the cell bracket, then choosing among the options in the Style submenu.

In *Mathematica* you are allowed to name variables, functions and other objects very flexibly using combinations of letters and digits, bearing in mind that *Mathematica* is case sensitive, and you cannot name an object with the same name as an object that is already predefined in the system. You also cannot begin a name with a digit. For example, the trig function "Cos" is predefined, but you could name a new version of the function "cos" if you wanted. *Mathematica* capitalizes the first letter of all of its predefined objects.

Arithmetical operations are written using +, -, * for multiplication, and / for division. *Mathematica* can in some cases interpret juxtaposition of variables separated by a space as multiplication, and also numbers adjacent to variables are viewed as products. It is safest to use the * to multiply. Exponentiation is done using the ^ symbol. After your input is typed, position the cursor anywhere in the cell, hold down the Shift key and press the Enter key to get *Mathematica* to evaluate the cell. Here are a few examples of calculations:

> **2 / 16**

$$\frac{1}{8}$$

```
2 * 5
```

10

```
t = 9; s = 5;
t s
t * s
2 t + 1
```

45

45

19

```
6 ^ 2
```

36

Notice that when you carriage return in an input cell, *Mathematica* understands that the next line is a new calculation. But if a line is followed by a semicolon, then output is suppressed. If you want to reuse a symbol name, such as *t* and *s* above for a different purpose, it is usually enough to just reassign its value, but you should also know about the Clear[symbol] function, which clears out previous instances of the given symbol.

    Mathematica is quite particular in the way that the various types of brackets are to be used: open parentheses () only surround algebraic expressions for purposes of grouping; square brackets [] only surround arguments of functions, both in definitions and evaluations of functions; and curly braces {} only surround lists. Here are a few correct usages of brackets. Observe that in the last calculation, the two lists are being added componentwise.

```
Expand[16 (x + 1)]
```

16 + 16 x

```
f[x_] := x²;
f[3]
```

9

```
{1, 4, 3} + {3, 2, 1}
```

{4, 6, 4}

Some of the most useful predefined constants and mathematical functions are Pi, E (the Euler number), Max, Min, (the maximum and minimum value in a list of objects), Abs (absolute value), Exp (exponentiation with base equal to the Euler number, also written as E^), Log (the natural logarithm), Sqrt (square root function), and Mod[n,k] (modular arithmetic remainder on division of n by k). Some functions and constants are illustrated next.

```
Mod[5, 3]
```

2

```
Log[E²]
```

2

```
Abs[-4]
```

4

```
Max[{1, 8, 7, 4}]
```

8

```
Exp[Log[2.5]]
```

2.5

```
Sin[Pi]
```

0

```
Sqrt[16]
```

4

There are tools to shorten and beautify expressions. Depending on what *Mathematica* palettes you have installed, you may be able to select Greek and mathematical symbols and expression templates by clicking on a button on the palette. The BasicTypesetting palette is particularly useful. Certain shortcut key combinations are also very helpful to know about: Holding down the CTRL key and typing the dash/underscore key produces a subscript expression $b_\square$; similarly holding down CTRL while typing the 6^ key gives a superscript, or exponential form $a^\square$. The CTRL / key combination produces a fraction template $\frac{\square}{\square}$. Greek letters can be produced from the keyboard with ESC combinations; for instance typing ESC, then p, then ESC again gives π, and the ESC a ESC combination gives α.

Functions and Graphing

Functions of a real variable are defined as in the $f(x) = x^2$ example above, or as below.

```
f1[x_] := 2 * Sqrt[x] + 1;
```

Notice that the argument of a function in such a definition is surrounded by square brackets, and is followed by the underscore symbol. The syntax := is used to indicate a definition; evaluation of the expression to the right of this sign is delayed until the function is called for a particular input value x. When a definition cell is given to *Mathematica*, no output is produced.

Plotting a real function of a real variable is very easy using the Plot command, which takes as its first argument the name of the function or an expression defining the function, and as its second argument a list such as $\{x, a, b\}$ indicating the name of the domain variable, and the left and right endpoints of the domain. For example:

```
Plot[f1[x], {x, 0, 4}, PlotStyle → Black]
```

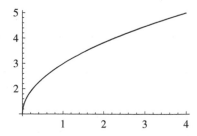

```
Plot[Sin[y2], {y2, 0, Pi}, PlotStyle → Black]
```

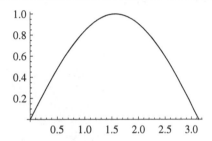

The Plot command, like numerous other commands in *Mathematica*, can be given optional arguments. Options are listed after the main arguments of the command, they have the form optionname->optionvalue, and they are separated by commas. A few of the most important options for Plot are: PlotRange->{c,d} to show only range values from *c* to *d*, or PlotRange->All for the whole graph; AxesLabel->{"xname","yname"} to label the axes, and PlotStyle->stylename to give a style such as a color to the plotting symbol. You can ask *Mathematica* for the options that can be used in a given command using the command Options[commandname]. For example, in the graph below we give names to the horizontal and vertical axes using AxesLabel, we force the graph to show values from −2 to 2 on the vertical scale using PlotRange, and we color the graph gray using PlotStyle.

```
Plot[Sin[2 * y2], {y2, 0, 2 * Pi},
  AxesLabel → {"y2", "ylabel"},
  PlotRange → {-2, 2}, PlotStyle → Gray]
```

Mathematica also has the capability to graph functions of two variables in three-dimensional space, and do corresponding contour plots, that is, curves in the plane defined by constant values of the function. The Plot3D and ContourPlot commands work much like Plot, with an additional variable, as shown below.

```
Plot3D[.5 * x² + .25 * y², {x, -1, 1}, {y, -1, 1},
  PlotStyle → Gray, Lighting → "Neutral"]
```

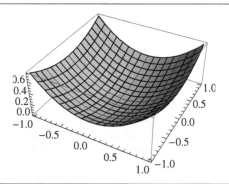

```
g2[x_, y_] := Cos[x * y];
ContourPlot[g2[x, y],
  {x, -1, 1}, {y, -1, 1}, ContourShading →
    {GrayLevel[0], GrayLevel[.2], GrayLevel[.3],
      GrayLevel[.4], GrayLevel[.5], GrayLevel[.6]}]
```

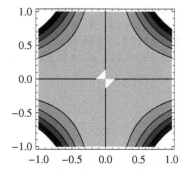

Algebraic and Numerical Commands

Mathematica also has numerous functions for symbolic algebraic manipulation and numerical approximation. The Simplify, Factor, Expand, Together, and Apart commands are fairly self-explanatory given their names. When applied to expressions, they attempt to rewrite them. Below are a few illustrations.

```
Simplify[(x² + 2 x + 1) - (x + 6) + 7 x² + 2]
```

$$-3 + x + 8 x^2$$

$$\text{Factor}\left[x^3 + 3 x^2 + 3 x + 1\right]$$

$$(1 + x)^3$$

$$\text{Expand}\left[(x + 1) (x - 2) (x + 4)\right]$$

$$-8 - 6 x + 3 x^2 + x^3$$

$$\text{Together}\left[\frac{x - 1}{x + 2} + \frac{x + 1}{x - 2}\right]$$

$$\frac{2 \left(2 + x^2\right)}{(-2 + x) (2 + x)}$$

$$\text{Apart}\left[\frac{2 x^2 + 4}{(x + 2) (x - 2)}\right]$$

$$2 + \frac{3}{-2 + x} - \frac{3}{2 + x}$$

There are functions that evaluate finite sums and products as well. Sums are computed as Sum[expression,{variable, lowerlimit, upperlimit}] and similarly for products, but symbolic templates are also available by clicking the appropriate button on the BasicMathInput palette and filling in the blanks. We illustrate both approaches below.

$$\text{Sum}\left[i^2, \{i, 1, 3\}\right]$$
$$\sum_{i=1}^{3} i^2$$

14

14

```
Product[2 j, {j, 1, 3}]
 3
 ∏  2 j
j=1
```

48

48

Mathematica can solve many equations and systems of equations in closed form using the Solve command, whose basic syntax is Solve[list of equations, list of variables]. The answer is reported as a list of so-called *rules* of the form {symbol->value}. A word is in order about syntax for equations in *Mathematica* however. *Mathematica* distinguishes carefully between three usages of the equal sign that humans usually blur together. Ordinary "=" is used when assigning a value to a variable immediately. From above, the ":=" notation is used for definitions that are not immediately evaluated. And finally, in logical equations we use the notation "==", that is, repeated equal signs. The following examples illustrate how to solve a simple quadratic equation for constant values, how to solve for one variable in terms of another in an equation, and how to solve a system of equations, expressed as a list.

```
Solve[x² - 6 x + 8 == 0, x]
```

$\{\{x \to 2\}, \{x \to 4\}\}$

```
Solve[ 2 y + x / x - 1 == 6, y]
```

$\{\{y \to \frac{1}{2} (-6 + 5 x)\}\}$

```
Solve[{x + y == 1, 2 x - y == 4}, {x, y}]
```

$\{\{x \to \frac{5}{3}, y \to -\frac{2}{3}\}\}$

Mathematica can solve many polynomial equations, but for other kinds of equations numerical techniques are necessary. The system provides a number of numerical operations. Some that you will find useful in this book are: N, NSolve, FindRoot, and NSum. The N[] function can apply to any numerical-valued expression to give an approximate decimal value, and it can be given an optional second argument to request desired digits of precision. It can be quite helpful in cases where *Mathematica* insists on giving closed form answers when a numerical estimation is far more helpful. Here are a few examples.

```
N[Cos[16]]
N[Cos[16], 10]
N[Pi, 20]
```

-0.957659

-0.9576594803

3.1415926535897932385

```
16 857 / 23 594
N[16 857 / 23 594]
```

$$\frac{16\,857}{23\,594}$$

0.714461

The NSolve command is a version of Solve that uses numerical techniques to look for solutions, including complex values, of polynomial equations. FindRoot is more general; its syntax is FindRoot[function, {variable, starting value}] and it uses iterative methods starting with the given starting value to locate zeros of the function close to that value. FindRoot works for a wide variety of non-linear equations. Examples are below.

```
NSolve[x^6 + 3 x^4 - 6 x - 12 == 0, x]
```

$\{\{x \to -1.07501\}, \{x \to -0.609288 - 1.39029\,i\},$
$\{x \to -0.609288 + 1.39029\,i\}, \{x \to 0.435812 - 1.79362\,i\},$
$\{x \to 0.435812 + 1.79362\,i\}, \{x \to 1.42196\}\}$

```
FindRoot[x^6 + 3 x^4 - 6 x - 12, {x, 1}]
```

$\{x \rightarrow 1.42196\}$

```
FindRoot[E^x - 3 x, {x, 1}]
```

$\{x \rightarrow 0.619061\}$

Finally, the NSum command is much like Sum, but it reports a decimal approximation to the result as below.

```
NSum[2.5 k² - 4.1 k + .8, {k, 1, 9}]
```

535.2

Calculus

Mathematica commands for calculus are abundant, but only a few are most relevant for this book. In continuous probability we find ourselves doing a great deal of integrating, so the most important of all are the Integrate[function, {x,a,b}] closed form definite integral and the similar NIntegrate command for numerical integration. (Integrate has a version for indefinite integration without limits of integration a and b, but we don't have as much cause to use it here.) The BasicMathInput palette has an integral template, which we also demonstrate in the examples below.

```
Integrate[x * Log[x], {x, 1, 2}]
∫₁² x * Log[x] dx
```

$-\dfrac{3}{4} + \text{Log}[4]$

$-\dfrac{3}{4} + \text{Log}[4]$

```
g[s_] := (6 s * Sin[s]) / E^s ;
NIntegrate[g[s], {s, 0, 1}]
```

0.878724

One can do multiple integrals by nesting, as follows:

```
Integrate[x * Integrate[y², {y, 1, 2}], {x, 0, 3}]
⌠³⌠²
⌡₀⌡₁ x * y² dy dx
```

$$\frac{21}{2}$$

$$\frac{21}{2}$$

Other useful commands for calculus problems are D[function, variable] for the derivative of a function with respect to a variable; Limit[function, variable->value] for the limit of a function at a value; and the commands Sum and NSum from above as applied to infinite series. The derivative command D also has variant forms: D[f, {x, n}] for the nth derivative of the function f relative to the variable x, and D[f, x1, x2,...] for the mixed partial of the several-variable function f with respect to the listed variables. Here are some examples of the use of D, Limit, and the sum commands.

```
Clear[f, t]
f[t_] := Cos[t] * Log[t];
D[f[t], t]
D[f[t], {t, 2}]
```

$$\frac{Cos[t]}{t} - Log[t]\, Sin[t]$$

$$-\frac{Cos[t]}{t^2} - Cos[t]\, Log[t] - \frac{2\, Sin[t]}{t}$$

```
D[x² * y², x, x, y]
(* this is the third order partial ∂/∂x ∂/∂x ∂/∂y *)
```

$$4\, y$$

```
Limit[x²-1 / x-1, x → 1]
```

2

$$\text{Sum}\left[\,(1\,/\,2)^{\,k}\,,\;\{k,\;1,\;\infty\}\right]$$
$$\sum_{k=1}^{\infty}\left(\frac{1}{2}\right)^{k}$$

1

1

$$\text{NSum}\left[\,(1\,/\,k)^{2}\,,\;\{k,\;1,\;\infty\}\right]$$
$$\sum_{k=1}^{\infty}\left(\frac{1}{k}\right)^{2}$$
$$\text{N}\left[\sum_{k=1}^{\infty}\left(\frac{1}{k}\right)^{2}\right]$$

1.64493

$$\frac{\pi^{2}}{6}$$

1.64493

Notice from the last example that for infinite series for which *Mathematica* is able to compute a closed form, the N[] function can still be applied to get a numerical approximation.

Lists

Lists in *Mathematica* are sequences of symbols, separated by commas and demarcated by braces {}. They can contain any type of object, and list members may themselves be lists. In probability we frequently store samples of data in lists, or simulate data into lists, which we then process in some way. Hence a facility with list manipulation in *Mathematica* is quite important for this book.

Some very simple operations on lists are: Length[list], which returns the number of elements in the list, Last[list] and First[list], which return the last and first element of the list, and Sort[list] which sorts the list into increasing (numerical or lexicographical) order. You should be able to easily verify the correctness of the output in all of the inputs below.

```
list1 = {"c", "a", "b"};
list2 = {1.4, 6.8, 4.3, 0.2};
Length[list1]
Last[list2]
First[list1]
Sort[list1]
Sort[list2]
```

3

0.2

c

{a, b, c}

{0.2, 1.4, 4.3, 6.8}

You can pick out individual elements of a list by referring to listname[[i]], which returns the ith list member (note that list subscripts start at 1). For example,

```
list1[[3]]
list2[[2]]
```

b

6.8

Lists can be defined as above by just enumerating their elements, but also patterned lists can be obtained by using the Table command. The standard form is Table[function of i, {i, imin, imax}], which lets the variable i successively take on values from imin to imax in increments of 1, and builds a list whose ith member is the function applied to i. We illustrate, and show some variants on the table iteration list below.

```
list3 = Table[j², {j, 1, 4}]
```

{1, 4, 9, 16}

```
list4 = Table[2 k, {k, 0, 1, .25}]
```

{0., 0.5, 1., 1.5, 2.}

```
list5 = Table[0, {6}]
```

{0, 0, 0, 0, 0, 0}

```
list6 = Table[i * j, {i, 1, 3}, {j, 2, 4}]
```

{{2, 3, 4}, {4, 6, 8}, {6, 9, 12}}

The list called list3 is an ordinary list of squares of integers from 1 through 4. In the definition of list4 we have added an extra value .25 to the table iteration list, which controls the size of the step taken from one value of the variable k to the next. So k will range through the values 0, .25, .50, .75, and 1.0 in this example. In list5 we abandon the list iteration variable altogether and ask for a list of zeros of length 6. The variable list6 is created by Table using two iteration variables i and j, producing a list of lists. First, i takes on the value 1, and *Mathematica* computes the products of 1 with $j = 2$, 3, and 4 and assembles those into a sublist to create the first entry of list6. Then i becomes 2, and the next sublist entry of list6 is the list of products of $i = 2$ with $j = 2$, 3, and 4. Similarly, the last sublist entry of list6 is the product of $i = 3$ with the values of j.

The last example leads us to nested lists and matrices in *Mathematica*. List members can themselves be lists, and this gives us a data structure to use to represent matrices. Each sublist is a row in the matrix. To display the matrix in the usual rectangular format, *Mathematica* provides the TableForm[matrix] and MatrixForm[matrix] commands. These are very similar, the main difference being in whether the output has parentheses surrounding the matrix entries. The arguments are lists, usually with nested sublists of equal length. For example,

```
TableForm[list6]
```

2 3 4
4 6 8
6 9 12

```
matrix7 = {{1, 1, 0, 0}, {0, 0, 1, 1}, {1, 0, 1, 0}};
MatrixForm[matrix7]
```

$$\begin{pmatrix} 1 & 1 & 0 & 0 \\ 0 & 0 & 1 & 1 \\ 1 & 0 & 1 & 0 \end{pmatrix}$$

References to individual matrix elements can be made by double subscripting; specifically the notation matrixname[[i, j]] refers to the element in row i, that is sublist i, and column j. For matrix7 we have:

```
matrix7[[1, 2]]
matrix7[[3, 4]]
```

1

0

There are a number of useful operations on lists and matrices that you should know about. The arithmetical operations +, −,*, and / can be applied to two lists of the same length, which applies the operation componentwise, as shown below.

```
{1, 0, 2} * {3, 4, 5}
{2, 4, 6} / {2, 2, 2}
```

{3, 0, 10}

{1, 2, 3}

The dot product of vectors or matrix product of two matrices for which the product is defined can be done with the period playing the role of the dot in the usual mathematical notation, as the next examples indicate.

```
{1, 0, 2}.{3, 4, 5}
matrix8 = {{1, 2}, {3, 4}};
matrix9 = {{0, 1}, {2, 0}};
MatrixForm[matrix8]
MatrixForm[matrix9]
MatrixForm[matrix8.matrix9]
```

13

$$\begin{pmatrix} 1 & 2 \\ 3 & 4 \end{pmatrix}$$

$$\begin{pmatrix} 0 & 1 \\ 2 & 0 \end{pmatrix}$$

$$\begin{pmatrix} 4 & 1 \\ 8 & 3 \end{pmatrix}$$

Matrices can be transposed with the Transpose[matrixname] command.

```
matrix8 = {{1, 2}, {3, 4}};
MatrixForm[matrix8]
matrix10 = Transpose[matrix8];
MatrixForm[matrix10]
```

$$\begin{pmatrix} 1 & 2 \\ 3 & 4 \end{pmatrix}$$

$$\begin{pmatrix} 1 & 3 \\ 2 & 4 \end{pmatrix}$$

Very often in this book we build lists by simulating values. Each new value must be appended to the end of the existing list. For this, *Mathematica* provides the AppendTo[listname, newvalue] command. This command modifies the given list by placing the new value at the end.

```
list2 = {1.4, 6.8, 4.3, 0.2};
AppendTo[list2, 3.1];
list2
```

{1.4, 6.8, 4.3, 0.2, 3.1}

Finally, for plotting numerical valued lists, we have the ListPlot[listname] command. It can be used in two ways: if its input is a one-dimensional list of numbers, say {a, b, c, ...}, it assumes that these are y-values to be paired up with x-values of {1, 2, 3, ...} and plotted as points in the $x - y$ plane. If the list that is input is already a list of pairs {x, y}, then ListPlot simply plots these points in the plane. Two useful options to know about for ListPlot are Joined->True, which connects the points with line segments, and PlotStyle->PointSize[value], which changes the size of the points. Here are some examples of ListPlot.

```
pointlist1 = {3, 6, 9, 12, 15};
ListPlot[pointlist1, PlotStyle → Black]
```

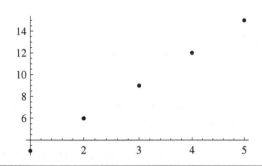

```
pointlist2 = {{1, 1}, {3, 2}, {4, 2}, {5, 3}};
ListPlot[pointlist2,
  PlotStyle → {Black, PointSize[.015]}]
```

```
pointlist2 = {{1, 1}, {3, 2}, {4, 2}, {5, 3}};
ListPlot[pointlist2, Joined → True,
  PlotStyle → {Black, PointSize[.015]}]
```

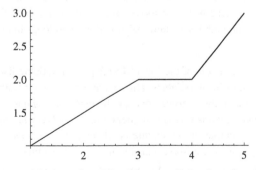

Notice from the last example that when Joined is set to True, the points do not plot.

Programming Tools

One of the pedagogical strategies that is highlighted in this book is to teach the concept of random variation by having the reader program simulations of various random phenomena. Programming is also an important problem-solving device to gain intuition about and to approximate solutions to problems that are not friendly to purely analytical approaches. So it is necessary for the reader to acquire some skill in writing short programs in *Mathematica*. When writing programs of any kind, the student should first clearly understand the problem and its assumptions, and then specify exactly what is to be input to the program and what is to be output. Good programming includes documentation of what the lines of code are intended to do, in any but the most short and simple programs. *Mathematica* provides the bracketing pair: (* *) for documentation; anything enclosed by this pair of symbols is considered a comment and ignored when the program is run.

A program is comprised essentially of header information including identification of input, a list of variables used internally by the program and not defined externally to it, and a sequence of statements producing some output. (This description really applies to a programming style called *imperative programming* that is used in this book. It must be acknowledged that many, if not most, *Mathematica* programmers prefer a functional programming style, in which problems are solved by sequences of function compositions, rather than statements executed one after another. I am of the opinion that for simulation programming, the imperative style is more straightforward.) In *Mathematica*, there is a function called Module that allows the programmer to group statements together and introduce local variables. When actually carrying out computations, the program needs the ability to react one way under one configuration of data and another way under a different configuration. So, *Mathematica* has boolean operators on logical statements, and *conditional functions* like If and Which to make the computation depend on the current state of the data. Also, most programming problems require repetitive operations to be done. Some problems require a fixed number of operations, such as replicating a phenomenon a fixed number of times, while others need to continue executing until some stopping condition goes into effect. For these repetitive code fragments, or *loops*, *Mathematica* includes the Do and While commands. We talk about modules, conditional functions, and loops in turn below.

Modules

A module in *Mathematica* is of the form Module[{local variable list}, statements]. The first argument, i.e., a list of local variables, allows the user to give names to temporary variables used in the computation that is to be done by the module, which are undefined outside of the module. The second argument is any sequence of valid *Mathematica* commands, separated by semicolons. The output of the module is whatever is computed in its last line. A typical template for using a module to solve a problem is to define a function

MyFunction[arg1_, arg2_, ...] := Module[{locals}, statements]

When the function is called, arguments are given to the module, whose values can be used as the program executes.

For example, here is a program to reverse the order of the first two elements of a list. We assume that the input list has at least two elements in it, otherwise the references to the 1st and 2nd elements will not be valid. In this module, a copy of the input list is made in the local variable listcopy. We then replace the first and second elements of the copy of the list by the second and first elements of the original list, and return the copy.

```
ReverseTwo[thelist_] :=
  Module[{listcopy},
    listcopy = thelist; (* Mathematica
      doesn't allow changes to input variables,
    so work on a copy of the input list *)
    listcopy[[1]] = thelist[[2]];
    (* swap into the first element *)
    listcopy[[2]] = thelist[[1]];
    (* swap into the second element *)
    listcopy]
```

```
ReverseTwo[{1, 2, 3}]
```

{2, 1, 3}

The function call above tests out the command on a small list, and the output is appropriate.

Conditionals and Logical Operators

The conditional functions If and Which in *Mathematica* enable the programmer to control the seqence of operations depending on conditions. If[boolean expression, st1, st2] returns st1 if the expression is True and st2 otherwise. The Which[expression1, st1, expression2, st2,...] function is a generalization of If that chooses to return st1 if expression1 is true, else st2 if expression2 is true, etc. Often, there are small code fragments that are executed, instead of simple values returned, depending on the boolean expression value. Here is a small question and answer program that responds accordingly if it finds a given element in a given list. The real work is done by a built-in boolean function called MemberQ[list, expression], which returns True if the expression is present in the list, and False if not.

```
MemberCheck[mylist_, element_] :=
  If[MemberQ[mylist, element],
    Print["Yes!"], Print["No"]]
```

Below, MemberCheck replies that 4 is not an element of the list {1, 2, 3}, and the letter "c" is an element of the list {a, b, c, d}

```
MemberCheck[{1, 2, 3}, 4]
MemberCheck[{"a", "b", "c", "d"}, "c"]
```

No

Yes!

Boolean expressions in *Mathematica* are usually logical equations or inequalities, such as $x >= 6$, $y == 3$, or $x + y == 4$. Such expressions can be combined by the logical operations Not, && (i.e., And), and || (i.e., Or). The following examples illustrate boolean expressions, Which, If, and logical operations. Notice that the logical "and" expressions (1<2) && (2<3) and And[(1<2), (2<3)] are the same, and similarly logical "or" expressions may either be written in binary form with the shorthand symbol || or in functional form with Or[cond1, cond2].

```
(1 < 2) && (2 < 3)
```

True

```
(* This is the functional form of the
 same compound boolean expression above *)
And[(1 < 2), (2 < 3)]
```

True

```
(* This is false because the
 second boolean expression is false *)
(1 < 2) && (3 < 2)
```

False

```
(1 < 2) || (3 < 2)
```

True

```
(* This is true because the
  first boolean expression is true *)
Or[(1 < 2), (3 < 2)]
```

True

```
x2 = 9;
Not[x2 > 10]
```

True

```
(* a piecewise-defined function *)
f[x_] := Which[x < 0, 0, 0 ≤ x ≤ 1, 2, x > 1, 4];
{f[-1], f[0], f[2]}
Plot[f[x], {x, -1, 2}, PlotStyle → Black]
```

{0, 2, 4}

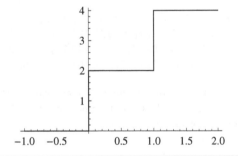

```
(* checking for bad input in a ratio *)
RatioCheck[num_, denom_] := If[Not[denom == 0],
    N[num / denom], Print["Error: division by zero"]];
RatioCheck[1, 4]
RatioCheck[1, 0]
```

0.25

Error: division by zero

Loops: Do, While

Two repetitive programming structures are available in *Mathematica*: the Do loop and the While loop. The syntax for Do is much like Table: Do[statements,{loopvar, initial value, final value}], or simply Do[statements, {numiterations}] if the statements in the loop do not depend on any indexing variable. Hence, Do is appropriate when it is known in advance how many times the sequence of statements is to be repeated. If a sequence of statements is to continue as long as some boolean condition remains true, then the While command is suitable: While[boolean expression, statements] checks whether the boolean expression is true, and if so, executes the statements. It then goes back and rechecks the boolean expression, and if it is true, the statements are evaluated again, etc. So it is up to the programmer to make sure that the statements give a way for the boolean expression to eventually turn false, otherwise the loop will not be able to end.

Here is a code fragment that, like Table, builds a list. In this case, starting with an empty list it constructs a list of $n = 20$ random, uniformly distributed numbers on [0, 1].

```
n = 20;
thelist = {};
Do[AppendTo[thelist, RandomReal[]], {n}];
thelist
```

```
{0.628989, 0.658139, 0.288552, 0.686715, 0.795205,
 0.283333, 0.0575438, 0.497726, 0.442312, 0.37098,
 0.00261852, 0.794867, 0.0854368, 0.512956, 0.138335,
 0.0765775, 0.555683, 0.130204, 0.548339, 0.472718}
```

By contrast, the next code fragment continues to simulate random real numbers on [1, 2] until their total is 20 or more (which can take no more than 20 passes through the While loop). It then returns the total of the numbers.

```
sum = 0;
While[sum < 20,
   sum = sum + RandomReal[{1, 2}]];
sum
```

```
20.4392
```

Graphics and Animation in *Mathematica*

It is rather easy to obtain graphics output from *Mathematica*; (we already have illustrated Plot, ListPlot, and Plot3D for example), but it is useful to know how to combine graphics by superposing or juxtaposing them. *Mathematica* also gives a way to animate graphics automatically, or to have the user control the process of animation. Finally, you might want to experiment with graphics primitives, which display basic geometric objects like lines, points, circles, and arrows, in order to form your own diagrams. We discuss these graphics functions in this subsection.

At times you want to show graphics output side-by-side for comparison purposes. To do this, give names to the graphics that you generate, and use the GraphicsRow[{graph1,-graph2,...}] command to do the display. (There is a similar GraphicsColumn command, but it is somewhat less useful.) Here we display the plots of two functions. In order to suppress the plotting of the two graphs prior to the side-by-side display, we place semicolons at the end of the two Plot commands.

```
g1 = Plot[Sin[x], {x, 0, 2 π}, PlotStyle → Black];
g2 = Plot[Sin[2 x], {x, 0, 2 π}, PlotStyle → Gray];
GraphicsRow[{g1, g2}]
```

Superposition of graphs is often possible from within the graphical command itself; for example, if we wanted to superpose instead of juxtapose the two function graphs above we could simply Plot the list of the two functions:

```
Plot[{Sin[x], Sin[2 x]},
    {x, 0, 2 π}, PlotStyle → {Black, Gray}]
```

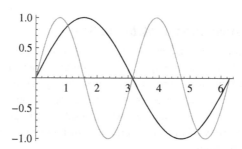

Alternatively, when the graphics have already been produced and named, or when the graphics must be generated from two different commands such as Plot and ListPlot, the Show[{list of graphs}] command can be used to superpose the graphs. Here we use it on the two sine curves:

```
Show[{g1, g2}]
```

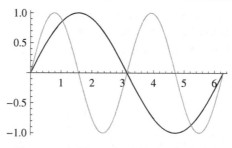

In a moment we will use Show to combine several kinds of primitive graphical objects.

Animation is a very powerful pedagogical device that can allow the user a deep understanding of how graphs change as their input parameters change. *Mathematica* has two ways of doing animations, using Animate and Manipulate. The former simply runs an animation, while the latter allows the user to control the animation. The basic syntax for both is illustrated by the two input cells below: the first argument to Animate or Manipulate is the graphics command, within which is a variable that makes the graph change (here it is the constant coefficient c of the linear term in the cubic function). The second argument is a range of values {variable, lowvalue, highvalue} that you want the graphics control variable to take. If this list contains a fourth member called increment, then animation frames will be generated by adding this increment to the most recent value of the graphics control variable.

```
Clear[x, c];
Animate[Plot[x³ - c * x + 1, {x, -2, 2},
    PlotRange → {-4, 4}, PlotStyle → Black],
   {c, -2, 2}, ControlPlacement → Top]
```

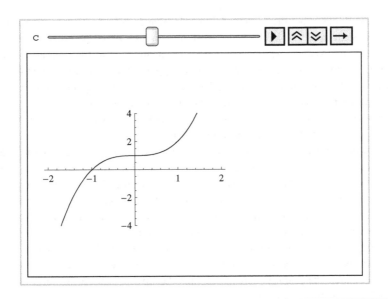

```
Clear[x, c];
Manipulate[Plot[x³ - c * x + 1, {x, -2, 2},
   PlotRange → {-4, 4}, PlotStyle → Black],
  {c, -2, 2}, ControlPlacement → Top]
```

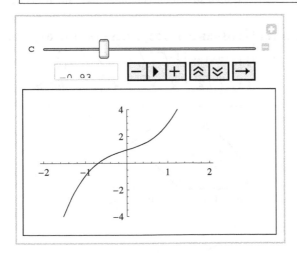

Notice two things about the commands: in order that each frame in the animation is plotted on a consistent scale, we have set the PlotRange option. Also, the control panels are a bit different. In the Animate panel, there is a start and stop button (the right arrow), buttons to speed up or slow down the animation, and a toggle button to run the animation forward, backward, or back and forth. In the Manipulate panel, the user is given a drag bar to position the animation to whatever frame is desired, and there are step forward and step backward buttons. So the ability of the user to exert more control makes Manipulate animations more interactive than animations produced with Animate.

Although it is not absolutely necessary for this book, you may want to use graphics primitives to create your own diagrams. These graphics primitives, such as Point, Line, Rectangle, Circle, and Arrow are used as arguments of the Graphics command in order to produce the diagram. Point[{x, y}] is a point with coordinates (x, y). Point can also be applied to a list of such coordinates to produce several points at once. Line[{list of points}] represents a connected line graph in which each point in the list is connected to its successor in the list. Rectangle[lower left point, upper right point] is a filled rectangle with the given two corner points. Circle[{x, y}, radius] produces a circle with center (x, y) and the given radius. The graphics primitive Arrow[{startpoint,endpoint}] stands for an arrow beginning and ending at the two given points.

Below we use Show to combine a unit circle with center at the origin with points at the origin and at $\left(\sqrt{2}\,/2,\ \sqrt{2}\,/2\right)$ on the circle, and with an arrow connecting these two points. Notice that all of the graphics primitive objects are in a list inside Graphics, and the first element of that list specifies that points are to be given a larger size for better visibility. The Axes->True option for Show is used so that the coordinate axes are displayed.

```
Show[Graphics[{PointSize[.02], Circle[{0, 0}, 1],
    Point[{0, 0}], Point[{√2 / 2, √2 / 2}],
    Arrow[{{0, 0}, {√2 / 2, √2 / 2}}]}], Axes → True]
```

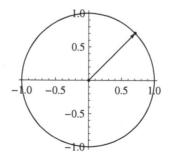

External Packages

At times you have need of commands that are not loaded automatically by *Mathematica* when its kernel starts. *Mathematica* has a few external packages of commands, including MultivariateStatistics` and *Combinatorica`*, that are important for the study of probability and statistics, and also this book comes with its own package, KnoxProb7`Utilities`. Packages can be loaded in with the Needs command. The syntax is simply Needs["package-name"], but you must remember that the package names follow a syntax in which directories are terminated by the backquote character. The commands below illustrate how two of the aforementioned packages are loaded. No output is generated by the Needs command unless an error is encountered in loading the package, but once the package is loaded, all of its commands and objects are available for use.

```
Needs["MultivariateStatistics`"]
```

```
Needs["KnoxProb7`Utilities`"]
```

If you attempt to use a symbol from a package before loading the package, and then try to load, you may find that your spurious use of the undefined symbol masks the true definition of the symbol in the package, and you get unexpected output or error messages. If this happens, you can correct matters by quitting the kernel, restarting it, and then making sure that you load the package before using the symbol.

Appendix B

Glossary of *Mathematica* Commands for Probability

Command Name   **Location**

AbsPctAtM[p, n, M, numreps]   Section 1.3

(* simulates numreps random walks with right step probability p, and returns the percentage of them that were absorbed at M *)

AssetPriceDistribution[numsteps]   Section 3.1

(* plots the distribution of the binomial branch process after a given number of steps *)

BernoulliDistribution[p]   kernel

(* an object representing the Bernoulli distribution with success parameter p *)

Binomial[n, k]   kernel

(* returns the binomial coefficient n choose k *)

BinomialDistribution[n, p]   kernel

(* an object representing the binomial distribution with n trials and success parameter p *)

CDF[dist, x]   kernel

(* returns the value of the cumulative distribution function of the given distribution at x *)

ChiSquareDistribution[r]   kernel

(* object representing the $\chi^2(r)$ distribution *)

Correlation[var1, var2]   kernel

(* returns the sample correlation between the two lists of data *)

Covariance[datalist]   kernel

(* returns the sample covariance matrix of the given list of data, which are formatted as a list of n-tuples, one component for each variable *)

DepartureTimes[arrtimelist, servtimelist]   Section 6.3

(* given the lists of arrival and service times in a queueing problem, returns list of customer departure times *)

DiscreteUniformDistribution[{a,b}]   kernel

(* an object representing the uniform distribution on $\{a, a + 1, ..., b\}$ *)

DotPlot[list]   KnoxProb7`Utilities`
(* draws a dot plot of the given list of data. It takes some of the options of ListPlot, although in the interest of getting a good dotplot it suppresses AspectRatio, AxesOrigin, Axes, PlotRange, and AxesLabel. It has four options of its own: VariableName, set to be a string with which to label the horizontal axis, NumCategories, initialized to 30, which gives the number of categories for stacking dots, DotSize, initially .02, and DotColor, initially RGBColor[0,0,0], that is black. *)

DrawIntegerSample[a,b, n]   Section 1.1, KnoxProb7`Utilities`
(* selects a sample of size n from the set of integers a, ..., b with optional Boolean arguments Ordered->True, and Replacement->False determining the four possible sampling scenarios. *)

ExpectedHighTank[n, θ]   Section 2.1
(* computes the expected value of the highest tank number in a random sample of size n from a uniformly distributed set of tank numbers in $\{1,2,...\theta\}$ *)

ExponentialDistribution[λ]   kernel
(* object representing the exp(λ) distribution *)

FRatioDistribution[m, n]   kernel
(* object representing the $F(m, n)$ distribution *)

Gamma[r]   kernel
(* returns the value of the gamma function at r *)

GammaDistribution[α, β]   kernel
(* object representing the gamma distribution with the given parameters *)

GeometricDistribution[p]   kernel
(* an object representing the geometric distribution with success parameter p *)

Histogram[datalist, numberofrectangles]   kernel
or Histogram[datalist, {rectanglewidth}]
(* plots a histogram of frequencies of a list of data, with a desired number of rectangles, or with rectangles of a desired width. An optional third argument has either of the values "Probability" or "ProbabilityDensity" depending on whether you want bar heights to be relative frequencies or relative frequencies divided by interval length. If the second argument is replaced by $\{a, b, dx\}$, then a histogram extending from a to b with rectangle widths dx is returned. *)

HypergeometricDistribution[n, M, N]   kernel
(* an object representing the hypergeometric distribution, where a sample of size n is taken from a population of size N, in which M individuals are of a certain type *)

KPermutations[list, k]   KnoxProb7`Utilities`
(* returns all permutations of k objects from the given list *)

KSubsets[list, k]   Combinatorica`
      (* lists all subsets of size k from the given list *)

MatrixPower[T, n]   kernel
      (* returns the nth power of the matrix T *)

Mean[datalist]   kernel
      (* returns the sample mean of the list of data *)

Mean[dist]   kernel
      (* returns the mean of the given distribution *)

Multinomial[x1, x2, ... , xk]   kernel
      (* returns the multinomial coefficient *)

MultinormalDistribution[meanvector, covariancematrix]   MultivariateStatistics`
      (* an object representing the bivariate normal distribution with the given two-component vector of means, and the given covariance matrix. Generalizes to many dimensions *)

MyRandomArray[initseed, n]   Section 1.3
      (* returns a list of uniform[0,1] random numbers of length n using the given initial seed value *)

NegativeBinomialDistribution[r, p]   kernel
      (* an object representing the negative binomial distribution with r successes required and success parameter p *)

NormalDistribution[μ, σ]   kernel
      (* an object representing the normal distribution with mean μ and variance σ^2 *)

OptValue[t, x, T, EP, b, a, r]   Section 6.5
      (* returns the present value at time t of a European call option on a stock whose current price is x. The exercise time is T, exercise price is EP, b and a are the up and down proportions for the binomial stock price model, and r is the riskless rate of interest *)

PDF[dist, x]   kernel
      (* returns the value of the probability mass function or density function of the given distribution at state x. Also defined in MultivariateStatistics` package in the form PDF[multivariatedist,{x,y}] it gives the joint density function value at the point $\{x, y\}$ *)

PlotContsProb[density, domain, between]   KnoxProb7`Utilities`
      (* plots the area under the given function on the given domain between the points in the list between, which is assumed to consist of two points in increasing order. Options are the options that make sense for Show, and ShadingStyle->RGBColor[1,0,0] which can be used to give a style to the shaded area region. *)

PlotStepFunction[F, endpoints, jumplist]   KnoxProb7`Utilities`

(* plots a step function on domain specified, with jumps at the points in jumplist, which is a list of sorted numbers. The step function is assumed to be right continuous, as a c.d.f. is. It accepts option DotSize→017 to change the size of the dots, StepStyle–RGBCol - or[0,0,0] to assign a style to the steps, and it inherits any options that make sense for Show. *)

PoissonDistribution[μ]   kernel

(* an object representing the Poisson distribution with parameter μ *)

ProbabilityHistogram[statelist, problist]   KnoxProb7`Utilities`

(* takes a list of states, assumed to be in order, an associated list of probabilities for those states, assumed to have the same length, and graphic options for bars as in Histogram, and displays a bar chart for the probability distribution. The user can control the colot of the bars with the BarColor option. *)

ProportionMInSample[n, k, m, numreps]   Section 1.4

(* returns the proportion of times among numreps replications that m was in a random sample without order or replacement of size k from $\{1, 2, ..., n\}$ *)

P4orfewer[λ]   Section 6.2

(* a function giving the probability of 4 or fewer arrivals of a Poisson process in the first unit of time, as a function of the arrival rate *)

Quantile[distribution, q]   kernel

(* returns the number x for which $P[X \le x] = q$ for X having the given continuous distribution *)

RandomInteger[distribution], RandomInteger[distribution,n] kernel

(* returns a random value, or n of them sampled from a given discrete distribution *)

RandomReal[distribution], RandomReal[distribution,n]   kernel

(* returns a random value, or n of them sampled from a given continuous distribu- tion. With an empty distribution argument it samples from the uniform distribution on [0,1] *)

RandomKPermutation[list, size]   KnoxProb7`Utilities`

(* returns a random permutation of the given size from the given list *)

RandomKSubset[list, k]   Combinatorica`

(* returns a randomly selected subset of size k of the given list *)

SeedRandom[seedvalue] or SeedRandom[ ]   kernel

(* if an integer seed value is supplied it resets the seed for the random number generator to that value, or if no argument is supplied it sets the seed randomly *)

ShowLineGraph[datalist]   Section 6.4

(* Utility that plots a path in 3-space in which the points in the datalist are triples, the first element of which is time. See Exercise 6.4-3. *)

SimArrivals[arrdist, numarrs]   Section 6.3

    (* simulates desired number of interarrival times from the given distribution in a queueing problem *)

SimContingencyTable[sampsize]   Section 1.6

    (* for the particular 2 by 2 contingency table in Example 3, it simulates a random sample from the population, categorizes it and returns the table of category frequencies *)

SimMeanSequence[dist, nummeans, m]   Section 3.3

    (* simulates a list of a desired number of running sample means from the given distribution, updating the list every *m* new observations *)

SimNormal[n, μ, σ]   Section 4.3

    (* simulates a list of *n* pseudo-random observations from the $N(\mu, \sigma^2)$ distribution *)

SimPairs[dist, numpairs]   Section 3.2

    (* two versions, the first simulates and plots the max and min of numpairs random pairs from the given distribution, and the second simulates and plots the random pairs themselves *)

SimSampleVariances[numvars, sampsize, μ, σ]   Section 4.6

    (* simulates a desired number of sample variances, where samples of the given size are taken from the $N(\mu, \sigma^2)$ distribution *)

SimServiceTimes[servicedist, numservices]   Section 6.3

    (* simulates desired number of service times from the given distribution in a queueing problem *)

SimStockAndOption[initprice, T, EP, b, a, r]   Section 6.5

    (*uses OptValue to simulate and plot a stock price and its call option value for times from 0 to *T*. initprice is the initial stock price, and the other parameters are as in OptValue *)

SimX[ ], SimY[ ]   Section 2.5

    (* return random *X* and *Y* observations from the discrete distributions in Example 3 *)

SimXYPairs[numpairs]   Section 2.5

    (* returns a list of a desired number of *X*, *Y* pairs from the discrete distributions in Example 3 *)

SimulateBrownianMotion[x, t, n, μ, σ2]   Section 6.4

    (* returns a list of simulated {time,state} pairs for a Brownian motion with intial state *x*, final time *t*, *n* subintervals of time, and drift and variance rates μ and σ2 *)

SimulateExp[n, λ]   Section 4.3

    (* simulates a list of *n* pseudo-random observations from the exp(λ) distribution *)

SimulateGrades[n]   Section 2.1

    (* simulates a random selection of *n* grades following the p.m.f. in formula (2) *)

SimulateMedians[nummeds, n, dist]   Section 4.4
(* simulates a desired number of medians of samples of size *n* from the given distribution *)

SimulateMinima[m, popsize, sampsize]   Section 1.1
(* uses DrawIntegerSample to simulate *m* observations of the minimum value among the sample of the given size from the set {1, 2, ..., popsize} *)

SimulateMysteryX[m]   KnoxProb7`Utilities`
(* see Exercise 1.1-11, simulates a list of *m* values of a mystery random variable *)

SimulateRandomWalk[p, n, M]   Section 1.3
(* simulates a list which is the sequence of states of a random walk with absorbing barriers at 0 and *M*, right move probability *p*, and initial position *n* *)

SimulateRange[numreps, a, b, sampsize]   Section 1.4
(* returns a list of a desired number of ranges, i.e., max-min, of samples of the given size taken in order and without replacement from {a, $a + 1$, ..., b} *)

SimulateRoulette[numreps]   Section 2.2
(* returns a list of numreps simulated roulette colors, red or other *)

SimulateSampleMeans[nummeans, distribution, sampsize]   Section 2.6
(* returns a list of a desired number of means of samples of the given size, taken from the given probability distribution *)

SimulateWaitingTimes[arrdist, servdist, numcustomers]   Section 6.3
(* simulates a list of waiting times for a single server queue with the given interarrival and service time distributions *)

StandardDeviation[datalist]   kernel
(* returns the sample standard deviation of the given list of data *)

StandardDeviation[dist]   kernel
(* returns the standard deviation of the given distribution *)

StepSize[p]   Section 1.3
(* returns 1 with probability *p*, else 0 *)

StudentTDistribution[r]   kernel
(* object representing the $t(r)$ distribution *)

Subsets[universe]   Combinatorica`
(* lists all subsets of any size from the given universe *)

UniformDistribution[{a, b}]   kernel
(* object representing the continuous uniform distribution on interval $[a, b]$ *)

Variance[datalist]   kernel
(* returns the sample variance of the given list of data *)

Variance[dist] kernel
 (* returns the variance of the given distribution *)

WeibullDistribution[α, β] kernel
 (* object representing the Weibull distribution with the given parameters *)

XYPairFrequencies[numpairs] Section 2.5
 (* in Example 3, returns a frequency table of X, Y pairs *)

Appendix C

Short Answers to Selected Exercises

Section 1.1

1. Approach 1: sample space {000, 001, ..., 999}. A typical simple event is a three digit number with leading 0's, such as 027 or 635. The probability of an event is its cardinality divided by 1000. Approach 2: sample space {−1, 499}. Give outcome −1 probability 999/1000 and outcome 499 probability 1/1000.

2. Sample space: {{A,B}, {A,C}, {A,D}, {B,C}, {B,D}, {C,D}}. P[A serves] = 3/6.

3. (a) P[25 ACT and B] = 10/179; P[C] = 56/179; P[23 ACT] = 57/179. (b) P[A given 24 ACT] = 6/43. (c) P[23 ACT given B] = 5/38 ≠ 57/179.

5. P[king on 1st] = 4/52. P[queen on 2nd] = 4/52.

6. $P[X = -1]$ = 999/1000 and $P[X = 499]$ = 1/1000. $E[X]$ = −1/2.

7. $P[X = 0]$ = 4/6 and $P[X = 1]$=2/6. $E[X]$ = 2/6.

8. $P[X = 0]$ = 1/4, $P[X = 1]$ = 2/4, $P[X = 2]$ = 1/4. $E[X]$ = 1.

9. $P[X = 23]$ = 57/179, $P[X = 24]$ = 43/179, $P[X = 25]$ = 39/179, $P[X = 26]$ = 40/179. $E[X]$ = 4358/179 ≈ 24.3.

13. The sample space has 9900 elements.

$$P[X = 0] = .6384; P[X = 1] = .3232; P[X = 2] = .0384$$

Section 1.2

2. The probability of a hit of any kind is 59/216 ≈ .273. This batter should have home run symbols in a set of squares whose probability totals to 11/216; for instance in squares red=1, whitesum=6 and red=3, whitesum=8, and red=5, whitesum=2 .

3. (a) Ω = {(r_1, r_2, r_3) where each $r_i \in$ {a, b, c, d}}. Examples: (a,c,d) (c,b,b). (c) P[at least 2 right] = 10/64.

4. (c) P[Bubba or Erma is in the sample] = 9/10.

6. P[10 is not in the sample] = 72/90.

7. (a) P[less than 20] ≈ .335; P[between 20 and 30] ≈ .152; P[between 30 and 40] ≈ .121; P[between 40 and 100] ≈ .313; P[over 100] = 10.5/133.9 ≈ .078. P[at least 30] ≈ .512.

(b) The probability that at least one of the two earned more than $100,000 is roughly .14976.

8. (b) P[3rd quadrant] = 1/4; (c) P[not in circle radius 2] = $1 - \pi/4$; (d) P[either 3rd quadrant or not in circle radius 2] = $1 - 3\pi/16$.

9. $P[(A \cup B)^c]$ = .26.

13. $P[\{\omega_1\}]$ = .2, $P[\{\omega_2\}]$ = .4, $P[\{\omega_3\}]$ = .4.

15. 15/24.

16. 1/13.

17. 13/30

18. .05.

19. 4.

20. 880.

21. 1/2.

Section 1.3

1. $P[\ (c, d)\ ] = (d - c)/(b - a) = P[\ [c, d]\ ]$ and $P[\ \{d\}\ ] = 0$.

2. Notice that because the first few seed values are small in comparison to the modulus, the first few random numbers are small.

9. The area under $y = 1/x$ is infinite and also there is no bounded rectangle in which to enclose the region. For $y = 1/\sqrt{x}$ the second problem remains but at least the area is finite.

10. One sees a rather symmetric hill shaped frequency distribution.

12. One usually gets a mean lifetime of about 13, and an asymmetrical histogram with most of its weight on the left and a peak around 12 or 13.

13. Starting at state 1: $p^2/(1 - p + p^2)$. Starting at state 2: $p/(1 - p + p^2)$.

Section 1.4

1.(a) There are 5! lineups. (b) The probability is $1/5$. (c) The probability is 4/5.

2. 544/625.

3. 3125.

	15.	0.252901
	16.	0.283604
	17.	0.315008
	18.	0.346911
	19.	0.379119
	20.	0.411438
	21.	0.443688
	22.	0.475695
5.	23.	0.507297
	24.	0.538344
	25.	0.5687
	26.	0.598241
	27.	0.626859
	28.	0.654461
	29.	0.680969
	30.	0.706316
	31.	0.730455

32.	0.753348
33.	0.774972
34.	0.795317
35.	0.814383

8. 2^n.

9. $C_{n,3} = n \cdot (n-1) \cdot (n-2)/6$.

11. The probability of a full house is about .00144. The probability of a flush is about .00197. The full house is more valuable.

12. 19,199,220,203,520,000.

14. (a) 243/490. (b) 10 bad packages suffices. (c) $n = 19$ is the smallest such n.

15. .515.

16. The probability that subject 1 is in group 1 is n_1/n.

17. 10080.

19. P[3 or fewer runs] = 1/429.

20. P[individual 1 in the first substrata of the first stratum] = 16/180.

Section 1.5

1. (a) 126/218. (b) 25/114. (c) 295/380.

3. .11425.

4. .34.

5. (a) P[all Republicans] = $4/35$. (b) P[at least 2 Rep | at least 1 Rep] = $132/204$.

6. 4/7.

9. 5/12.

10. 9/64.

11. The proportion should approach 1/4.

13. P[lands in 1] = $1/8$; P[lands in 2] = $3/8$; P[lands in 3] = $3/8$; P[lands in 4] = $1/8$.

15. P[at least C | score between 10 and 20] = $68/74$.

17. .1728.

18. .20.

19. .0140845.

20. .4.

Section 1.6

1. P[no more than 1 right] $\approx$ .376.

2. The two events are independent.

4. A and B cannot be independent.

8. It is impossible to make A and B independent.

9. (b) P[HHH] = $(.49)^3$, P[HTH] = $(.49)^2(.51)$ = P[THH] = P[HHT], P[TTH] = $(.51)^2(.49)$ = P[THT] = P[HTT], P[TTT] = $(.51)^3$.

10. P[current flows from A to B] $= \left(1 - (1 - p)^2\right)\left(1 - (1 - q)^3\right)$.

11. P[system fails by time 15] $= .125$.

15. $.08704$.

17. $.328416$.

Section 2.1

1. The values of $f_3(x)$ sum to 1 exactly; hence it is valid. The c.d.f. of this distribution is

$$F_3(x) = \begin{cases} 0 & \text{if } x < 1 \\ .16 & \text{if } 1 \le x < 2 \\ .47 & \text{if } 2 \le x < 3 \\ .65 & \text{if } 3 \le x < 4 \\ .75 & \text{if } 4 \le x < 5 \\ 1 & \text{if } x \ge 5 \end{cases}$$

2. (a) The table below gives the p.m.f.

x	2	3	4	5	6	7	8	9	10	11	12
$f(x)$	$\frac{1}{36}$	$\frac{2}{36}$	$\frac{3}{36}$	$\frac{4}{36}$	$\frac{5}{36}$	$\frac{6}{36}$	$\frac{5}{36}$	$\frac{4}{36}$	$\frac{3}{36}$	$\frac{2}{36}$	$\frac{1}{36}$

(b) The p.m.f. of the maximum random variable is

x	1	2	3	4	5	6
$f(x)$	$\frac{1}{36}$	$\frac{3}{36}$	$\frac{5}{36}$	$\frac{7}{36}$	$\frac{9}{36}$	$\frac{11}{36}$

5. If $m \in \{1, 2, ..., 20\}$, $F(m) = 1 - \left(1 - \frac{m}{20}\right)^5$ and $f(m) = \left(1 - \frac{m-1}{20}\right)^5 - \left(1 - \frac{m}{20}\right)^5$.

7. For m in $\{5, 6, ..., 20\}$,

$$P[X = m] = \frac{\binom{m-1}{4}}{\binom{20}{5}}.$$

9. $\mu = 2.4$, $\sigma^2 = .64$, third moment $= .168$.

10. $\sigma_1^2 = .50$, $\sigma_2^2 = .25$.

11. For the distribution on the left, the third moment is 0, and for the distribution on the right, it is 2.25.

13. $E[X^3] - 3\mu E[X^2] + 2\mu^3$.

16. $E[X] = 0$; $\text{Var}(X) = n(n + 1)/3$.

17. $E[2X - 6] = 7$, $\text{Var}(2X - 6) = 31$.

18. (a) game 1: mean 2.0, variance 9.0; game 2: mean 2.25, variance 54.1875. (b) game 1.

20. The most likely population size is 26.

21. The mean number defective is .7. The variance is about .5918.

24. $220.

25. 45%.

Section 2.2

3. $E\left[(X - p)^3\right] = 2\,p^3 - 3\,p^2 + p$.

6. .382281.

7. $E[X^2] = np(1 - p) + (np)^2$.

8. The expected number of cabs is 36, and the probability that at least 30 cabs arrive is .889029.

9. P[at least 300 advances] = .999999. The mean and variance are 350 and 105, respectively.

12. The expected number of hits is 150. The standard deviation of batting average is .0205.

14. The probability that the promoters must issue at least one refund is .0774339. The expected total refund is $10.21.

16. 1/9.

18. .096.

19. .0488.

20. .469154.

Section 2.3

1. The necessary probability p is approximately .0001.

2. $F(x) = P[X \le x] = 1 - (1 - p)^{x+1}$.

3. $P[X > m + n \mid X > n] = (1 - p)^m$.

4. .23999.

6. (a) .537441 (b) .518249 (c) .38064 (d) mean = 30, standard deviation = $\sqrt{420} \approx 20.5$.

8. p is about .285.

11. expected time = $16\frac{2}{3}$, variance = 38.9, probability of between 15 and 30 jumps around .554.

14. The expected number of games played is 70, the variance is 46.66... and the standard deviation is 6.83.

15. 1/25.

16. $2694.40.

Section 2.4

1. (a) .0888353; (b) .440907; (c) .686714.

2.

	F(x;1)	F(x;2)	F(x;3)	F(x;4)
0	0.367879	0.135335	0.0497871	0.0183156
1	0.735759	0.406006	0.199148	0.0915782
2	0.919699	0.676676	0.42319	0.238103
3	0.981012	0.857123	0.647232	0.43347
4	0.99634	0.947347	0.815263	0.628837
5	0.999406	0.983436	0.916082	0.78513
6	0.999917	0.995466	0.966491	0.889326
7	0.99999	0.998903	0.988095	0.948866
8	0.999999	0.999763	0.996197	0.978637
9	1.	0.999954	0.998898	0.991868
10	1.	0.999992	0.999708	0.99716

3. μ = sample mean = 3.625.

4. $E[X^3] = \mu^3 + 3\mu^2 + \mu$.

5. $E[X(X-1)(X-2)\cdots(X-k)] = \mu^{k+1}$.

7. .0103.

8. The probability that a randomly selected piece has either 1 or 2 defects is about .578624.

9. .0426209.

10. (a) $P[X > 12] = .424035$; (b) $P[X > 15 \mid X > 12] = .366914$.

11. $P[X_1 = 2, \; X_2 = 3, \; X_3 = 5] = .0124468$.

12. $P[X_s = j \cap X_{t+s} = k] = \dfrac{e^{-\lambda s}(\lambda s)^j}{j!} \dfrac{e^{-\lambda t}(\lambda t)^{k-j}}{(k-j)!}$.

13. $\lambda = 3$ makes them most likely.

14. 2.

15. \$698.90.

Section 2.5

1. $q(0 \mid 0) = 4/31$, $q(1 \mid 0) = 6/31$, $q(2 \mid 0) = 10/31$, $q(3 \mid 0) = 8/31$, $q(4 \mid 0) = 3/31$
$p(0 \mid 2) = 10/41$, $p(1 \mid 2) = 7/41$, $p(2 \mid 2) = 10/41$, $p(3 \mid 2) = 9/41$, $p(4 \mid 2) = 5/41$.

2. The X marginal is $p(1) = 4/16$, $p(2) = 3/16$, $p(3) = 3/16$, $p(4) = 6/16$. The Y marginal is $q(1) = 9/32$, $q(2) = 9/32$, $q(3) = 7/32$, $q(4) = 7/32$.

4. (a) $P[X = 0] = 4/10$, $P[X = 1] = 3/10$, $P[X = 2] = 2/10$, $P[X = 3] = 1/10$.

(b) $P[Y = 0] = 4/10$, $P[Y = 1] = 3/10$, $P[Y = 2] = 2/10$, $P[Y = 3] = 1/10$.

(c) $P[X = 0 \mid Y = 1] = 1/3, P[X = 1 \mid Y = 1] = 1/3$, $P[X = 2 \mid Y = 1] = 1/3$.

(d) $P[Y = 0 \mid X = 1] = 1/3, P[Y = 1 \mid X = 1] = 1/3$, $P[Y = 2 \mid X = 1] = 1/3$.

5. $p(y \mid x = 3) = \begin{pmatrix} 30 \\ y \end{pmatrix} \begin{pmatrix} 25 \\ 5-y \end{pmatrix} / \begin{pmatrix} 55 \\ 5 \end{pmatrix}$.

10. P[hits safely] = .264.

11. $P[X = 1 \mid Y = 1] = .993$, $P[X = 0 \mid Y = 1] = .007$;
$P[X = 1 \mid Y = 0] = .486$, $P[X = 0 \mid Y = 0] = .514$.

13. $f(z) = \frac{1}{25} \cdot \left(\frac{4}{5}\right)^z \cdot (z + 1)$, $z = 0, 1, 2, \ldots$.

14. $f(x_1, x_2) = 1/4$. Similarly the other two joint marginals $f(x_1, x_3)$ and $f(x_2, x_3)$ are equal to 1/4 for all pairs. $f(x_1) = 1/2$ and the other two marginals are identical.

16. 1/64.

17. $83/144 \approx .58$.

Section 2.6

2. (a) $E[X + Y] = 3$; (b) $E[2X - Y] = 3$.

3. $E[X_1 + X_2 + X_3] = 8$; $E[X_1 + X_2 + X_3 + X_4 + X_5] = 20$.

4. mean = 24; variance = 144.

6. When the random variables are independent, $\text{Var}(\sum_{i=1}^{n} c_i X_i) = \sum_{i=1}^{n} c_i^2 \text{Var}(X_i)$.

8. Both the covariance and correlation equal 1/2.

9. $\text{Cov}(X, Y) = n\, p(1 - p)$; correlation = 1.

13. A sample size of 217 is sufficient.

15. 2/5.

16. 113/16.

17. $E[Y \mid X = 1] = 10/4$; $E[Y \mid X = 2] = 17/6$.

18. $h(0) = 0$, $h(1) = 1/2$, $h(2) = 5/3$, $h(3) = 14/4$; $E[Y^2] = 2$.

19. The conditional mean and variance are 3/7 and 12/49, respectively.

22. 19,300.

23. 8.8.

24. .9856.

Section 3.1

2. (a) $\mathcal{H} = \{\Omega, \emptyset, \{2, 3\}, \{1, 4, 5\}, \{1, 5\}, \{2, 3, 4\}\}$;

(b) $\mathcal{H} = \{\Omega, \emptyset, \{1, 2\}, \{3, 4, 5\}, \{1, 2, 3\}, \{4, 5\}, \{1, 2, 4, 5\}, \{3\}\}$.

5. $\mathcal{H} = \{\{x\}, \Omega - \{x\}, \Omega, \emptyset\}$.

7. $c = 1$. $P[\,(2, 4) \cup (8, \infty)\,] = 3/8$.

8. $c = 1$. $P[\,[2, e]\,] = 2 - \log(4)$. $P[\,[1, 2]\,] = \log(4) - 1$.

9. $c = 1/\ln(2)^2$; $P[\,[1.5, 2] \times [0, 1.5]\,] = .242781$.

10. Exact probability = .109375; approximate probability = .158655.

Section 3.2

1. 1/2; 1/2.

2. .952637.

3. $C = k$, $F(x) = 1 - e^{-kx}$, $x \geq 0$.

4. For $x \in [a, b]$, $F(x) = \frac{x-a}{b-a}$; $P\left[X > \frac{a+b}{2}\right] = 1/2$.

6. $P[T > 10] = 2/e$.

7. $P[1.5 \leq X \leq 3.5] = 3/4$.

9. $x = \beta\left(\frac{\alpha-1}{\alpha}\right)^{1/\alpha}$.

10. Y is uniformly distributed on $[-1, 1]$.

11. .752796; $30111.80.

12. $P[V \geq 40] = .622146$; v^* is about 54.6.

13. 1/48.

14. $P[V - U \geq 1/4] = 9/16$.

16. $P[Y > X] = .432332$.

18. $P[V > 3/4 \mid U = 1/2] = 1/2$.

20. $P[Y > 5/4 \mid X = 3/2] = .716737$; $P[X < 3/2 \mid Y = 6/5] = .546565$.

21. (a) 2/27; (b) 8/27.

24. 5/8.

25. .416667.

26. .578125.

Section 3.3

1. $\text{Var}(X) = (b - a)^2/12$.

3. $\mu = 0$; $\text{Var}(X) = 1/6$.

5. $E[T] = 20$, $E[T^2] = 600$, $E[T^3] = 24\,000$.

6. (a) $P[\,|X - \mu| < \sigma] = .5774$; (b) $P[\,|X - \mu| < 2\sigma] = 1$; (c) $P[\,|X - \mu| < 3\sigma] = 1$

8. The expected profit is $2387.

9. The covariance is $-9/100$, the correlation is $-9/31$.

11. 1/36

12. $E[X + 6Y] = 7/(2\log(1024/729))$.

16. $E[2XY] = 1/3$.

19. $E[Y \mid X = x] = \frac{x^2/2 + 1/4}{(x^2 + 1/3)}$.

20. 1.70667.

21. 5.72549.

22. 1/12.

Section 4.1

5. $1/(\sigma\sqrt{2\pi})$.

7. (a) .841345; (b) .97725; (c) .99865.

8. About .026.

9. About .0001.

11. (a) .182; (b) .8680; (c) .0069; (d) .1251.

12. .1111.

14. (a) About .32; (b) About .95.

15. -1.28155, -0.841621, -0.524401, -0.253347, 0., 0.253347, 0.524401, 0.841621, 1.28155.

17. $711.61.

18. (a) .308538; (b) .598706.

Section 4.2

2. The probability is roughly .218.

3. (a) .351828; (b) .0102; (c) 17.3616.

4. (a) 96.0; (b) 99.7; (c) 100.3.

6. There does appear to be a good linear relationship between the logged variables. For the logged SO2 variable the mean estimate is 3.19721 and the standard deviation is 1.49761. For the logged mortality variable the mean estimate is 6.84411 and the standard deviation is 0.0662898. The correlation estimate is 0.414563. The conditional variance and standard deviation come out to be .00363912 and .0603251, respectively.

8. (a) .551128; (b) .53446.

9. .307806.

11. (a) .692077; (b) .956638.

13. If you divide $f(x, y)$ by $p(x)$ for each fixed x the resulting function will integrate to 1.

14. ρ.

16. $\dfrac{1}{2\pi\sqrt{1-\rho^2}}\exp\left[\dfrac{-1}{2(1-\rho^2)}\left(z^2 - 2\rho z w + w^2\right)\right]$.

Section 4.3

1. (a) The density of Y is $g_Y(y) = \frac{3}{2}y^{1/2}$, $y \in [0, 1]$.   (b) The density of Y is $g_Y(y) = 9\,y^8$, $y \in [0, 1]$.

2. $g_U(u) = \dfrac{1}{\sqrt{2\pi}}u^{-1/2}\,e^{-u/2}$, $u \geq 0$.

6. The probability that C exceeds 20 is about .655.

7. $M(t) = \left(e^{tb} - e^{ta}\right)/(t(b-a))$.

8. $M(t) = p/(1 - (1-p)e^t)$, $t < -\log(1-p)$.

9. $M(t) = (8/27)\,e^{1\cdot t} + (19/27)$.

11. The Bernoulli m.g.f. is $M(t) = p\,e^t + (1 - p)$.

12. The m.g.f. of the Poisson(μ) distribution is $M(t) = e^{\mu(e^t - 1)}$.

13. $P[\overline{X} < .250] \approx .53$; $P[\overline{X} > .280] \approx .0086$.

14. (a) The probability that the average time is more than 1.7 is about .057; (b) the probability that the total time is less than 16 is about .785.

15. $F(v) = (\ln(v/10\,000) - .04)/.04$.

16. $f(r) = 5/(2\,r^2)$.

17. 10560.

Section 4.4

2. (a) .755859; (b) .755859; (c) .527344.

4. $f(y) = \frac{3}{2}\,y^2 - \frac{3}{4}\,y^3,\ y \in (-1,\ 1)$.

6. .167772.

7. .03125.

8. .653671.

9. $1 - e^{-3/2}$.

10. $1 - e^{-13/6}$.

12. $h(x) = \frac{7}{6}\,x - 1$.

13. 2.22656.

14. $2,025.

Section 4.5

1. (a) .65697; (b) .135353; (c) .125731.

4. The probability $P[T > .5]$ is .342547.

5. The probability that the project will not be finished is .199689.

7. $E[T] = 10$, and $P[T \geq 10] = e^{-1}$.

8. (a) 2.13773; (b) .919174.

9. $F(x) = 1 - e^{-\lambda x} - \lambda\,x\,e^{-\lambda x}$.

10. .714943; .691408.

11. β is around .748.

13. The jump times are 1.1, 3.4, 4.9, and 9.1; the states are 0, 1, 2, and 3; estimate of λ: .43956.

14. .336664.

16. 173.287.

17. .4.

Section 4.6

1. (a) .34378; (b) .132062; (c) 6.7372; (d) $a = 3.24697$, $b = 20.4832$.

2. The m.g.f. of a $\chi^2(r)$ random variable is $M(t) = (1 - 2\,t)^{-r/2}$.

3. Let a and b be points for the $\chi^2(n-1)$ distribution for which $P[X < a] = P[X > b] = .025$. The confidence interval is $(n-1)\,S^2/b \le \sigma^2 \le (n-1)\,S^2/a$.

4. $[191.271, 396.663]$.

5. It is nearly certain that the standard deviation exceeds 2.1.

6. .228418.

7. If $\mu = -6$ is true, it is only about .01 likely to observe such a small sample mean, so we have strong evidence that the true μ is smaller than -6.

9. (a) .0234488; (b) .0616232; (c) $t = 1.31946$.

10. The data are not sufficiently unusual to cast doubt on the hypothesis $\mu = 98.7$.

11. Write $t_{.05}$ for the 95th percentile of the $t(n-1)$ distribution, i.e., the point such that $P[T \le t_{.05}] = .95$. The confidence interval is $\overline{X} \pm t_{.05}\,\dfrac{s}{\sqrt{n}}$.

15. Some graphical experimentation produces $r = 14$.

16. (a) .49425; (b) .467569; (c) $a = .379201$, $b = 2.53424$.

17. We do not have strong evidence against equality of variances.

20. The 90% confidence interval is $[.437126, 1.7022]$.

Section 4.7

1. $\dfrac{1}{(2\pi)^{3/2}\sqrt{3.84}}\, e^{-(1/2)\left((x_1-2.5)^2 + 1.04167\,x_2{}^2 - 2\,(.104167)\,x_2(x_3-1.8) + .260417\,(x_3-1.8)^2\right)}$

3. U and V are independent $N(0, 2)$ random variables.

4. The joint distribution is trivariate normal with the mean $(-1, -1, 0)$ and covariance matrix

$$\begin{pmatrix} 14.2 & 7.3 & -6.9 \\ 7.3 & 15.4 & 8.1 \\ -6.9 & 8.1 & 15 \end{pmatrix}$$

5. The marginals are $N(2, 3)$, $N(1, 1)$, and $N(0, 4)$.

6. $\chi^2(1)$.

9. The $\chi^2(r_1 + r_2)$ distribution.

Section 5.1

1. We find that $n = 25$ is the smallest number of rolls such that the desired probability exceeds 1/2.

5. (a) In all three cases, the Chebyshev upper bound is $1/4$. When $X \sim \Gamma(3, 4)$, the actual probability is about .044; (b) When $X \sim$ uniform$(0, 1)$ the actual probability is 0; (c) When $X \sim N(2, 4)$ the actual probability is about .0455.

7. A sample of size 2560 suffices.

Section 5.2

2. $f(x) = \sqrt{k}\ e^{-kx^2/2} \big/ \sqrt{2\pi}$.

6. Approximately .846.

7. The probability that the sample mean could have come out as large as 550 or larger is about .006. Since this probability is so small, we have significant evidence against $\mu = 540$.

8. .818595.

9. .769861.

Section 6.1

1. The distribution of X_3 is $\{5/96, 5/64, 67/192, 25/48\}$. The limiting distribution is $\{3/71, 6/71, 24/71, 38/71\}$.

2. The distribution of X_2 given $X_0 = 4$ is $\{0, 0, .16, .48, .36\}$. The limiting distribution is $\{.1, .25, .25, .25, .15\}$.

5. The chain is not regular.

7. The limiting distribution is $\{1/3, 1/3, 1/3\}$. The smallest power is $n = 8$.

8. The limiting distribution is $\left\{\dfrac{1-q}{2-(p+q)}, \dfrac{1-p}{2-(p+q)}\right\}$. Among many solutions is $p = 1/2$, $q = 3/4$.

9. The conditional distribution of X_4 given $X_0 = 3$ is $\{53/96, 43/192, 43/192\}$. The limiting distribution is $\{5/9, 2/9, 2/9\}$.

10. The solutions are $f(2) = .588235$, $f(3) = .411765$.

12. 0, $p^2\big/(1-p+p^2)$, $p^2\big/(1-p+p^2)$, and 1. The point of most rapid increase is about .653.

14. .6561.

15. $g(n) = \dbinom{n-1}{3}.05^4\,.95^{n-4}$, $n = 4,\ 5,\ 6,\ \dots$

Section 6.2

1. (a) .0464436; (b) .0656425.

2. .241165.

3. .0116277.

4. To the hundredths place $\lambda = 1.67$.

7. 5/6.

10. N_t^I has the Poisson distribution with parameter $(5/4)\,t$.

11. $(125/8)\,t$.

Section 6.3

1. $\rho = 2/3$.

2. The probability of at least 2 in the system is 4/9. The probability that a customer goes directly into service is 1/3.

5. $E[W] = E[W_q] + 1/\mu$.

6. The c.d.f. of W_q is $F_q(s) = 1 - e^{-\mu(1-\rho)s} + (1 - \rho)e^{-\mu s}$.

7. The long-run probability that a customer will take at least 2 minutes is around .716.

8. If C is the weighting factor on service rate, and D is the weighting factor on waiting time, then the optimal $\mu = \lambda + \sqrt{D/C}$.

11. $f_1(t) = \frac{\lambda}{\lambda+\mu} - \frac{\lambda}{\lambda+\mu} e^{-(\lambda+\mu)t}$.

12. The distribution of the time until the next customer is served is $\exp(n\,\mu)$. The differential equation is $f_k'(t) = -(\lambda + \mu k)\, f_k(t) + \lambda\, f_{k-1}(t) + \mu\,(k + 1)\, f_{k+1}(t)$, $k \geq 1$.

Section 6.4

1. (a) .226919; (b) .295677; (c) .429893.

4. $C(t, s) = \sigma^2 t$.

6. About 5.1 years.

7. $2\left(1 - \int_{-\infty}^{a} \frac{1}{\sqrt{2\pi\sigma^2 t}} e^{-x^2/2\sigma^2 t}\, dx\right)$

9. $G(y) = 2 \int_{-\infty}^{y} \frac{1}{\sqrt{2\pi t}} e^{-x^2/2t}\, dx - 1$, $y > 0$; $g(y) = \frac{2}{\sqrt{2\pi t}} e^{-y^2/2t}$, $y > 0$; $E[Y_t] = \sqrt{2t/\pi}$.

10. .344509.

13. .011981.

Section 6.5

1. You would be indifferent between the first two stocks.

2. The optimal proportion q^* in the stock is around 26%.

4. The optimal solutions are

$$q_1^* = \frac{\mu_1 - r}{2a\sigma_1^2}, \quad q_2^* = \frac{\mu_2 - r}{2a\sigma_2^2}$$

5. The optimal proportion in the stock is about -13%; so we see in this case that because the non-risky asset performs so much better, it is wise to borrow on the stock and invest all proceeds in that asset.

7. $p = .5$. In two time units, the stock will either go up to 34.99 with probability 1/4, or go to 32.40 with probability 1/2, or it will stay at 30 with the remaining probability 1/4. The option value is about 87.6 cents.

10. The portfolio weights are

$$N = \frac{\max\{(1+b)\,S - E, 0\} - \max\{(1+a)\,S - E, 0\}}{S(b-a)}, \quad B = \frac{N\,S(1+b) - \max\{(1+b)\,S - E, 0\}}{(1+r)}$$

12. \$445,777,000.

13. The amount to be saved is about \$798,500.

References

Abell, Martha L. and James P. Braselton, *Mathematica by Example, 4th ed.*, Elsevier/Academic Press, Burlington, MA, 2009.

Billingsley, Patrick, *Probability and Measure*. John Wiley & Sons, New York, 1979.

Burton, David M., *The History of Mathematics: An Introduction*. Allyn & Bacon Inc., Boston, 1985.

Chung, Kai Lai, *Elementary Probability Theory with Stochastic Processes*. Springer-Verlag, New York, 1979.

Çinlar, Erhan. *Introduction To Stochastic Processes*. Prentice-Hall, Englewood Cliffs, NJ, 1975.

Cooke, Roger, *The History of Mathematics: A Brief Course*. John Wiley & Sons, New York, 1997.

Feller, William, *An Introduction to Probability Theory and Its Applications, Vol. I and II, 3rd ed.*, John Wiley & Sons, New York, 1968.

Gross, Donald and Carl M. Harris, *Fundamentals of Queueing Theory, 2nd ed.*, John Wiley & Sons, New York, 1985.

Hacking, Ian, *The Emergence of Probability*. Cambridge University Press, London, 1975.

Hoel, Paul G., Sidney C. Port, and Charles J. Stone, *Introduction to Probability Theory*. Houghton Mifflin, Boston, 1971.

Hogg, Robert V. and Allen T. Craig, *Introduction to Mathematical Statistics, 4th ed.*, Macmillan, New York, 1978.

Karlin, Samuel and Howard M. Taylor, *A First Course in Stochastic Processes, 2nd ed.*, Academic Press, New York, 1975.

Katz, Victor J., *A History of Mathematics: An Introduction*. HarperCollins, New York, 1993.

Lamberton, Damien and Bernard Lapeyre, *Introduction to Stochastic Calculus Applied to Finance*. Chapman & Hall, London, 1996.

Loève, Michel, *Probability Theory I, 4th ed.*, Springer-Verlag, New York, 1977.

Parzen, Emanuel, *Modern Probability Theory and Its Applications*. John Wiley & Sons, New York, 1960.

Roman, Steven, *Introduction to the Mathematics of Finance*. Springer, New York, 2004.

Ross, Sheldon M., *Stochastic Processes*. John Wiley & Sons, New York, 1983.

Rubinstein, Reuven, *Simulation and the Monte Carlo Method*. John Wiley & Sons, New York, 1981.

Saaty, Thomas L., *Elements of Queueing Theory with Applications*. Dover Publications, New York, 1961.

Sharpe, William, *Portfolio Theory and Capital Markets*. McGraw-Hill, New York, 1970.

Snell, J. Laurie, *Introduction to Probability*. Random House, New York, 1988.

Society of Actuaries and Casualty Actuarial Society. Exam P Sample Questions, 2005. http://www.soa.org/files/pdf/P-09-05ques.pdf.

Index

(*Mathematica* commands in bold)

9 780367 385194